Axial Flux Permanent Magnet Brushless Machines

Jacek F. Gieras
Rong-Jie Wang
Maarten J. Kamper

Axial Flux Permanent Magnet Brushless Machines

Second Edition

Springer

Prof. Jacek F. Gieras
Hamilton Sundstrand, Applied Research
Rockford, Illinois, U.S.A.
University of Technology and Life Sciences
Bydgoszcz, Poland
jgieras@ieee.org

Prof. Maarten J. Kamper
Laboratory for Electrical Machines
and Drives
Department of Electrical
and Electronic Engineering
University of Stellenbosch, South Africa
kamper@sun.ac.za

Dr. Rong-Jie Wang
Laboratory for Electrical Machines
and Drives
Department of Electrical
and Electronic Engineering
University of Stellenbosch, South Africa
rwang@sun.ac.za

ISBN 978-1-4020-6993-2 e-ISBN 978-1-4020-8227-6

Library of Congress Control Number: 2008925238

Printed on acid-free paper.

9 8 7 6 5 4 3 2 1

springer.com

Preface

Growing interest in new topologies of permanent magnet (PM) brushless machines has prompted the authors to update the first edition of *Axial flux permanent magnet brushless machines*, which was originally published in 2004. In the second edition new sections (non-overlap concentrated coil windings, rotor dynamics, miniature axial flux PM motors, in wheel motors), new examples of applications and more numerical examples have been added. On request of engineers involved in axial flux PM motors technology, a CD with all numerical examples solved in Mathcad has been attached.

The importance of PM brushless machines technology and its impact on energy conversion systems are receiving increasing attention each year. While d.c. brush machines production is shrinking, PM brushless machines are replacing d.c. brush and sometimes induction motors in consumer electronics, kitchen and bath equipment, public life, instrumentation and automation system, clinical engineering, industrial electromechanical drives, automobile manufacturing industry, electric and hybrid electric vehicles, marine vessels, toys, more electric aircrafts and many other applications on larger scale. New applications have emerged in distributed generation systems (wind turbine generators, high-speed microturbine generators), miniature power supplies, flywheel energy storages, aircraft and rotorcraft actuators, missile fin actuators, naval integrated motor-propellers (rim driven thrusters). The role of axial flux PM brushless machines is increasing especially in applications where integration of motors with other mechanical parts is imperative.

We believe that the new edition will provide the most up-to-date knowledge on the analysis, design, control and applications of axial flux PM brushless machines. It will increase awareness and stimulate innovations in this field.

Jacek F. Gieras
Rong-Jie Wang
Maarten J. Kamper

Contents

1

Introduction

1.1 Scope

The term *axial flux permanent magnet* (AFPM) machine in this book relates only to permanent magnet (PM) machines with *disc type rotors*. Other AFPM machine topologies, e.g. *transverse flux machines*, have not been considered. In principle, the electromagnetic design of AFPM machines is similar to its radial flux PM (RFPM) counterparts with cylindrical rotors. However, the mechanical design, thermal analysis and assembly process are more complex.

1.2 Features

The AFPM machine, also called the *disc-type* machine, is an attractive alternative to the cylindrical RFPM machine due to its pancake shape, compact construction and high power density. AFPM motors are particularly suitable for electrical vehicles, pumps, fans, valve control, centrifuges, machine tools, robots and industrial equipment. The large diameter rotor with its high moment of inertia can be utilised as a flywheel. AFPM machines can also operate as small to medium power generators. Since a large number of poles can be accommodated, these machines are ideal for low speed applications, as for example, electromechanical traction drives, hoists or wind generators.

The unique disc-type profile of the rotor and stator of AFPM machines makes it possible to generate diverse and interchangeable designs. AFPM machines can be designed as single air gap or multiple air gaps machines, with slotted, slotless or even totally ironless armature. Low power AFPM machines are frequently designed as machines with slotless windings and surface PMs.

As the output power of the AFPM machines increases, the contact surface between the rotor and the shaft in proportion to the power becomes smaller. Careful attention must be given to the design of the rotor-shaft mechanical joint as this is usually the cause of failures of disc type machines.

Table 1.1. Specifications of double-sided disc type AFPM brushless servo motors manufactured by *E. Bautz GmbH*, Weiterstadt, Germany

Quantity	S632D	S634D	S712F	S714F	S802F	S804F
Rated power, W	680	940	910	1260	1850	2670
Rated torque, Nm	1.3	1.8	2.9	4.0	5.9	8.5
Maximum torque, Nm	7	9	14	18	28	40
Standstill torque, Nm	1.7	2.3	3.5	4.7	7.0	10.0
Rated current, A	4.0	4.9	4.9	6.6	9.9	11.9
Maximum current, A	21	25	24	30	47	56
Standstill current, A	5.3	6.3	5.9	7.8	11.7	14.0
Rated speed, rpm	5000	5000	3000	3000	3000	3000
Maximum speed, rpm	6000	6000	6000	6000	6000	6000
Armature constant, V/1000 rpm	23	25	42	42	42	50
Torque constant, Nm/A	0.35	0.39	0.64	0.64	0.64	0.77
Resistance, Ω	2.5	1.8	2.4	1.5	0.76	0.62
Inductance, mH	3.2	2.8	5.4	4.2	3.0	3.0
Moment of inertia, $kgm^2 \times 10^{-3}$	0.08	0.12	0.21	0.3	0.6	1.0
Mass, kg	4.5	5.0	6.2	6.6	9.7	10.5
Diameter of frame, mm	150	150	174	174	210	210
Length of frame, mm	82	82	89	89	103	103
Power density, W/kg	151.1	188.0	146.8	190.9	190.7	254.3
Torque density, Nm/kg	0.289	0.36	0.468	0.606	0.608	0.809

In some cases, rotors are embedded in power-transmission components to optimise the number of parts, volume, mass, power transfer and assembly time. For electric vehicles (EVs) with built-in wheel motors the payoff is a simpler electromechanical drive system, higher efficiency and lower cost. Dual-function rotors may also appear in pumps, elevators, fans and other types of machinery, bringing new levels of performance to these products.

Most applications use the AFPM machine as a d.c. brushless motor. Encoders, resolvers or other rotor position sensors are thus a vital part of brushless disc motors.

Table 1.1 shows specifications of AFPM brushless servo motors rated up to 2.7 kW, manufactured by *E. Bautz GmbH*, Weiterstadt, Germany.

1.3 Development of AFPM Machines

The history of electrical machines reveals that the earliest machines were axial flux machines (M. Faraday, 1831, anonymous inventor with initials P.M., 1832, W. Ritchie, 1833, B. Jacobi, 1834). However, shortly after T. Davenport (1837) claimed the first patent [P1] for a radial flux machine, conventional radial flux machines have been widely accepted as the mainstream configuration for electrical machines [33, 53].

The first primitive working prototype of an axial flux machine ever recorded was M. Faraday's disc (1831) - see Numerical Example 1.2. The disc type construction of electrical machines also appears in N. Tesla's patents, e.g. U.S. patent No. 405 858 [P2] entitled *Electro-Magnetic Motor* and published in 1889 (Fig. 1.1). The reasons for shelving the axial flux machine were multi-fold and may be summarised as follows:

- strong axial (normal) magnetic attraction force between the stator and rotor;
- fabrication difficulties, such as cutting slots in laminated cores and other methods of making slotted stator cores;
- high costs involved in manufacturing the laminated stator cores;
- difficulties in assembling the machine and keeping the uniform air gap.

Although, the first PM excitation system was applied to electrical machines as early as the 1830s, the poor quality of hard magnetic materials soon discouraged their use. The invention of Alnico in 1931, barium ferrite in the 1950s and especially the rare-earth neodymium-iron-boron (NdFeB) material (announced in 1983) have made a *comeback* of the PM excitation system possible.

It is generally believed that the availability of high energy PM materials (especially *rare earth* PMs) is the main driving force for exploitation of novel PM machine topologies and has thus revived AFPM machines. Prices of rare-earth PMs have been following a descending curve in the last decade of the 20th century with a sharp decline in the last three years. A recent market survey shows that the NdFeB PMs can now be purchased in the Far East for less than U.S.$ 20 per kilogram. With the availability of more affordable PM materials, AFPM machines may play a more important role in the near future.

1.4 Types of Axial Flux PM Machines

In principle, each type of a radial flux machine should have its corresponding axial flux (disc type) version. In practice, disc type machines are limited to the following three types:

- PM d.c. commutator machines;

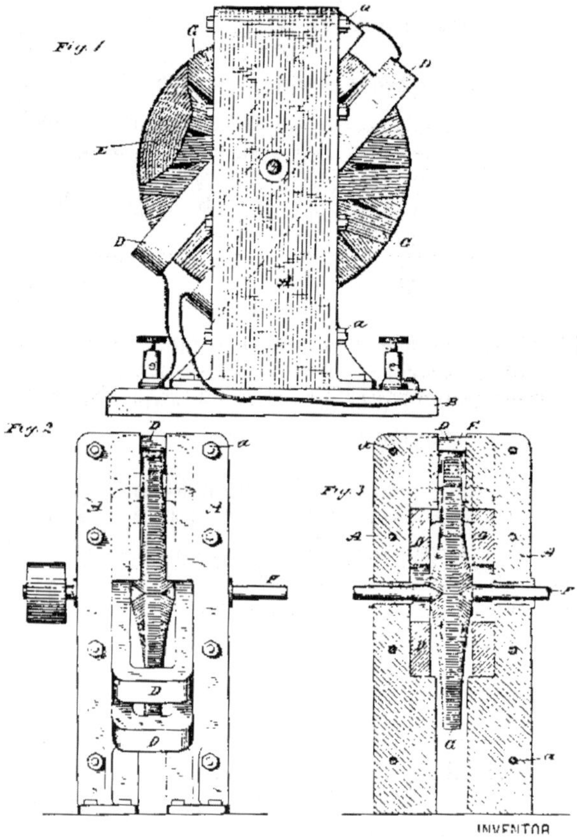

Fig. 1.1. Electro-magnetic motor with disc rotor according to N. Tesla's patent No. 405 858, 1889 [P2].

- PM brushless d.c. and synchronous machines;
- induction machines

Similar to its RFPM counterpart, the AFPM d.c. commutator machine uses PMs to replace the electromagnetic field excitation system. The rotor (armature) can be designed as a *wound rotor* or *printed winding* rotor.

In the wound rotor, the armature winding is made of copper wires and moulded with resin. The commutator is similar to that of the conventional type, i.e. it can be either a cylindrical or radial commutator.

The *disc-type printed armature winding motor* is shown in Fig. 1.2. The rotor (armature) does not have a ferromagnetic core and its winding is similar to the wave winding of conventional d.c. commutator machines. The coils are stamped from pieces of sheet copper and then welded, forming a wave winding.

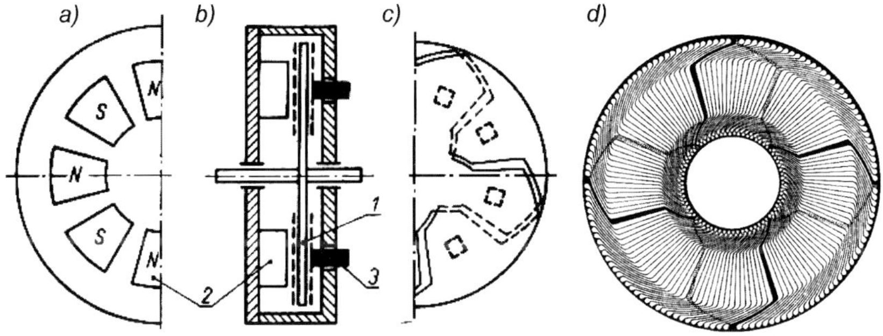

Fig. 1.2. AFMPM 8-pole d.c. commutator motor with printed rotor winding: (a) stator with PMs, (b) cross section, (c) rotor (armature) windings and brushes, (d) construction of $2p = 8$ winding with 145 bars. 1 — rotor with double-sided printed winding, 2 — PMs, 3 — brushes.

When this motor was invented by J. Henry Baudot [18], the armature was made using a similar method to that by which printed circuit boards are fabricated. Hence, this is called the printed winding motor. The magnetic flux of a d.c. printed winding commutator motor with a large air gap can be produced using cost effective Alnico magnets with high remanence.

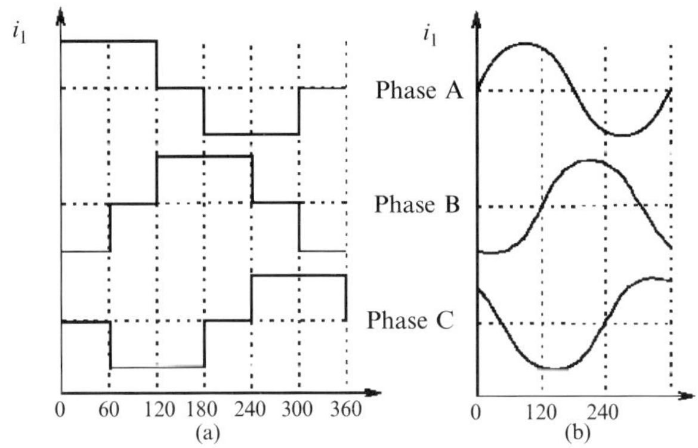

Fig. 1.3. Current waveforms for AFPM brushless machines: (a) square-wave machine, (b) sinewave machine.

AFPM d.c. commutator motors are still a versatile and economical choice for certain industrial, automotive and domestic applications such as fans, blowers, small EVs, power tools, appliances, etc.

Practically, d.c. brushless and a.c. synchronous machines have almost the same structure, though their theory and operation principles are somewhat different [106, 122, 187]. The main difference is in the shape of the operation current waveform (Fig. 1.3), i.e.:

- the d.c. brushless machine generates a trapezoidal EMF waveform and is operated with a rectangular line current waveform (also called a *square-wave* machine);
- the a.c. synchronous machine generates a sinusoidal EMF waveform and is operated with sinewave currents (also called a *sinewave* machine).

It is difficult to manufacture a laminated rotor with cage winding for a disc-type induction machine [160]. If the cage winding is replaced with a non-magnetic high conductivity (Cu or Al) homogenous disc or steel disc coated with copper layer, the performance of the machine drastically deteriorates. Therefore, there is little interest in disc type induction machines so far [160, 267].

1.5 Topologies and Geometries

From a construction point of view, brushless AFPM machines can be designed as single-sided or double-sided, with or without armature slots, with or without armature core, with internal or external PM rotors, with surface mounted or interior PMs and as single stage or multi-stage machines.

In the case of double-sided configurations, either the external stator or external rotor arrangement can be adopted. The first choice has the advantage of using fewer PMs at the expense of poor winding utilisation while the second one is considered as a particularly advantageous machine topology [38]. The diverse topologies of AFPM brushless machines may be classified as follows:

- single-sided AFPM machines
  - with slotted stator (Fig. 1.4a)
  - with slotless stator
  - with salient-pole stator
- double-sided AFPM machines
  - with internal stator (Fig. 1.4b)
    - with slotted stator
    - with slotless stator
      - with iron core stator
      - with coreless stator (Fig. 1.4d)
      - without both rotor and stator cores
    - with salient pole stator (Fig. 1.5)
  - with internal rotor (Fig. 1.4c)
    - with slotted stator
    - with slotless stator
    - with salient pole stator (Fig. 1.6)
- multi-stage (multidisc) AFPM machines (Fig. 1.7)

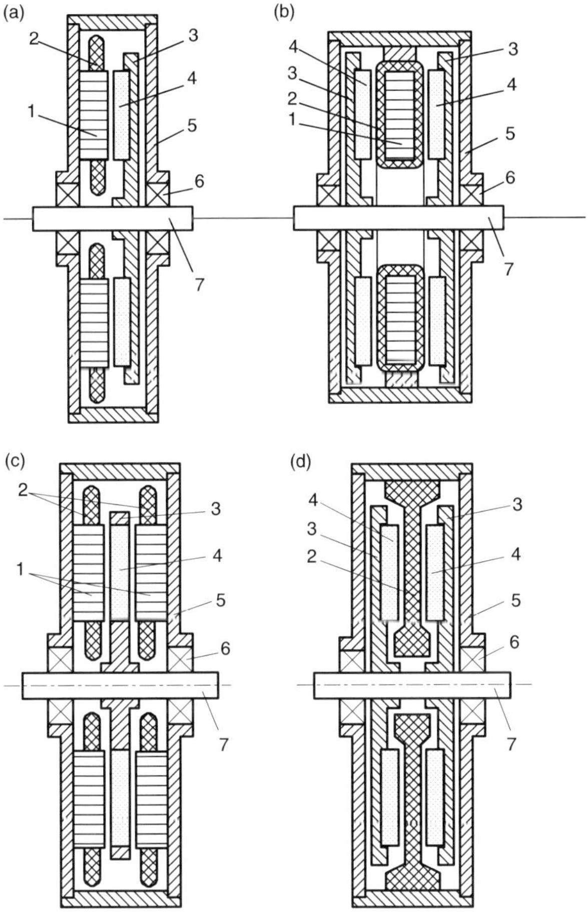

Fig. 1.4. Basic topologies of AFPM machines: (a) single-sided slotted machine, (b) double-sided slotless machines with internal stator and twin PM rotor, (c) double sided machine with slotted stator and internal PM rotor, (d) double-sided coreless motor with internal stator. 1 — stator core, 2 — stator winding, 3 — rotor, 4 — PM, 5 — frame, 6 — bearing, 7 — shaft.

The air gap of the slotted armature AFPM machine is relatively small. The mean magnetic flux density in the air gap decreases under each slot opening due to increase in the reluctance. The change in the mean magnetic flux density caused by slot openings corresponds to a fictitious increase in the air gap [121]. The relation between fictitious g' and physical air gap g is expressed with the aid of Carter coefficient $k_C > 1$, i.e.,

$$g' = gk_C \tag{1.1}$$

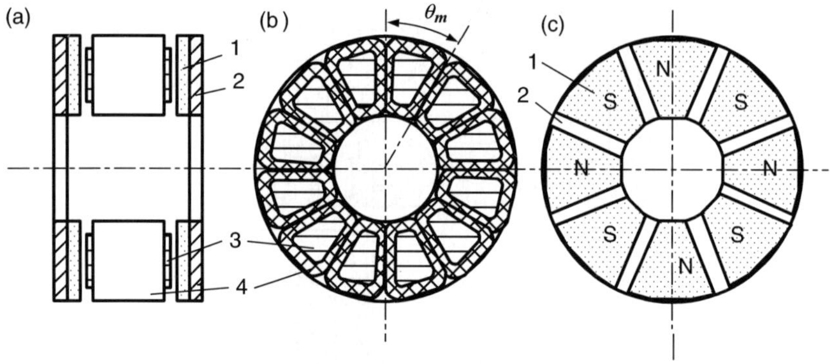

Fig. 1.5. Double-sided AFPM brushless machine with internal salient-pole stator and twin external rotor [185]: (a) construction; (b) stator; (c) rotor. 1 — PM, 2 — rotor backing steel disc, 3 — stator pole, 4 — stator coil.

$$k_C = \frac{t_1}{t_1 - \gamma g} \tag{1.2}$$

$$\gamma = \frac{4}{\pi}\left[\frac{b_{14}}{2g}\arctan\left(\frac{b_{14}}{2g}\right) - \ln\sqrt{1+\left(\frac{b_{14}}{2g}\right)^2}\right] \tag{1.3}$$

where t_1 is the average slot pitch and b_{14} is the width of slot opening.

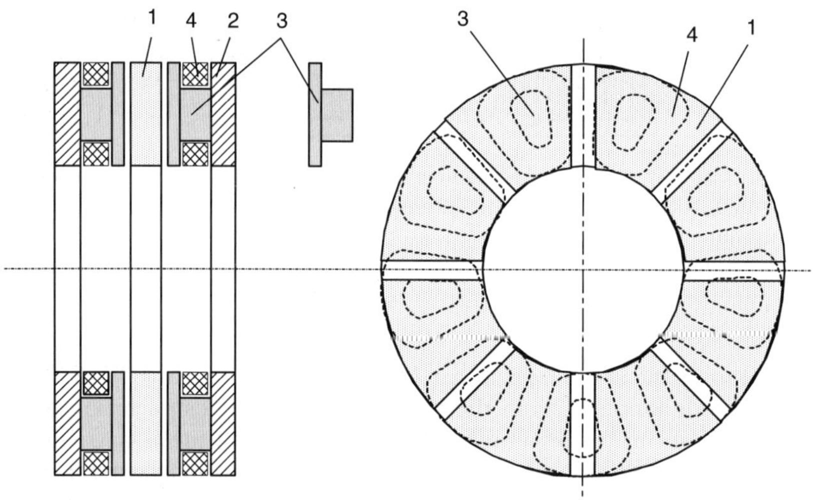

Fig. 1.6. Double-sided AFPM brushless machine with three-phase, 9-coil external salient-pole stator and 8-pole internal rotor. 1 — PM, 2 — stator backing ferromagnetic disc, 3 — stator pole, 4 — stator coil.

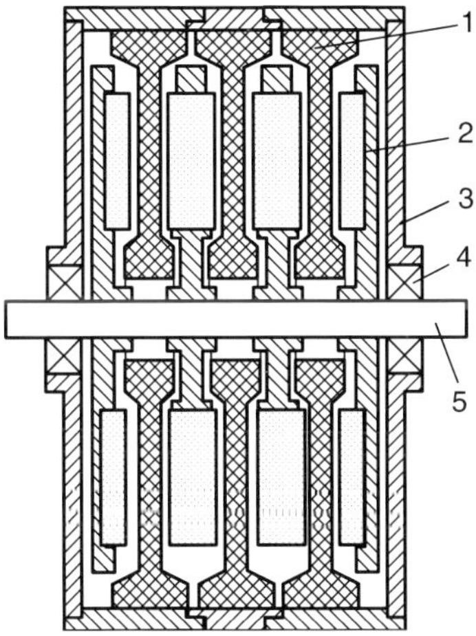

Fig. 1.7. Coreless multidisc AFPM machine with three coreless stators and five PM rotor units: 1 — stator winding, 2 — rotor unit, 4 — PM, 3 — frame, 4 — bearing, 5 — shaft.

For AFPM machines with *slotless windings* the air gap is much larger and equal to the mechanical clearance plus the thickness of all non-magnetic materials (winding, insulation, potting, supporting structure) that is passed by the main magnetic flux. Since there are no slots, Carter coefficient $k_C = 1$. Compared to a conventional slotted winding, the slotless armature winding has advantages such as simple stator assembly, elimination of the cogging torque and reduction of rotor surface losses, magnetic saturation and acoustic noise. The disadvantages include the use of more PM material, lower winding inductances sometimes causing problems for inverter-fed motors and significant eddy current losses in slotless conductors [49].

In the double-sided, salient-pole AFPM brushless machine shown in Fig. 1.5, the stator coils with concentrated parameters are wound on axially laminated poles. To obtain a three-phase self-starting motor, the number of the stator poles should be different from the number of the rotor poles, e.g. 12 stator poles and 8 rotor poles [173, 174, 185]. Fig. 1.6 shows a double-sided AFPM machine with external salient pole stators and internal PM rotor. There are nine stator coils and eight rotor poles for a three-phase AFPM machine.

Depending on the application and operating environment, slotless stators may have ferromagnetic cores or be completely coreless. *Coreless stator* configurations eliminate any ferromagnetic material from the stator (armature) system, thus making the associated eddy current and hysteresis core losses nonexisting. This type of configuration also eliminates axial magnetic attraction forces between the stator and rotor at zero-current state. It is interesting that slotless AFPM machines are often classified according to their winding arrangements and coil shapes, namely, *toroidal*, *trapezoidal* and *rhomboidal* forms [38, 49, 87].

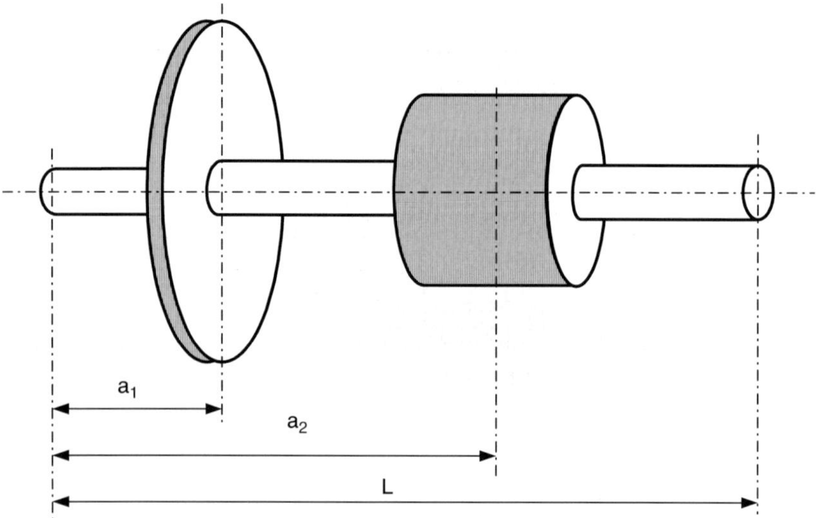

Fig. 1.8. Solid cylindrical shaft loaded with two masses m_1 and m_2.

1.6 Rotor Dynamics

All spinning shafts, even in the absence of external load, deflect during rotation. Fig. 1.8 shows a shaft with two rotating masses m_1 and m_2. The mass m_1 can represent a disc-type rotor of an AFPM machine while the mass m_2 can represent a load. The mass of the shaft is m_{sh}. The combined mass of the rotor, load and shaft can cause deflection of the shaft that will create resonant vibration at certain speed called *whirling* or *critical speed*. The frequency when the shaft reaches its critical speed can be found by calculating the frequency at which transverse vibration occurs.

The critical speed in rev/s of the ith rotating mass can be found as [227]

$$n_i = \frac{1}{2\pi}\sqrt{\frac{g}{\sigma_i}} \tag{1.4}$$

where $g = 9.81$ m/s^2 is the acceleration due to gravity and δ_i is the static deflection at the ith position of the rotor due to ith rotor only, i.e.,

$$\sigma_i = \frac{m_i g a_i^2 (L - a_i)^2}{3 E_i I_i L} \tag{1.5}$$

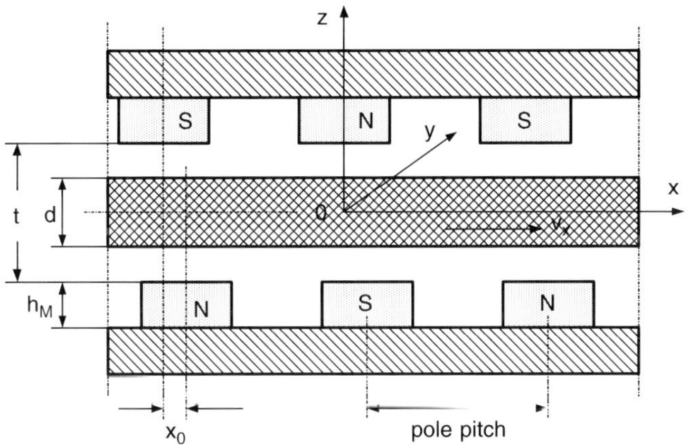

Fig. 1.9. Twin-stator double-sided AFPM machine in Cartesian coordinate system.

In the above equation E_i is the modulus of elasticity (for steel $E = 200 \times 10^9$ Pa), I_i is the area moment of inertia of cross-sectional area, L is the length of the shaft and a_i is the location of the ith rotor from the left end of the shaft (Fig. 1.8). The area moment of inertia can be found as

$$I_i = \frac{\pi D_i^4}{64} \tag{1.6}$$

The resultant angular critical speed $\Omega_{cr} = 2\pi n_{cr}$ for the shaft loaded with i number of rotors ($\Omega_i = 2\pi n_i$) can be found on the basis of Dunkerley's equation [80] as

$$\frac{1}{\Omega_{cr}^2} = \sum_i \frac{1}{\Omega_i^2} \tag{1.7}$$

or Rayleigh's equation [227]

$$\Omega_{cr} = \sqrt{\frac{g \sum_i (m_i \sigma_i)}{\sum_i (m_i \sigma_i^2)}} \tag{1.8}$$

The shaft is also considered a rotor with mass m_{sh} concentrated at $0.5L$ where L is the length of the shaft (bearing–to–bearing).

Dunkerley's empirical method uses the frequencies that each individual load creates when each load acts alone and then combines them to give an approximation for the whole system [80]. Thus eqn (1.7) is an approximation to the *first natural frequency* of vibration of the system which is assumed nearly equal to the critical speed of rotation. Rayleigh's method is based on the fact that the maximum kinetic energy must be equal to maximum potential energy for a conservative system under free vibration [227].

1.7 Axial Magnetic Field Excited by PMs

A double-sided AFPM machine with twin PM rotor in xyz rectangular coordinate system is shown in Fig. 1.9. Assuming that the radius of curvature is higher than the pole pitch and the centre axes of the opposite rotor poles are shifted by a linear distance x_0, the normal component of the magnetic flux density on the surface of the rotor can be described in the stationary xyz coordinate system by the following equations:

- at $z = 0.5d$

$$B_{mz1}(x,t) = B_0 \sum_{\nu=1}^{\infty} b_\nu \cos(\omega_\nu t \mp \beta_\nu x - \frac{\pi}{2}) \tag{1.9}$$

- at $z = -0.5d$

$$B_{mz2}(x,t) = B_0 \sum_{\nu=1}^{\infty} b_\nu \cos\left[\omega_\nu t \mp \beta_\nu (x - x_0) - \frac{\pi}{2}\right] \tag{1.10}$$

where B_0 is the value of the normal component in the center axis of the North pole, and

$$\beta_\nu = \nu \frac{\pi}{\tau} \tag{1.11}$$

$$\omega_\nu = \beta_\nu v = \beta_\nu \pi D n \tag{1.12}$$

$$b_\nu = \frac{4}{\tau}\left[\frac{c_p}{c_p^2+\beta_\nu^2}\sinh\alpha - 6\left(\frac{1}{\beta_\nu}\right)^4 \frac{1}{b_t^3}\cosh\alpha + 3\left(\frac{1}{\beta_\nu}\right)^2 \frac{2}{\tau - b_p}\cosh\alpha\right]$$

$$\times \sin(\nu\frac{\pi}{2})\sin(\nu\frac{\pi b_t}{\tau})$$

$$+\frac{4}{\tau}\left[\frac{\beta_\nu}{c_p^2+\beta_\nu^2} + 6\left(\frac{1}{\beta_\nu}\right)^3 \frac{1}{b_t^2} - \frac{1}{\beta_\nu}\right]\cosh\alpha \sin\left(\nu\frac{\pi}{2}\right)\cos\left(\beta_\nu b_t\right) \tag{1.13}$$

$$b_t = \frac{\tau - b_p}{2} \qquad c_p = 2\frac{\alpha}{b_p}$$

In the above eqns (1.9), (1.10), (1.11), (1.12), and (1.13) $v = v_x$ is the linear speed of the rotor in the x-direction, $n = \pi D/v$ is the rotational speed in rev/s and the parameter α depends on the shape of the distribution of the normal component of the magnetic flux density (Fig. 1.10). For the flat-topped

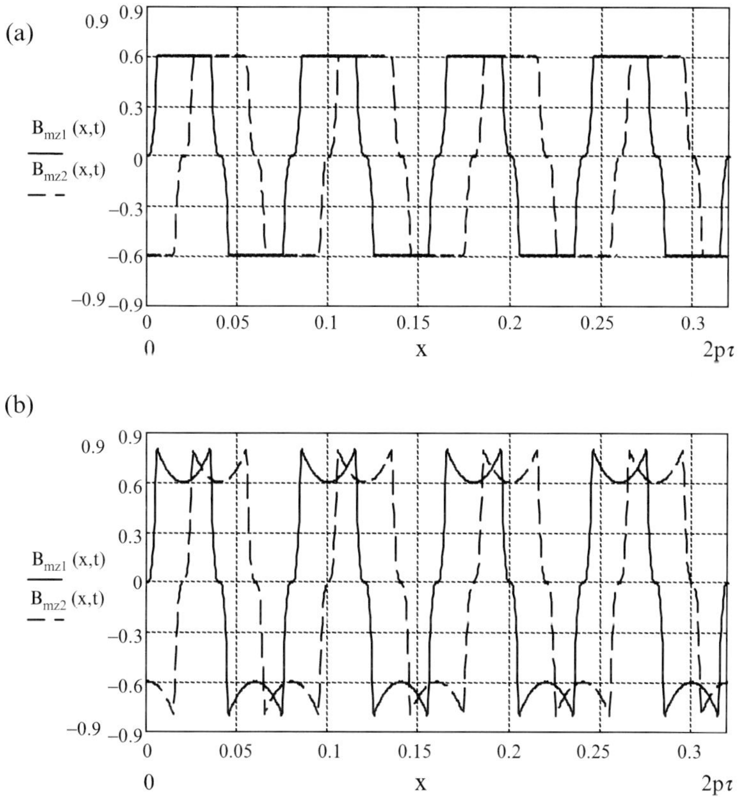

Fig. 1.10. Distribution of the normal components of the magnetic flux density according to eqns (1.9), (1.10) and (1.13) for $\tau = 0.04$ m, $b_p = 0.03$ m, $B_0 = 0.6$ T, $t = 0$, $x_0 = 0.5\tau$: (a) $\alpha = 0$, (b) $\alpha = 0.8$.

curve $\alpha = 0$ and for the concave curve (armature or eddy-current reaction) $0 < \alpha \leq 1$. The coefficient b_ν according to eqn (1.13) has been derived in [103].

The average diameter D of the PM excitation system and corresponding average pole pitch are:

$$D = 0.5(D_{out} + D_{in}) \qquad \tau = \frac{\pi D}{2p} \tag{1.14}$$

where D_{in} is the inner diameter of PMs, D_{out} is the outer diameter of PMs and $2p$ is the number of poles.

The electromagnetic field analysis in AFPM brushless machines has been discussed in e.g., [99, 100, 281, 282].

1.8 PM Eddy-Current Brake as the Simplest AFPM Brushless Machine

A double-sided, PM excited *eddy-current brake* with high conductivity non-magnetic disc-type rotor is one of the simplest brushless AFPM machines (Fig. 1.9). In an eddy-current brake the PM excitation system is stationary and the conductive rotor rotates at the speed n. Eqns (1.9) to (1.14) are valid since the stationary PM excitation system and rotating electric circuit (armature) are equivalent to the rotating PMs and the stationary electric circuit.

It is assumed that the eddy currents in the non-magnetic conductive disc flow only in the radial direction, i.e., in the y-direction (Fig. 1.9). Thus, the magnetic vector potential $\mathbf{A}$ in the disc is described by the following scalar equation (2D analysis):

$$\frac{\partial^2 A_{my\nu}}{\partial x^2} + \frac{\partial^2 A_{my\nu}}{\partial z^2} = \alpha_\nu A_{my\nu} \tag{1.15}$$

where

$$\alpha_\nu = \sqrt{j\omega_\nu \mu_0 \mu_r \sigma} = (1+j)k_\nu \tag{1.16}$$

$$k_\nu = \sqrt{\frac{\omega_\nu \mu_0 \mu_r \sigma}{2}} \tag{1.17}$$

In eqns (1.16) and (1.17) the electric conductivity σ depends on the disc temperature. The relative magnetic permeability of paramagnetic (Al) or diamagnetic (Cu) materials $\mu_r \approx 1$. The angular frequency for higher space harmonics is according to eqn (1.12) or

$$\omega_\nu = 2\pi f_\nu; \qquad f_\nu = \nu f; \qquad \nu = 1, 3, 5, \ldots \tag{1.18}$$

General solution to eqn (1.15) can be written, for example, as

$$A_{my} = \sum_{\nu=1}^{\infty} \sin(\omega_\nu t + \beta_\nu x - \frac{\pi}{2})[A_{1\nu}\exp(-\kappa_\nu z) + A_{1\nu}\exp(\kappa_\nu z)] \tag{1.19}$$

where

$$\kappa_\nu = \sqrt{\alpha_\nu^2 + \beta_\nu^2} = (a_{R\nu} + a_{X\nu})k_\nu \tag{1.20}$$

$$a_{R\nu} = \frac{1}{\sqrt{2}}\sqrt{\sqrt{4 + \left(\frac{\beta_\nu}{k_\nu}\right)^4} + \left(\frac{\beta_\nu}{k_\nu}\right)^2} \tag{1.21}$$

$$a_{X\nu} = \frac{1}{\sqrt{2}}\sqrt{\sqrt{4 + \left(\frac{\beta_\nu}{k_\nu}\right)^4} - \left(\frac{\beta_\nu}{k_\nu}\right)^2} \tag{1.22}$$

Since the currents in the disc flow only in the radial direction y, $E_{mx} = 0$ $E_{mz} = 0$ and $B_{my} = 0$ for $-0.5d \leq z \leq 0.5d$. Using the magnetic vector potential $\nabla \times \mathbf{A} = \mathbf{B}$ and the second Maxwell's equation $\nabla \times \mathbf{E} = -\partial\mathbf{B}/\partial t$, the remaining electric and magnetic components in the disc can be found as

$$E_{my} = -j\omega_\nu A_{my}$$

$$= -\sum_{\nu=1}^{\infty} j\omega_\nu \sin(\omega_\nu t + \beta_\nu x - \frac{\pi}{2})[A_{1\nu}\exp(-\kappa_\nu z) + A_{2\nu}\exp(\kappa_\nu z)] \tag{1.23}$$

$$B_{mx} = -\frac{\partial A_{my}}{\partial z}$$

$$= \sum_{\nu=1}^{\infty} \kappa_\nu \sin(\omega_\nu t + \beta_\nu x - \frac{\pi}{2})[A_{1\nu}\exp(-\kappa_\nu z) - A_{2\nu}\exp(\kappa_\nu z)] \tag{1.24}$$

$$B_{mz} = \frac{\partial A_{my}}{\partial x}$$

$$= \sum_{\nu=1}^{\infty} \beta_\nu \cos(\omega_\nu t + \beta_\nu x - \frac{\pi}{2})(A_{1\nu}\exp(-\kappa_\nu z) + A_{2\nu}\exp(\kappa_\nu z)) \tag{1.25}$$

The integration constants $A_{1\nu}$ and $A_{2\nu}$ can be found on the basis of equality of normal components of the magnetic flux density in the air and in the disc at $z = 0.5d$ and $z = -0.5d$, i.e.

- at $z = 0.5d$

$$\beta_\nu \cos(\omega_\nu t + \beta_\nu x - \frac{\pi}{2})[A_{1\nu}\exp(-\kappa_\nu d/2) + A_{2\nu}\exp(\kappa_\nu d/2)]$$

$$= B_0 b_\nu cos(\omega_\nu t + \beta_\nu x - \frac{\pi}{2})$$

- at $z = -0.5d$

$$\beta_\nu \cos[\omega_\nu t + \beta_\nu(x - x_0) - \frac{\pi}{2}][A_{1\nu}\exp(-\kappa_\nu d/2) + A_{2\nu}\exp(\kappa_\nu d/2)]$$

$$= B_0 b_\nu \cos[\omega_\nu t + \beta_\nu(x - x_0) - \frac{\pi}{2}]$$

There is only backward-rotating magnetic field in the air gap of an eddy-current brake, so that the terms $\beta_\nu x$ and $\beta_\nu(x - x_0)$ in eqns (1.9) and (1.10) are with the + sign. Thus,

$$A_{1\nu} = A_{2\nu} = \frac{1}{\beta_\nu} B_0 b_\nu \frac{\sinh(\kappa_\nu d/2)}{\sinh(\kappa_\nu d)} = \frac{1}{2}\frac{1}{\beta_\nu} B_0 b_\nu \frac{1}{\cosh(\kappa_\nu d/2)} \sinh(\kappa_\nu d) \quad (1.26)$$

because $\sinh(2x) = 2 \sinh x \cosh x$. Putting eqn (1.26) into (1.23), (1.24) and (1.25), the particular solution to eqn (1.15) for E_{my}, B_{mx} and B_{mz} components are

$$E_{my} = -\sum_{\nu=1}^{\infty} j\omega_\nu \frac{B_0 b_\nu}{\beta_\nu} \frac{1}{\cosh(\kappa_\nu d/2)} \sin(\omega_\nu t + \beta_\nu x - \frac{\pi}{2}) \cosh(\kappa_\nu z) \quad (1.27)$$

$$B_{mx} = \sum_{\nu=1}^{\infty} \kappa_\nu \frac{B_0 b_\nu}{\beta_\nu} \frac{1}{\cosh(\kappa_\nu d/2)} \sin(\omega_\nu t + \beta_\nu x - \frac{\pi}{2}) \sinh(\kappa_\nu z) \quad (1.28)$$

$$B_{mz} = \sum_{\nu=1}^{\infty} B_0 b_\nu \frac{1}{\cosh(\kappa_\nu d/2)} \cos(\omega_\nu t + \beta_\nu x - \frac{\pi}{2}) \cosh(\kappa_\nu z) \quad (1.29)$$

The surface wave impedance for the νth space harmonic is calculated on the basis of eqns (1.27) and (1.28)

$$z_\nu = r_\nu + jx_\nu = -\mu_0\mu_r \left[\frac{E_{my}}{B_{mx}}\right]_{z=0.5d} = \frac{\alpha_\nu^2}{\sigma}\frac{1}{\kappa_\nu} \coth(\kappa_\nu \frac{d}{2})$$

$$= [(B_{R\nu}A_{R\nu} - B_{X\nu}A_{X\nu}) + j(B_{X\nu}A_{R\nu} - B_{R\nu}A_{X\nu})]\frac{k_\nu}{\sigma} \quad (1.30)$$

where

$$A_{R\nu} = \frac{\sinh(a_{R\nu}k_\nu d)}{\cosh(a_{R\nu}k_\nu d) - \cos(a_{X\nu}k_\nu d)} \quad (1.31)$$

$$A_{X\nu} = \frac{-\sin(a_{X\nu}k_\nu d)}{\cosh(a_{R\nu}k_\nu d) - \cos(a_{X\nu}k_\nu d)} \quad (1.32)$$

$$B_{R\nu} = \frac{a_{X\nu}}{0.5(a_{R\nu}^2 + a_{X\nu}^2)} \qquad B_{R\nu} = \frac{a_{R\nu}}{0.5(a_{R\nu}^2 + a_{X\nu}^2)} \quad (1.33)$$

The impedance of the disc for the νth space harmonic

$$Z_\nu = z_\nu \frac{0.5(D_{out} - D_{in})}{\tau/\nu} k_{z\nu} = \frac{\alpha_\nu^2}{\sigma}\frac{1}{\kappa_\nu}\frac{0.5(D_{out} - D_{in})}{\tau/\nu} k_{z\nu} \coth(\kappa_\nu \frac{d}{2}) \quad (1.34)$$

where D_{out} is the outer diameter of PMs, D_{in} is the inner diameter of PMs, τ is the average pole pitch (1.14) and $k_{z\nu}$ is the impedance increase factor due to circumferential currents (in the x direction). The impedance increase factor for the νth harmonic is [69]

$$k_{z\nu} = 1 + \frac{1}{\nu^2} \frac{\tau}{0.5(D_{out} - D_{in})} \tag{1.35}$$

1.9 AFPM Machines versus RFPM Machines

In pace with the application of new materials, innovation in manufacturing technology and improvements in cooling techniques, further increase in the power density (output power per mass or volume) of the electrical machine has been made possible. There is an inherent limit to this increase for conventional radial flux PM (RFPM) machines because of [30, 53, 106, 165, 192]:

- the bottle-neck feature for the flux path at the root of the rotor tooth in the case of induction and d.c. commutator machines or brushless machines with external rotors (Fig. 1.11);
- much of the rotor core around the shaft (rotor yoke) is hardly utilised as a magnetic circuit;
- heat from the stator winding is transferred to the stator core and then to the frame — there is poor heat removal through the stator air gap, rotor and shaft without forced cooling arrangements.

These limitations are inherently bound with radial flux structures and cannot be removed easily unless a new topology is adopted. The AFPM machine, recognised as having a higher power density than the RFPM machine, is more compact than its radial flux counterpart [29, 53, 106, 165].

Moreover, since the inner diameter of the core of an AFPM machine is usually much greater than the shaft diameter (see Fig. 1.4), better ventilation and cooling can be expected. In general, the special properties of AFPM machines, which are considered advantageous over RFPM machines in certain applications, can be summarised as follows [52, 106]:

- AFPM machines have much larger diameter to length ratio than RFPM machines;
- AFPM machines have a planar and somewhat adjustable air gap;
- capability of being designed to possess a higher power density with some saving in core material;
- the topology of an AFPM machine is ideal to design a modular machine in which the number of the same modules is adjusted to power or torque requirements;
- the larger the outer diameter of the core, the higher the number of poles that can be accommodated, making the AFPM machines a suitable choice for high frequency or low speed operations.

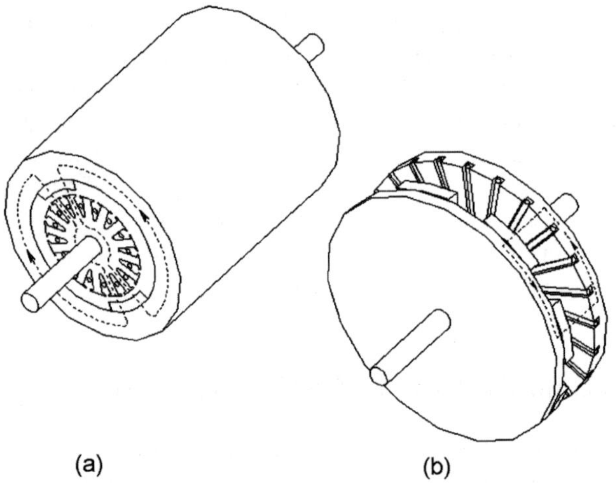

Fig. 1.11. Topologies of (a) RFPM machine (b) AFPM machine.

Consequently, AFPM type machines are particularly suitable for servo, traction, distributed generation and special-purpose applications where their properties offer distinct advantages over their conventional RFPM counterparts.

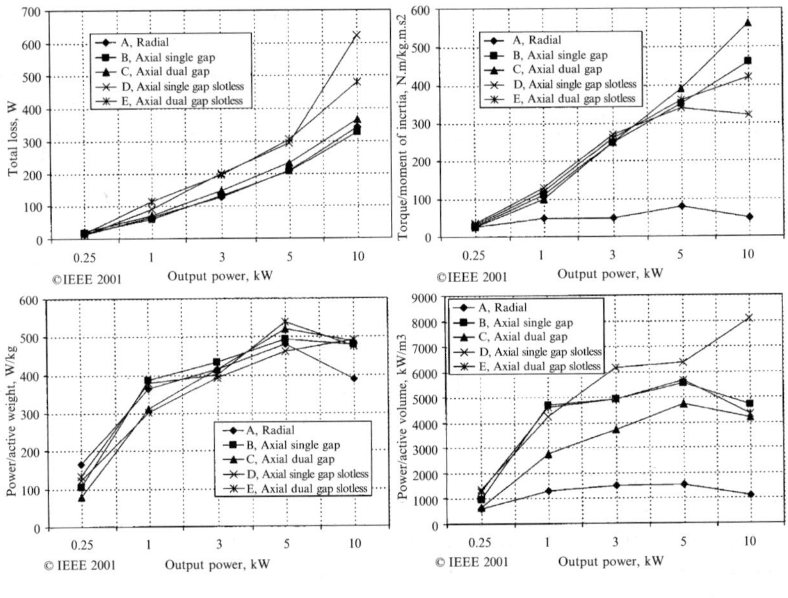

Fig. 1.12. Performance comparison of RFPM and AFPM machines [239].

The quantitative comparison between traditional RFPM machine and AFPM machine is always difficult as it may raise the question of this comparison's fairness. Some published work dealt with quantitative investigations of RFPM and AFPM machine configurations in terms of sizing and power density equations [8, 53, 132, 256, 278]. Fig. 1.12 gives the performance comparison between a conventional RFPM machine and a number of AFPM machines of different configurations at five different power levels [239], which shows that the AFPM machine has a smaller volume and less active material mass for a given power rating than the RFPM machine.

1.10 Power Limitation of AFPM Machines

The power range of AFPM disc-type brushless machines is now from a fraction of a Watt to sub-MW. As the output power of the AFPM machine increases, the contact surface between the rotor and shaft becomes smaller in comparison with the rated power. It is more difficult to design a high mechanical integrity rotor-shaft mechanical joint in the higher range of the output power. A common solution to the improvement of the mechanical integrity of the rotor-shaft joint is to design a multidisc (multi stage) machine (Fig. 1.7).

Since the scaling of the torque capability of the AFPM machine as the cube of the diameter (see eqn (2.101)) while the torque of a RFPM machines scale as the square of the diameter times the length, the benefits associated with axial flux geometries may be lost as the power level or the geometric ratio of the length to diameter of the motor is increased [189]. The transition occurs near the point where the radius equals twice the length of RFPM machine. This may be a limiting design consideration for the power rating of a single-stage disc machine as the power level can always be increased by simply stacking of disc machines on the same shaft and in the same enclosure.

Numerical Example 1.1

Find the critical speed of rotation of the system consisting of a disc-type rotor (steel disc with PMs), steel shaft and driven wheel. Moduli of elasticity, specific mass densities, diameters and widths (lengths) are as follows:

(a) $E_1 = 200 \times 10^9$ Pa, $\rho_1 = 7600$ kg/m^3, $D_1 = 0.3$ m, $w_1 = 0.01$ m for the disc-type rotor;
(b) $E_2 = 200 \times 10^9$ Pa, $\rho_3 = 7650$ kg/m^3, $D_2 = 0.15$, $w_2 = 0.1$ m for the driven wheel;
(c) $E_{sh} = 210 \times 10^9$ Pa, $\rho_{sh} = 7700$ kg/m^3, $D_{sh} = 0.0245$ m, $L = 0.6$ m for the shaft.

The location of the rotor from the left end of the shaft is $a_1 = 0.2$ m and the location of the driven wheel from the same end of the shaft is $a_2 = 0.35$ m (Fig. 1.8). The acceleration of gravity is 9.81 m/s^2.

Solution

Mass of rotor

$$m_1 = \rho_1 \frac{\pi D_1^2}{4} w_1 = 7600 \frac{\pi 0.3^2}{4} \times 0.01 = 5.372 \text{ kg}$$

Mass of driven wheel

$$m_2 = \rho_2 \frac{\pi D_2^2}{4} w_2 = 7650 \frac{\pi 0.15^2}{4} \times 0.1 = 13.519 \text{ kg}$$

Mass of shaft

$$m_{sh} = \rho_{sh} \frac{\pi D_{sh}^2}{4} L = 7700 \frac{\pi 0.0245^2}{4} \times 0.6 = 2.178 \text{ kg}$$

Area moment of inertia of the rotor according to eqn (1.6)

$$I_1 = \frac{\pi 0.3^4}{64} = 3.976 \times 10^{-4} \text{ m}^4$$

Area moment of inertia of the driven wheel to eqn (1.6)

$$I_2 = \frac{\pi 0.15^4}{64} = 2.485 \times 10^{-5} \text{ m}^4$$

Area moment of inertia of the shaft according to eqn (1.6)

$$I_{sh} = \frac{\pi 0.0245^4}{64} = 1.769 \times 10^{-8} \text{ m}^4$$

Static deflection of the shaft at position of rotor due to rotor only as given by eqn (1.5)

$$\sigma_1 = \frac{5.372 \times 9.81 \times 0.2^2 (0.6 - 0.2)^2}{3 \times 200 \times 10^9 \times 3.976 \times 10^{-4} \times 0.6} = 2.356 \times 10^{-9} \text{ m}$$

Static deflection of the shaft at position of driven wheel due to driven wheel only as given by eqn (1.5)

$$\sigma_2 = \frac{13.519 \times 9.81 \times 0.35^2 (0.6 - 0.35)^2}{3 \times 200 \times 10^9 \times 2.485 \times 10^{-5} \times 0.6} = 1.135 \times 10^{-7} \text{ m}$$

Static deflection of shaft due to shaft only as given by eqn (1.5)

$$\sigma_{sh} = \frac{2.178 \times 9.81 \times 0.3^2 (0.6 - 0.3)^2}{3 \times 210 \times 10^9 \times 1.769 \times 10^{-8} \times 0.6} = 2.59 \times 10^{-5} \text{ m}$$

where the mid point of the shaft is $0.5L = 0.5 \times 0.6 = 0.3$ m. Thus, critical speeds according to eqn (1.4) are

- Critical speed of the disc-type rotor

$$n_1 = \frac{1}{2\pi}\sqrt{\frac{9.81}{2.356 \times 10^{-9}}} = 10\ 269.2 \text{ rev/s} = 616\ 150.9 \text{ rpm}$$

- Critical speed of the driven wheel

$$n_2 = \frac{1}{2\pi}\sqrt{\frac{9.81}{1.235 \times 10^{-7}}} = 1479.7 \text{ rev/s} = 88\ 780 \text{ rpm}$$

- Critical speed of the shaft

$$n_{sh} = \frac{1}{2\pi}\sqrt{\frac{9.81}{2.59 \times 10^{-5}}} = 97.97 \text{ rev/s} = 5878.4 \text{ rpm}$$

Critical angular speeds for the disc-type rotor is $\Omega_1 = 2\pi 10\ 269.3 = 64\ 523.2$ rad/s, for the driven wheel is $\Omega_2 = 2\pi 1479.7 = 9297$ rad/s and for the shaft is $\Omega_{sh} = 2\pi 97.97 = 615.6$ rad/s. According to Dunkerley equation

$$x = \frac{1}{\Omega_1^2} + \frac{1}{\Omega_2^2} + \frac{1}{\Omega_{sh}} = \frac{1}{64\ 523.2} + \frac{1}{9297} + \frac{1}{615.6} = 2.651 \times 10^{-6} \text{ s}^2/\text{rad}^2$$

Critical angular speed of the system as given by eqn (1.7)

$$\Omega_{cr} = \frac{1}{\sqrt{x}} = \frac{1}{\sqrt{2.651 \times 10^{-6}}} = 614.21 \text{ rad/s}$$

Critical speed of rotation of the system according to Dunkerley equation

$$n_{cr} = \frac{\Omega_{cr}}{2\pi} = \frac{614.21}{2\pi} = 97.75 \text{ rev/s} = 5865.3 \text{ rpm}$$

Critical speed of rotation of the system according to Rayleigh's method - eqn (1.8)

$$n_{cr} =$$

$$\frac{1}{2\pi}\sqrt{\frac{9.81(5.372 \times 2.356 \times 10^{-9} + 13.519 \times 1.135 \times 10^{-7} + 2.178 \times 2.59 \times 10^{-5})}{5.372 \times (2.356 \times 10^{-9})^2 + 13.519 \times (1.135 \times 10^{-7})^2 + 2.178 \times (2.59 \times 10^{-5})^2}}$$

$$= 99.3 \text{ rev/s} = 5958.15 \text{ rpm}$$

The results according to Dunkerley and Rayleigh are similar.

Numerical Example 1.2

A copper disc with its dimensions as shown in Fig. 1.13 rotates with the speed of 12 000 rpm between U-shaped laminated pole shoes of a PM. A sliding contact consisting of two brushes is used to collect the electric current generated by this primitive homopolar generator: one brush slides on the external diameter $D_{out} = 0.232$ m and the second brush is located directly below one of the poles at the distance of $0.5D_{in} = 0.03$ m from the axis of the disc. The remanent magnetic flux density of the NdFeB PM is $B_r = 1.25$ T, coercivity is $H_c = 950\,000$ A/m and height $2h_M = 0.016$ m. The thickness of the disc is $d = 0.005$ m, one-sided air gap is $g = 0.0015$ m and the width of the pole is $b_p = 20$ mm. The relative magnetic permeability of the laminated core is $\mu_r = 1000$ and the conductivity of the disc is $\sigma = 57 \times 10^6$ S/m at 20°C. The length of the flux path in the laminated core is $l_{Fe} = 0.328$ m. Find:

(a) the magnetic flux density in the air gap;
(b) the EMF between brushes;
(c) the current, if the line resistance is $R_l = 0.001\ \Omega$ and load resistance is $R_L = 0.02\ \Omega$.

The magnetic flux fringing effect in the air gap, the variation of magnetic permeability with the magnetic field intensity in the laminated core and brush voltage drop are neglected.

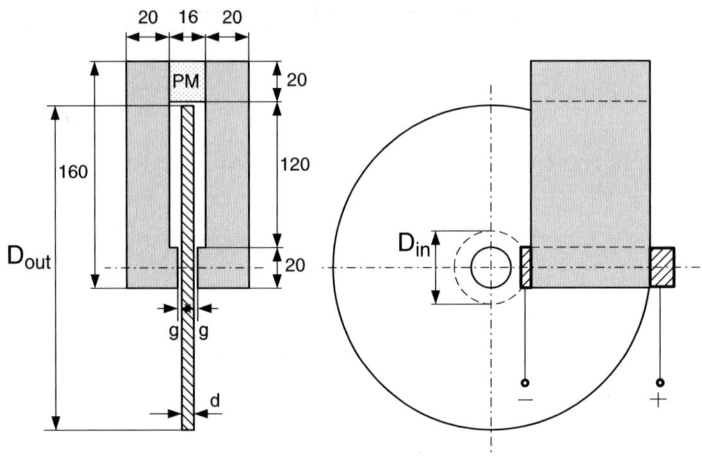

Fig. 1.13. Faraday's disc: a nonmagnetic conductive disc rotating in a stationary magnetic field according to *Numerical example 1.2*.

Solution

This is a homopolar type d.c. generator known as Faraday's disc and can be used as a current source, e.g. for electrolysis.

(a) Magnetic flux density in the air gap

The relative recoil magnetic permeability

$$\mu_{rrec} = \frac{1}{\mu_0}\frac{\Delta B}{\Delta H} = \frac{1}{0.4\pi \times 10^{-6}}\frac{1.25 - 0}{950000 - 0} = 1.047$$

The magnetic flux density in the air gap and saturation factor of the magnetic circuit can be found on the basis of Kirchhoff's magnetic voltage law, i.e.

$$\frac{B_r}{\mu_0\mu_{rrec}}2h_M = \frac{B_g}{\mu_0\mu_{rrec}}2h_M + \frac{B_g}{\mu_0}2g + H_{Fe}l_{Fe}$$

$$\frac{B_r}{\mu_0\mu_{rrec}}2h_M = \frac{B_g}{\mu_0\mu_{rrec}}2h_M + \frac{B_g}{\mu_0}2gk_{sat}$$

where the saturation factor of the magnetic circuit

$$k_{sat} = 1 + \frac{l_{Fe}}{2\mu_r(g + 0.5d)} = 1 + \frac{l_{Fe}}{2 \times 1000(0.0015 + 0.5 \times 0.005)} = 1.042$$

Thus, the air gap magnetic flux density is

$$B_g = \frac{B_r}{1 + (g + 0.5d)k_{sat}\mu_{rrec}/h_M}$$

$$= \frac{1.25}{1 + (0.0015 + 0.5 \times 0.005)1.042 \times 1000/0.008} = 0.809 \text{ T}$$

(b) EMF between brushes

Since $\mathbf{E}/l = \mathbf{v} \times \mathbf{B_g}$ or $dE = B_g v dr$ where $v = v_x = 2\pi r n$ is the linear speed, r is the radius of the disc and n is the rotational speed in rev/s, $dE = B_g(2\pi r n)dr$. Thus

$$E = \int_{0.5D_{in}}^{0.5D_{out}} B_g(2\pi r n)dr = \frac{\pi n B_g}{4}\left(D_{out}^2 - D_{in}^2\right)$$

$$= \frac{\pi 12000}{60}\frac{0.809}{4}\left(0.232^2 - 0.06^2\right) = 6.38 \text{ V}$$

(c) Current, if the line resistance is $R_l = 0.001\ \Omega$ and load resistance is $R_L = 0.02\ \Omega$.

Neglecting the current fringing effect in the disc, the resistance of the disc for induced current at 20°C is

$$R_d \approx \frac{0.5(D_{out} - D_{in})}{\sigma d b_p} = \frac{0.5(0.232 - 0.06)}{57 \times 10^6 \times 0.005 \times 0.02} = 0.0000151\ \Omega$$

The current is

$$I = \frac{E}{R_d + R_l + R_L} = \frac{6.38}{0.0000151 + 0.001 + 0.02} = 303.6\ \text{A}$$

The terminal voltage is $V = I \times R_L = 303.6 \times 0.02 = 6.1$ V and the line voltage drop is $\Delta V = I \times R_l = 303.6 \times 0.001 = 0.304$ V.

Numerical Example 1.3

Find the impedance of the aluminum disc of a double-sided eddy current brake at ambient temperature 20°C. The inner diameter of the PMs is $D_{in} = 0.14$ m, the outer diameter of the PMs is $D_{out} = 0.242$, thickness of the disc $d = 3$ mm, number of poles $2p = 16$ and speed $n = 3000$ rpm. Assume the conductivity of the aluminum $\sigma = 30 \times 10^6$ S/m at 20°C and its relative magnetic permeability $\mu_r \approx 1$.

Solution

The average diameter of the disc

$$D = 0.5(D_{in} + D_{out}) = 0.5(0.14 + 0.232) = 0.191\ \text{m}$$

The average pole pitch

$$\tau = \frac{\pi D}{2p} = \frac{\pi 0.191}{16} = 0.037\ \text{m}$$

The frequency and angular frequency of the current in the disc for fundamental space harmonic $\nu = 1$

$$f = pn = 8\frac{3000}{60} = 400\ \text{Hz} \qquad \omega = 2\pi f = 2\pi 400 = 2513.3\ 1/\text{s}$$

The attenuation factor for the fundamental space harmonic $\nu = 1$ of the electromagnetic field in the disc according to eqn (1.17) is

$$k = \sqrt{\frac{2513.3 \times 0.4\pi 10^{-6} \times 1.0 \times 30 \times 10^6}{2}} = 217.66\ 1/\text{m}$$

The coefficients according to eqns (1.11), (1.21), (1.22), (1.31), (1.32) and (1.33) for $\nu = 1$, are respectively

$$\beta = \frac{\pi}{0.037} = 84.865\ 1/\mathrm{m}$$

$$a_R = \frac{1}{\sqrt{2}}\sqrt{\sqrt{4 + \left(\frac{84.865}{217.66}\right)^4} + \left(\frac{84.865}{217.66}\right)^2} = 1.038$$

$$a_X = \frac{1}{\sqrt{2}}\sqrt{\sqrt{4 + \left(\frac{84.865}{217.66}\right)^4} - \left(\frac{84.865}{217.66}\right)^2} = 0.964$$

$$A_R = \frac{\sinh(1.038 \times 217.66 \times 0.003)}{\cosh(1.038 \times 217.66 \times 0.003) - \cos(0.964 \times 217.66 \times 0.003)} = 1.699$$

$$A_X = \frac{-\sin(0.964 \times 217.66 \times 0.003)}{\cosh(1.038 \times 217.66 \times 0.003) - \cos(0.964 \times 217.66 \times 0.003)} = -1.369$$

$$B_R = \frac{0.964}{0.5(1.038^2 + 0.964^2)} = 0.961$$

$$B_X = \frac{1.038}{0.5(1.038^2 + 0.964^2)} = 1.035$$

The surface wave impedance for $\nu = 1$ according to eqn (1.30)

$$r = [0.961 \times 1.699 - 1.035 \times (-1.369)]\frac{217.66}{57 \times 10^6} = 2.2123 \times 10^{-5}\ \Omega$$

$$x = [1.035 \times 1.699 + 0.961 \times (-1.369)]\frac{217.66}{57 \times 10^6} = 3.216 \times 10^{-6}\ \Omega$$

$$z = 2.2123 \times 10^{-5} + j3.216 \times 10^{-6}\ \Omega$$

The impedance increase factor due to circumferential currents (in the x-direction) according to eqn (1.35) is

$$k_{z\nu} = 1 + \frac{1}{1^2}\frac{0.037}{0.5(0.242 - 0.14)} = 1.733$$

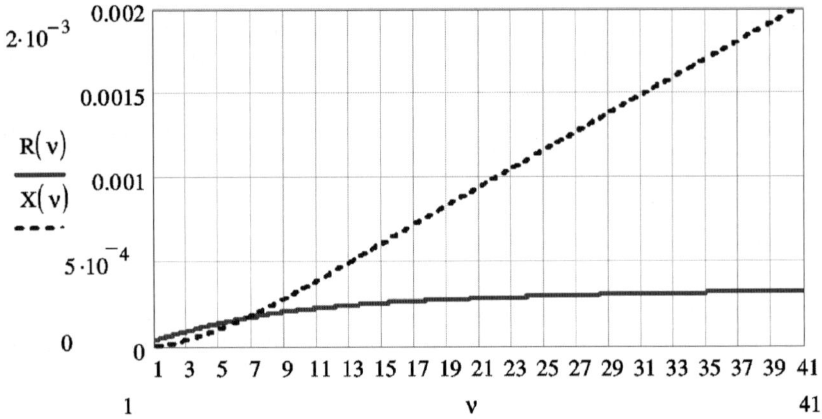

Fig. 1.14. Resistance and reactance of the aluminum disc as a function of higher space harmonics ν according to *Numerical example 1.3*.

The disc impedance per pole for the fundamental space harmonic $\nu = 1$ according to eqn (1.34)

$$Z = (2.2123 \times 10^{-5} + j3.216 \times 10^{-6})\frac{0.242 - 0.14}{0.037}1.366$$

$$= 5.231 \times 10^{-5} + j7.605 \times 10^{-6}\ \Omega$$

$$R = 5.231 \times 10^{-5}\ \Omega \qquad X = 7.605 \times 10^{-6}\ \Omega$$

The disc resistance R_ν and reactance X_ν for higher space harmonics $\nu > 1$ is plotted in Fig. 1.14.

Numerical Example 1.4

Estimate the cost of materials of a 3-phase, 250 kW, 2300 rpm brushless AFPM machine. The internal stator does not have any ferromagnetic core. The twin-type external rotor (Fig. 1.4d) consists of sintered NdFeB PMs (40 poles) glued to two solid steel discs. The mass of the stator copper conductors is $m_{Cu} = 18.7$ kg, mass of the stator winding insulation (including encapsulation) is $m_{ins} = 1.87$ kg, mass of PMs is $m_{PM} = 32.1$ kg, mass of the twin rotor core is $m_{rc} = 50.8$ kg (two discs) and mass of the shaft is $m_{sh} = 48.5$ kg. The cost of materials in U.S. dollars per kilogram is: copper conductor $c_{Cu} = 10.50$, insulating materials $c_{ins} = 8.50$, sintered NdFeB magnets $c_{PM} = 24.00$ and carbon steel $c_{steel} = 1.40$. The cost of frame, end discs (bells) and bearings is $C_f = \$164.00$. The cost of components that are independent of the machine geometry (encoder, terminal leads, terminal board, nameplate) is $C_0 = \$140.00$.

Coefficients taking into account the manufacturing and utilisation of steel for the shaft and rotor core (steel discs) are as follows:

- total volume of the steel bar to the volume of the shaft, $k_{ush} = 1.94$
- coefficient taking into account the cost of machining of the shaft, $k_{msh} = 2.15$
- total volume of the steel plate to the volume of the rotor discs, $k_{ur} = 2.1$
- coefficient taking into account the cost of machining of the rotor steel discs, $k_{mr} = 1.8$

Solution

The cost of the copper winding

$$C_w = m_{Cu} c_{Cu} = 18.7 \times 10.50 = \$196.35$$

The cost of the insulation and encapsulation

$$C_{ins} = m_{ins} c_{ins} = 1.87 \times 8.50 = \$15.90$$

The cost of PMs

$$C_{PM} = m_{PM} c_{PM} = 32.1 \times 24.00 = \$770.40$$

The cost of the rotor core (two steel discs)

$$C_{rc} = k_{ur} k_{mr} m_{rc} c_{steel} = 2.1 \times 1.8 \times 50.8 \times 1.40 = \$268.83$$

The cost of the shaft

$$C_{sh} = k_{ush} k_{msh} m_{sh} c_{steel} = 1.94 \times 2.15 \times 48.5 \times 1.40 = \$283.21$$

Total cost of materials

$$C = C_f + C_0 + C_w + C_{ins} + C_{PM} + C_{rc} + C_{sh}$$

$$= 164.00 + 140.00 + 196.35 + 15.90 + 770.40 + 268.93 + 283.21 = \$1838.69$$

2

Principles of AFPM Machines

In this chapter the basic principles of the AFPM machine are explained in details. Considerable attention is given to the magnetic circuits, windings, torque production, losses, equivalent circuits, sizing procedure, armature reaction and performance characteristics of AFPM machines.

2.1 Magnetic Circuits

2.1.1 Single-Sided Machines

The *single-sided construction* of an axial flux machine is simpler than the double-sided one, but the torque production capacity is lower. Fig. 2.1 shows typical constructions of single-sided AFPM brushless machines with surface PM rotors and laminated stators wound from electromechanical steel strips. The single-sided motor according to Fig. 2.1a has a standard frame and shaft. It can be used in industrial, traction and servo electromechanical drives. The motor for hoist applications shown in Fig. 2.1b is integrated with a sheave (drum for ropes) and brake (not shown). It is used in gearless elevators [113].

2.1.2 Double-Sided Machines With Internal PM Disc Rotor

In the *double-sided machine with internal PM disc rotor*, the armature winding is located on two stator cores. The disc with PMs rotates between two stators. An eight-pole configuration is shown in Fig. 2.2. PMs are embedded or glued in a nonmagnetic rotor skeleton. The nonmagnetic air gap is large, i.e. the total air gap is equal to two mechanical clearances plus the thickness of a PM with its relative magnetic permeability close to unity. A double-sided machine with parallel connected stators can operate even if one stator winding is broken. On the other hand, a *series connection is preferred* because it can provide equal but opposing axial attractive forces.

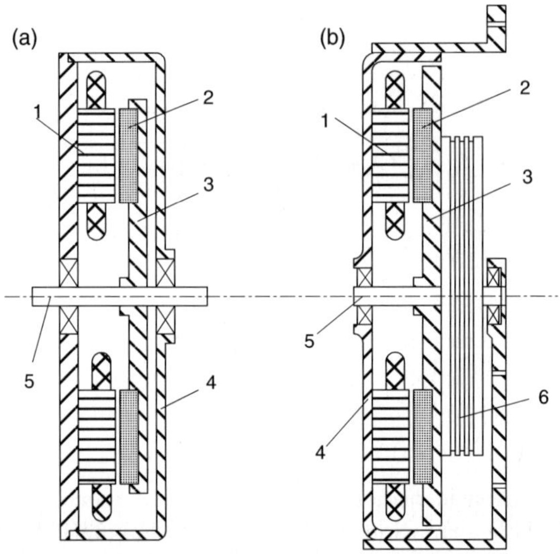

Fig. 2.1. Single sided disc type machines: (a) for industrial and traction electromechanical drives, (b) for hoist applications. 1 — laminated stator, 2 — PM, 3 — rotor, 4 — frame, 5 — shaft, 6 — sheave.

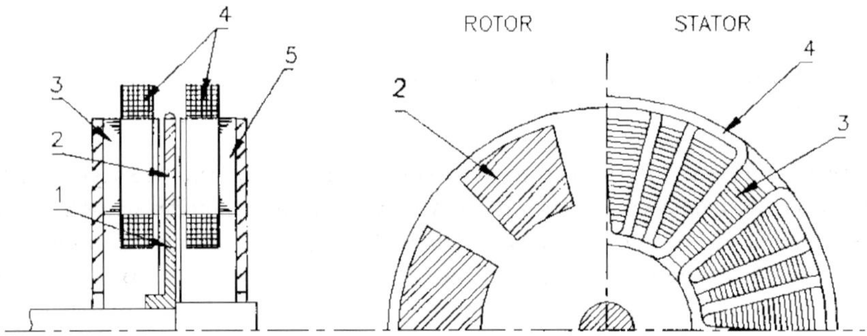

Fig. 2.2. Configuration of double-sided AFPM brushless machine with internal disc rotor: 1 — rotor, 2 — PM, 3 — stator core, 4 — stator winding.

A practical three-phase, 200 Hz, 3000 rpm, double-sided AFPM brushless motor with built-in brake is shown in Fig. 2.3 [157]. The three-phase winding is Y-connected, while phase windings of the two stators are in series. This motor is used as a flange-mounted servo motor. The ratio $X_{sd}/X_{sq} \approx 1.0$ so the motor can be analysed as a cylindrical non-salient rotor synchronous machine [128, 155, 157].

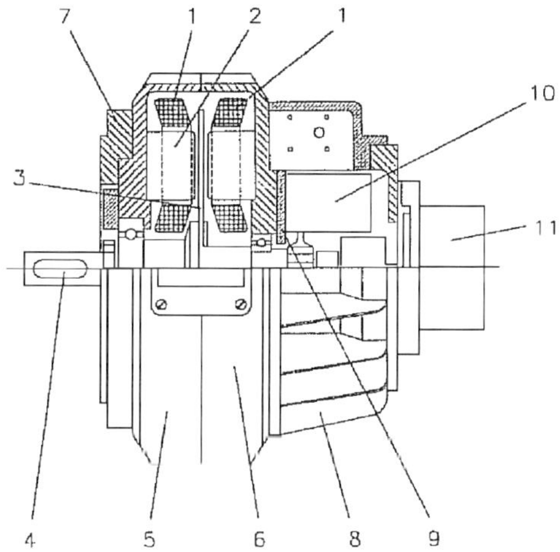

Fig. 2.3. Double-sided AFPM brushless servo motor with built-in brake and encoder: 1 — stator winding, 2 — stator core, 3 — disc rotor with PMs, 4 — shaft, 5 — left frame, 6 — right frame, 7 — flange, 8 — brake shield, 9 — brake flange, 10 — electromagnetic brake, 11 — encoder or resolver. Courtesy of *Slovak University of Technology STU*, Bratislava and *Electrical Research and Testing Institute*, Nová Dubnica, Slovakia.

2.1.3 Double-Sided Machines With Internal Ring-Shaped Core Stator

A *double-sided machine* with *internal ring-shaped stator core* has a poly-phase slotless armature winding (toroidal type) wound on the surface of the stator ferromagnetic core [101, 173, 244, 276]. In this machine the *ring-shaped stator core* is formed either from a continuous steel tape or sintered powders. The total air gap is equal to the thickness of the stator winding with insulation, mechanical clearance and the thickness of the PM in the axial direction. The double sided rotor simply called *twin rotor* with PMs is located at two sides of the stator. The configurations with internal and external rotors are shown in Fig. 2.4. The three phase winding arrangement, magnet polarities and flux paths in the magnetic circuit are shown in Fig. 2.5.

The AFPM machines designed as shown in Fig. 2.4a can be used as a propulsion motor or combustion engine synchronous generator. The machine with external rotor, as shown in Fig. 2.4b, has been designed for hoist applications. A similar machine can be designed as an electric car wheel propulsion motor. Additional magnets on cylindrical parts of the rotor are sometimes added [173] or U-shaped magnets can be designed. Such magnets embrace the armature winding from three sides and only the internal portion of the winding does not produce any electromagnetic torque.

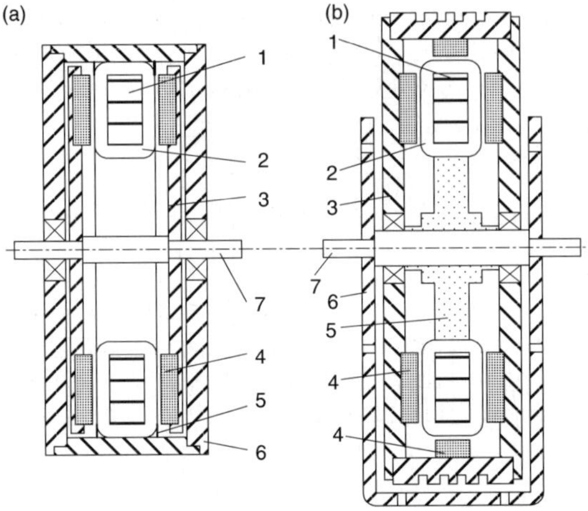

Fig. 2.4. Double-sided machines with one slotless stator: (a) internal rotor, (b) external rotor. 1 — stator core, 2 — stator winding, 3 — steel rotor, 4 — PMs, 5 — resin, 6 — frame, 7 — shaft.

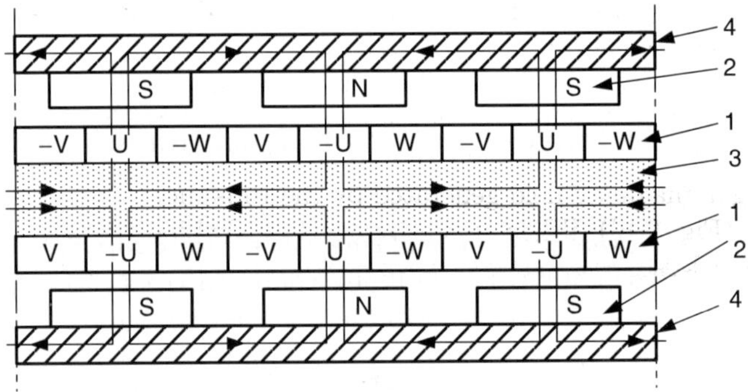

Fig. 2.5. Three phase winding, PM polarities and magnetic flux paths of a double-sided disc machine with one internal slotless stator. 1 — winding, 2 — PM, 3 — stator yoke, 4 — rotor yoke.

Owing to the large air gap the maximum flux density does not exceed 0.65 T. To produce this flux density a large volume of PMs is usually required. As the permeance component of the flux ripple associated with the slots is eliminated, the cogging torque is practically absent. The magnetic circuit is unsaturated (slotless stator core). On the other hand, the machine structure lacks the necessary robustness [244]. Both buried magnet and surface magnet rotors can be used.

There are a number of applications for medium and large power axial flux motors with external PM rotors, especially in electrical vehicles [101,

276]. Disc-type motors with external rotors have a particular advantage in low speed high torque applications, such as buses and shuttles, due to their large radius for torque production. For small electric cars the electric motor mounted directly into the wheel is recommended [101].

2.1.4 Double-Sided Machines With Internal Slotted Stator

The ring-type stator core can also be made with slots (Fig. 2.6). For this type of motor, slots are progressively notched into the steel tape as it is passed from one mandrel to another and the polyphase winding is inserted [276]. In the case of the slotted stator the air gap is small ($g \leq 1$ mm) and the air gap magnetic flux density can increase to 0.85 T [101]. The magnet volume is less than 50% that of the previous design, shown in Figs 2.4 and 2.5.

2.1.5 Double-Sided Machines With Internal Coreless Stator

AFPM machines with *coreless stators* have the stator winding wound on a non-magnetic and non-conductive supporting structure or mould. The stator core losses, i.e. hysteresis and eddy current losses do not exist. The losses in PMs and rotor solid steel disc are negligible. This type of design offers higher efficiency at zero cogging torque. In order to maintain a reasonable level of flux density in the air gap, a much larger volume of PMs in comparison with laminated stator core AFPM machine is required. The stator winding is placed in the air gap magnetic field generated by the PMs mounted on two opposing rotor discs (Fig. 1.4d). When operating at relatively high frequency, significant eddy current losses in the stator winding conductors may occur [263].

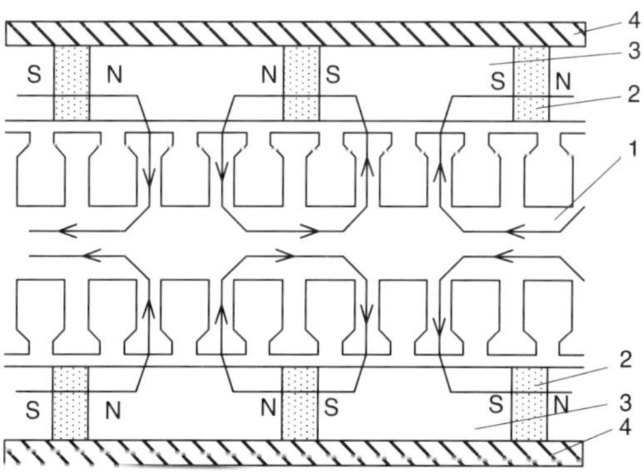

Fig. 2.6. Double-sided machine with one internal slotted stator and buried PMs. 1 — stator core with slots, 2 — PM, 3 — mild steel core (pole), 4 — nonmagnetic rotor disc.

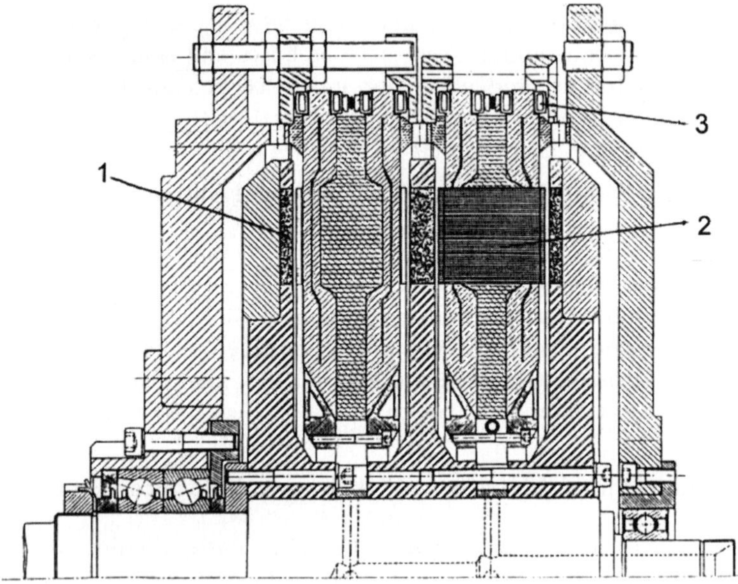

Fig. 2.7. Double stator, triple rotor AFPM brushless motor with water cooling system [59]. 1 — PM, 2 — stator core, 3 — stator winding.

2.1.6 Multidisc Machines

There is a limit on the increase of motor torque that can be achieved by enlarging the motor diameter. Factors limiting the single disc design are:

(a) axial force taken by bearings;
(b) integrity of mechanical joint between the disc and shaft;
(c) disc stiffness.

A more reasonable solution for large torques are *double* or *triple disc motors*.

There are several configurations of *multidisc motors* [5, 6, 7, 59, 81, 82]. Large multidisc motors rated at least 300-kW have a water cooling system (Fig. 2.7) with radiators around the winding end connections [59]. To minimize the winding losses due to skin effect, variable cross section conductors may be used so that the cross section of conductors is bigger in the slot area than in the end connection region. Using a variable cross section means a gain of 40% in the rated power [59]. However, variable cross section of conductors means an increased cost of manufacturing. Owing to high mechanical stresses a titanium alloy is recommended for disc rotors. Multidisc AFPM brushless machines are also proposed for applications such as light-weight high speed generators [P66, 222], propeller-driven stratospheric aircraft [81], and brakes for landing gears of aicrafts [P146, P148].

2.2 Windings

2.2.1 Three-Phase Windings Distributed in Slots

In a single-layer winding, only one coil side is located in a single slot. The number of all coils is $s_1/2$ and the number of coils per phase is $n_c = s_1/(2m_1)$ where s_1 is the number of stator slots and m_1 is the number of phases. In a double-layer winding two sides of different coils are accommodated in each slot. The number of all coils is s_1 and the number of coils per phase is $n_c = s_1/m_1$. The number of slots per pole is

$$Q_1 = \frac{s_1}{2p} \tag{2.1}$$

where $2p$ is the number of poles. The number of slots per pole per phase is

$$q_1 = \frac{s_1}{2pm_1} \tag{2.2}$$

The number of conductors per coil can be calculated as

- for a single-layer winding

$$N_c = \frac{a_p a_w N_1}{n_c} = \frac{a_p a_w N_1}{s_1/(2m_1)} = \frac{a_p a_w N_1}{pq_1} \tag{2.3}$$

- for a double-layer winding

$$N_c = \frac{a_p a_w N_1}{n_c} = \frac{a_p a_w N_1}{s_1/m_1} = \frac{a_p a_w N_1}{2pq_1} \tag{2.4}$$

where N_1 is the number of turns in series per phase, a_p is the number of parallel current paths and a_w is the number of parallel conductors. The number of conductors per slot is the same for both single-layer and double-layer windings, i.e.

$$N_{sl} = \frac{a_p a_w N_1}{pq_1} \tag{2.5}$$

The full coil pitch measured in terms of the number of slots is $y_1 = Q_1$ where Q_1 is according to eqn (2.1). The short coil pitch can be expressed as

$$y_1 = \frac{w_c(r)}{\tau(r)} Q_1 \tag{2.6}$$

where $w_c(r)$ is the coil pitch measured in units of length at a given radius r and $\tau(r)$ is the pole pitch measured at the same radius. The *coil pitch–to–pole pitch* ratio is independent on the radius, i.e.

$$\beta = \frac{w_c(r)}{\tau(r)} \tag{2.7}$$

The *distribution factor* of a polyphase winding for the fundamental space harmonic $\nu = 1$ is defined as the ratio of the *phasor sum—to—arithmetic sum of EMFs induced in each coil* and expressed by the following equation:

$$k_{d1} = \frac{\sin(\pi/2m_1)}{q_1 \sin[\pi/(2m_1 q_1)]} \tag{2.8}$$

The *pitch factor* for the fundamental space harmonic $\nu = 1$ is defined as the ratio of the *phasor sum—to—arithmetic sum of the EMFs per coil side* and expressed as:

$$k_{p1} = \sin\left(\beta\frac{\pi}{2}\right) \tag{2.9}$$

The *winding factor* for fundamental is the product of the distribution factor (eqn (2.8)) times pitch factor (eqn (2.9)), i.e.

$$k_{w1} = k_{d1} k_{p1} \tag{2.10}$$

The angle in electrical degrees between neighbouring slots is

$$\gamma = \frac{360^o}{s_1} p \tag{2.11}$$

Fig. 2.8 shows a single layer winding distributed in $s_1 = 36$ slots of a three-phase, $2p = 6$ AFPM machine.

2.2.2 Toroidal Winding

Toroidal stator windings are used in twin-rotor double-sided AFPM machines (Fig. 2.4). The toroidal stator winding of a three-phase, six-pole AFPM machine with twin external rotor is shown in Fig. 2.9. Each phase of the winding has an equal number of coils connected in opposition so as to cancel the possible flux circulation in the stator core. Those coils are evenly distributed along the stator core diametrically opposing each other so the only possible number of poles are $2, 6, 10, \ldots$ etc. The advantages of the toroidal stator winding [50, 245] are short end connection, simple stator core and easy design of any number of phases.

2.2.3 Coreless Stator Winding

Coreless stator windings are used in twin-rotor double-sided AFPM machines (Fig. 1.4d). For the ease of construction, the stator winding normally consists of a number of single layer trapezoidally shaped coils. The assembly of the stator is made possible by bending the ends of the coils by a certain angle, so that the active conductors lie evenly in the same plane and the end windings

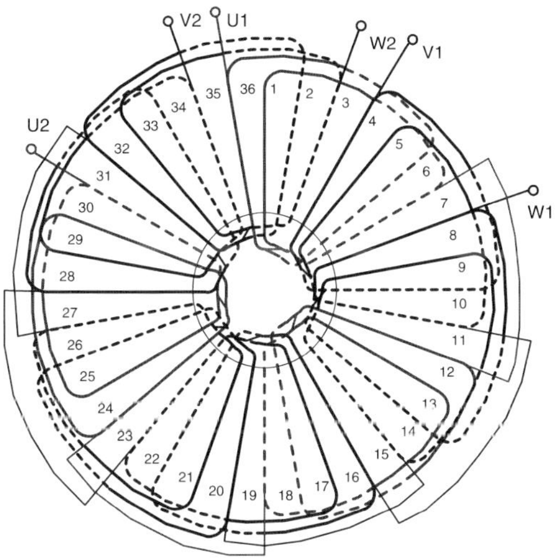

Fig. 2.8. Single-layer winding of an AFPM machine with $m_1 = 3$, $2p = 6$, $s_1 = 36$, $y_1 = Q_1 = 6$ and $q_1 = 2$.

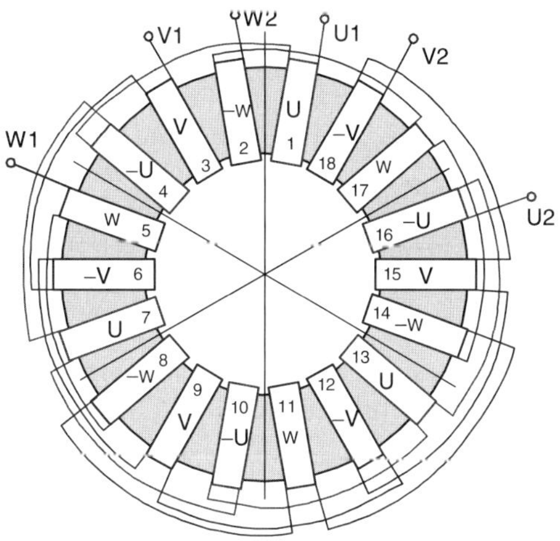

Fig. 2.9. Toroidal winding of a three-phase, six-pole, 18-coil AFPM machine with twin external rotor.

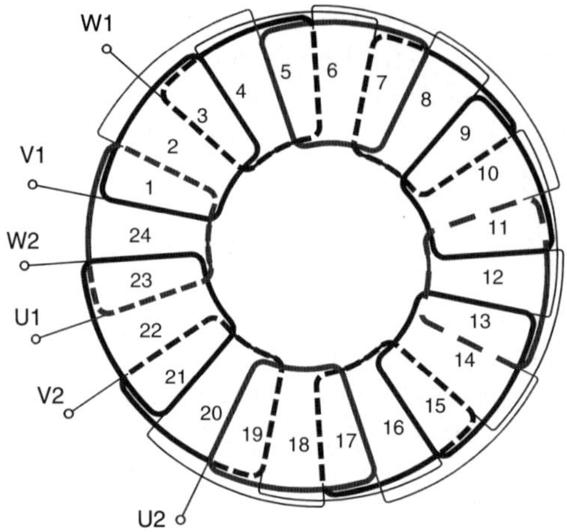

Fig. 2.10. Coreless winding of a three-phase, eight-pole AFPM machine with twin external rotor.

nest closely together. The windings are held together in position by using a composite material of epoxy resin and hardener. Fig. 2.10 shows the coreless stator winding of a three-phase, eight-pole AFPM machine. Obviously, the relations used in the slotted stator windings can be directly used for coreless trapezoidal stator winding with the exception that the term "slot" is replaced by the "coil side". Another coil profile that has been used in coreless stator AFPM machines is the *rhomboidal* coil. It has shorter end connections than the trapezoidal coils. The inclined arrangement of the coil's active sides makes it possible to place water cooling ducts inside the stator. The main drawback of rhomboidal coils is the reduction of the torque.

2.2.4 Non-Overlap (Concentrated Coil) Windings

The term *non-overlap windings* defined here refers to the windings of which the coils do not overlap. As with overlap windings non-overlap windings can be of single or double layer, concentrated or distributed, integral or fractional and air-cored or iron-cored. AFPM machines with non-overlap windings can be of single-sided or double-sided construction (Figs. 1.5, 1.6, 2.12).

In double-layer, non-overlap slotted iron-cored windings, two coils are sharing a slot (two coil layers per slot), which means that all teeth are wound. Consequently these windings are sometimes called *tooth windings*. In air-cored, double-layer non-overlap windings, the stator coils lie side-by-side against each other (see e.g. Fig. 1.5b). Note that the term *slot* can also be visualized when applied to air-cored windings.

In single-layer, non-overlap slotted iron-cored windings, only one coil layer is in a slot and therefore every second tooth is wound (*alternate teeth wound*);

air-cored, single-layer non-overlap windings have their coils displaced from each other as shown e.g. in Fig. 2.11. In double-layer, non-overlap windings, the number of stator coils is always equal to the number of stator slots, i.e. $Q_c = n_c m_1 = s_1$, where n_c is the number coils per phase. In single-layer non-overlap windings the number of stator coils is equal to half the number of stator slots i.e. $Q_c = s_1/2$.

In concentrated, non-overlap windings, there is only one coil in a coil group or coil phase belt ($z = 1$), but in distributed non-overlap windings two or more coils are distributed and connected in series to form a coil group ($z = 2, 3, \ldots$). In integral non-overlap windings, the coils in a coil group are distributed evenly, but in fractional non-overlap windings they are not.

As with overlap windings, a non-overlap winding can also be divided into a number of sections F that repeats itself (negative or positive periodicity) in the machine every unit or section of poles. Such a pole section, thus, consists of a set of phase coils that completes an m_1-phase stator coil layout. The number of pole sections F can be determined by taking the greatest common divisor (GCD) of the number of poles and the number of coils, i.e.

$$F = \mathrm{GCD}(2p, Q_c) \tag{2.12}$$

If the number of poles per pole section, $2p/F$, is uneven, then negative periodic boundary conditions can be applied in finite element analysis. If $2p/F$ is even, then positive boundary conditions must be applied. With F known from (2.12), the number of coils in a coil group can be determined by

$$z = \frac{Q_c}{m_1 F} = \frac{n_c}{F} \tag{2.13}$$

There are a set of rules for the pole and slot combinations that make non-overlap windings valid. These are:

- the number of poles must be even;
- the number of slots must be a multiple of the number of phases and must be even in the case of single layer windings;
- the number of coils and slots are equal in double layer windings; in single layer windings the number of coils is equal to half the number of slots;
- the number of coils in a coil group (z) must be an integer;
- the number of slots can not be equal to the number of poles.

Electromagnetic analysis of AFPM machines with non-overlap stator windings are the same as those with overlap windings provided that the winding factors are calculated according to the following equations: the distribution factor for the fundamental space harmonic is calculated by [240]

$$k_{d1} = \frac{\sin(\pi/2m_1)}{z\sin[(\pi/(2m_1 z)]} \tag{2.14}$$

For single and double layer slotted iron-cored non-overlap windings, the pitch factor for the fundamental space harmonic is calculated by using eqn (2.9) or else

$$k_{p1} = \sin(\frac{\pi}{2}\frac{2p}{s_1}) = \sin(\theta_m/2) \tag{2.15}$$

where θ_m is the slot pitch angle (also corresponding to the coil span angle in the case of double layer windings shown in Fig. 1.5b or corresponding to the coil pitch angle in the case of single layer windings as shown in Fig. 2.11) given by

$$\theta_m = \frac{2\pi p}{s_1} \tag{2.16}$$

For air-cored, non-overlap windings, the pitch factor for the fundamental space harmonic is derived in [146]. For *single layer* non-overlap windings the pitch factor is given by

$$k_{p1} = \sin(\theta_m/2)\frac{\sin(\theta_{re}/2)}{\theta_{re}/2} \tag{2.17}$$

while for *double layer*, non-overlap windings, the pitch factor is given by

$$k_{p1} = \sin[(\theta_m - \theta_{re})/2]\frac{\sin(\theta_{re}/2)}{\theta_{re}/2} \tag{2.18}$$

where θ_{re} is the electrical angle corresponding to the coil-layer width. For example, for a double layer, air-cored non-overlap winding with $m_1 = 3$, $p = 12$, $s_1 = Q_c = 18$ (thus $n_c = 6$) and $\theta_{re} = 0.1\,\theta_m$, the number of winding sections is $F = 6$, the number of coils distributed in a coil group is $z = 1$, the angular coil span is 1.333π, the distribution factor is $k_{d1} = 1$ and the pitch factor is $k_{p1} = \sin[(1.333 - 0.1333)\pi/2]sin(0.1333\pi/2)/(0.1333\pi/2) = 0.944$. Fig. 1.5b shows a double layer, three-phase, 12-coil non-overlap stator winding of a double-sided AFPM machine with $2p = 8$ rotor poles. Figures 1.6 and 2.11 show a single layer, 9-coil ($Q_c = 9$, $s_1 = 2Q_c = 18$) non-overlap winding, while Fig. 2.12 shows a double layer 9-coil ($s_1 = Q_c = 9$) non-overlap stator winding. The difference in the number of the stator coils and rotor poles provides the starting torque for motors and the reduction of torque pulsations.

2.3 Torque Production

Since the dimensions of AFPM machines are functions of the radius, the electromagnetic torque is produced over a continuum of radii, not just at a constant radius as in cylindrical machines.

The *pole pitch* $\tau(r)$ and *pole width* $b_p(r)$ of an axial flux machine are functions of the radius r, i.e.

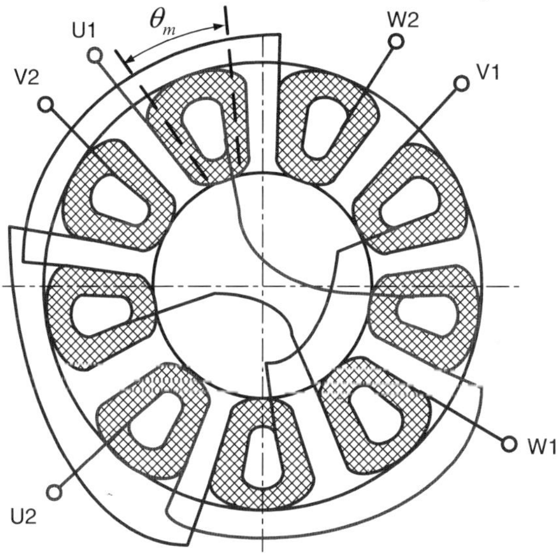

Fig. 2.11. Connection diagram of a three-phase, nine-coil winding of an AFPM brushless machine.

Fig. 2.12. Three-phase, nine-coil stator of a single-sided AFPM brushless machine with salient-pole stator. Photo-courtesy of *Mii Technologies*, Lebanon, NH, U.S.A.

$$\tau(r) = \frac{2\pi r}{2p} = \frac{\pi r}{p} \tag{2.19}$$

$$b_p(r) = \alpha_i \tau(r) = \alpha_i \frac{\pi r}{p} \tag{2.20}$$

where α_i is the ratio of the average B_{avg} to peak value B_{mg} of the magnetic flux density in the air gap, i.e.

$$\alpha_i = \frac{B_{avg}}{B_{mg}} \qquad or \qquad \alpha_i = \frac{b_p(r)}{\tau(r)} \tag{2.21}$$

Both the pole pitch $\tau(r)$ and pole width $b_p(r)$ are functions of the radius r, the parameter α_i is normally independent of the radius.

The *line current density* is also a function of the radius r. Thus the peak value of the line current density is

$$A_m(r) = \frac{m_1\sqrt{2}N_1 I_a}{p\tau(r)} = \frac{m_1\sqrt{2}N_1 I_a}{\pi r} \tag{2.22}$$

The *tangential force* acting on the disc can be calculated on the basis of Ampere's equation

$$\mathbf{dF}_x = I_a(\mathbf{dr} \times \mathbf{B}_g) = A(r)(\mathbf{dS} \times \mathbf{B}_g) \tag{2.23}$$

where $I_a\mathbf{dr} = A(r)\mathbf{dS}$, $A(r) = A_m(r)/\sqrt{2}$ according to eqn (2.22), $\mathbf{dr}$ is the radius element, $\mathbf{dS}$ is the surface element and $\mathbf{B_g}$ is the vector of the normal component (perpendicular to the disc surface) of the magnetic flux density in the air gap. An AFPM disc-type machine provides $\mathbf{B}_g$ practically independent of the radius r.

Assuming the magnetic flux density in the air gap B_{mg} is independent of the radius r, $dS = 2\pi r dr$ and $B_{avg} = \alpha_i B_{mg}$ according to eqn (2.21), the electromagnetic torque on the basis of eqn (2.23) is

$$dT_d = rdF_x = r[k_{w1}A(r)B_{avg}dS] = 2\pi\alpha_i k_{w1}A(r)B_{mg}r^2 dr \tag{2.24}$$

The line current density $A(r)$ is the electric loading per one stator active surface in the case of a typical stator winding distributed in slots (double-sided stator and internal rotor) or electric loading of the whole stator in the case of an internal toroidal type or coreless stator.

2.4 Magnetic Flux

For sinusoidal distribution of the magnetic flux density waveform excited by PMs, the *average magnetic flux density* is

$$B_{avg} = \frac{1}{\pi/p - 0}\int_0^{\pi/p} B_{mg}\sin(p\alpha)d\alpha = -\frac{p}{\pi}B_{mg}\left[\frac{1}{p}\cos(p\alpha)\right]_0^{\pi/p}$$

$$= -\frac{1}{\pi}B_{mg}[\cos\pi - \cos 0] = \frac{2}{\pi}B_{mg} \tag{2.25}$$

Since the surface element per pole is $2\pi r dr/(2p)$, the magnetic flux excited by PMs per pole for a nonsinusoidal magnetic flux density waveform $B_{avg} = \alpha_i B_{mg}$ is

$$\Phi_f = \int_{R_{in}}^{R_{out}} \alpha_i B_{mg}\frac{2\pi}{2p} r dr = \alpha_i B_{mg}\frac{\pi}{p}\left[\frac{r^2}{2}\right]_{R_{in}}^{R_{out}}$$

$$= \alpha_i B_{mg} \frac{\pi}{2p} (R_{out}^2 - R_{in}^2) \tag{2.26}$$

where B_{mg} is the *peak value of the magnetic flux density* in the air gap, p is the number of pole pairs, $R_{out} = 0.5D_{out}$ is the outer radius of the PMs and $R_{in} = 0.5D_{in}$ is the inner radius of the PMs.

It is convenient to use the *inner–to–outer PM radius* or *inner–to–outer PM diameter* ratio, i.e.

$$k_d = \frac{R_{in}}{R_{out}} = \frac{D_{in}}{D_{out}} \tag{2.27}$$

Thus,

$$\Phi_f = \alpha_i B_{mg} \frac{\pi}{2p} [(0.5D_{out})^2 - (0.5D_{in})^2] = \alpha_i B_{mg} \frac{\pi}{8p} D_{out}^2 (1 - k_d^2) \tag{2.28}$$

The same equation for a cylindrical type machine is [106]

$$\Phi_f = \frac{2}{\pi} \tau L_i B_{mg} \tag{2.29}$$

where τ is the pole pitch and L_i is the effective length of the stack.

The *permeance of the air gap* in the d-axis at the radius r is given by $dG_g = \mu_0 \frac{1}{g'} \alpha_i \frac{\pi r}{p} dr$ or

$$G_g = \frac{\mu_0}{g'} \alpha_i \frac{\pi}{p} \int_{R_{in}}^{R_{out}} r\,dr = \frac{\mu_0}{g'} \alpha_i \frac{\pi}{p} \left[\frac{r^2}{2} \right]_{R_{in}}^{R_{out}} = \mu_0 \frac{1}{g'} \alpha_i \frac{\pi}{2p} (R_{out}^2 - R_{in}^2)$$

$$= \lambda_g \alpha_i \frac{\pi}{2p} (R_{out}^2 - R_{in}^2) \tag{2.30}$$

where $\lambda_g = \mu_0 / g'$ is the permeance per unit surface and g' is the equivalent air gap.

2.5 Electromagnetic Torque and EMF

The *average electromagnetic torque* developed by AFPM motor according to eqns (2.22) and (2.24) is

$$dT_d = 2\alpha_i m_1 I_a N_1 k_{w1} B_{mg} r dr$$

If the above equation is integrated from $D_{out}/2$ to $D_{in}/2$ with respect to r, the average electromagnetic torque may be written as

$$T_d = \frac{1}{4}\alpha_i m_1 I_a N_1 k_{w1} B_{mg}(D_{out}^2 - D_{in}^2)$$
$$= \frac{1}{4}\alpha_i m_1 N_1 k_{w1} B_{mg} D_{out}^2 (1 - k_d^2) I_a \tag{2.31}$$

where k_d is according to eqn (2.27). Putting eqn (2.28) into eqn (2.31) the average torque is

$$T_d = 2\frac{p}{\pi} m_1 N_1 k_{w1} \Phi_f I_a \tag{2.32}$$

To obtain the *rms* torque for sinusoidal current and sinusoidal magnetic flux density, eqn (2.32) should be multiplied by the coefficient $\pi\sqrt{2}/4 \approx 1.11$, i.e.

$$T_d = \frac{m_1}{\sqrt{2}} p N_1 k_{w1} \Phi_f I_a = k_T I_a \tag{2.33}$$

where the *torque constant*

$$k_T = \frac{m_1}{\sqrt{2}} p N_1 k_{w1} \Phi_f \tag{2.34}$$

In some publications [92, 244] the electromagnetic force on the rotor is simply calculated as the product of the magnetic and electric loading $B_{avg}A$ and active surface of PMs $S = \pi(R_{out}^2 - R_{in}^2)$, i.e. $F_x = \pi B_{avg} A(R_{out}^2 - R_{in}^2)$, where A is the *rms* line current density at the inner radius R_{in}. For a double-sided AFPM machine $S = 2\pi(R_{out}^2 - R_{in}^2)$. Thus, the average electromagnetic torque of a double-sided AFPM machine is

$$T_d = F_x R_{in} = 2\pi B_{avg} A(R_{out}^2 - R_{in}^2) R_{in} = 2\pi B_{avg} A R_{out}^3 (k_d - k_d^3) \tag{2.35}$$

Taking the first derivative of the electromagnetic torque T_d with respect to k_d and equating it to zero, the maximum torque is for $k_d = 1/\sqrt{3}$. Industrial practice shows that the maximum torque is for $k_d \neq 1/\sqrt{3}$.

The *EMF at no load* can be found by differentiating the first harmonic of the magnetic flux waveform $\Phi_{f1} = \Phi_f \sin\omega t$ and multiplying by $N_1 k_{w1}$, i.e.

$$e_f = N_1 k_{w1} \frac{d\Phi_{f1}}{dt} = 2\pi f N_1 k_{w1} \Phi_f \cos\omega t$$

The magnetic flux Φ_f is expressed by eqns (2.26) and (2.28). The *rms* value is obtained by dividing the peak value $2\pi f N_1 k_{w1} \Phi_f$ of the EMF by $\sqrt{2}$, i.e.

$$E_f = \pi\sqrt{2} f N_1 k_{w1} \Phi_f = \pi\sqrt{2} p N_1 k_{w1} \Phi_f n_s = k_E n_s \tag{2.36}$$

where the *EMF constant* (armature constant) is

$$k_E = \pi\sqrt{2} p N_1 k_{w1} \Phi_f \tag{2.37}$$

The same form of eqn (2.36) can be obtained on the basis of the developed torque $T_d = m_1 E_f I_a/(2\pi n_s)$ in which T_d is according to eqn (2.33). For the toroidal type winding the winding factor $k_{w1} = 1$.

2.6 Losses and Efficiency

2.6.1 Stator Winding Losses

The *stator (armature) winding resistance* per phase for the d.c. current is

$$R_{1dc} = \frac{N_1 l_{1av}}{a_p a_w \sigma s_a} \tag{2.38}$$

where N_1 is the number of armature turns per phase, l_{1av} is the average length of turn, a_p is the number of parallel current paths, a_w is the number of parallel conductors, σ_1 is the electric conductivity of the armature conductor at given temperature (for a copper conductor $\sigma_1 \approx 57 \times 10^6$ S/m at 20°C and $\sigma_1 \approx 47 \times 10^6$ S/m at 75°C), and s_a is the conductor cross section. The average length of the armature turn is

$$l_{1av} = 2L_i + l_{1in} + l_{1out} \tag{2.39}$$

in which l_{1in} is the length of the *inner end connection* and l_{1out} is the length of the *outer end connection*.

For a.c. current and stator winding distributed in slots (ferromagnetic core) the stator winding resistance should be divided into the resistance of bars (radial portions of conductors) R_{1b} and resistance of the end connections R_{1e}, i.e.

$$R_1 = R_{1b} + R_{1e} = \frac{N_1}{a\sigma s_a}(2L_i k_{1R} + l_{1in} + l_{1out}) \approx k_{1R} R_{1dc} \tag{2.40}$$

where k_{1R} is the skin-effect coefficient for the stator (armature) resistance. For a double-layer winding and $w_c(r) = \tau(r)$ or $\beta = 1$ [161]:

$$k_{1R} = \varphi_1(\xi_1) + \left[\frac{m_{sl}^2 - 1}{3} - \left(\frac{m_{sl}}{2} \sin\frac{\gamma}{2}\right)^2\right] \Psi_1(\xi_1) \tag{2.41}$$

where

$$\varphi_1(\xi_1) = \xi_1 \frac{\sinh 2\xi_1 + \sin 2\xi_1}{\cosh 2\xi_1 - \cos 2\xi_1} \tag{2.42}$$

$$\Psi_1(\xi_1) = 2\xi_1 \frac{\sinh \xi_1 - \sin \xi_1}{\cosh \xi_1 + \cos \xi_1} \tag{2.43}$$

$$\xi_1 = h_c \sqrt{\pi f \mu_o \sigma_1 \frac{b_{1con}}{b_{11}}} \tag{2.44}$$

and m_{sl} is the number of conductors per slot arranged above each other in two layers (this must be an even number), γ is the phase angle between the currents of the two layers, f is the input frequency, b_{1con} is the width of all the conductors in a slot, b_{11} is the slot width and h_c is the height of a conductor in the slot. If there are n_{sl} conductors side by side at the same height of the slot, they are taken as a single conductor carrying n_{sl}-times greater current.

In general, for a three-phase winding $\gamma = 60^{\circ}$ and

$$k_{1R} = \varphi_1(\xi_1) + \left(\frac{m_{sl}^2 - 1}{3} - \frac{m_{sl}^2}{16} \right) \Psi_1(\xi_1) \tag{2.45}$$

For a chorded winding ($w_c(r) < \tau$ (r)) and $\gamma = 60^{\circ}$

$$k_{1R} \approx \varphi_1(\xi_1) + \left[\frac{m_{sl}^2 - 1}{3} - \frac{3(1 - w_c/\tau)}{16} m_{sl}^2 \right] \Psi_1(\xi_1) \tag{2.46}$$

The skin-effect coefficient k_{1R} for hollow conductors is given, for example, in [161].

(a) If $m_{sl} = 1$ and $\gamma = 0$, the skin-effect coefficient $k_{1R} = \varphi_1(\xi_1)$ (the same as for a cage winding).
(b) If $\gamma = 0$, the currents in all conductors are equal and

$$k_{1R} \approx \varphi_1(\xi_1) + \frac{m_{sl}^2 - 1}{3} \Psi_1(\xi_1) \tag{2.47}$$

For small motors with round armature conductors fed from power frequencies of 50 or 60 Hz,

$$R_1 \approx R_{1dc} \tag{2.48}$$

The *armature winding losses* are

$$\Delta P_{1w} = m_1 I_a^2 R_1 \approx m_1 I_a^2 R_{1dc} k_{1R} \tag{2.49}$$

Since the skin effect is only in this part of the conductor which is located in the slot — eqn (2.40), the armature winding losses should be multiplied by the coefficient

$$\frac{k_{1R} + l_{1in}/(2L_i) + l_{1out}/(2L_i)}{1 + l_{1in}/(2L_i) + l_{1out}/(2L_i)} \tag{2.50}$$

rather than by k_{1R}.

2.6.2 Stator Core Losses

The magnetic flux in the stator (armature) core is nonsinusoidal. The rotor PM excitation system produces a trapezoidal shape of the magnetic flux density waveform. The stator windings are fed from switched d.c. sources with PWM or square wave control. The applied voltage thus contains many harmonics which are seen in the stator flux.

The *eddy current losses* can be calculated using the following classical formula

$$\Delta P_{eFe} = \frac{\pi^2}{6}\frac{\sigma_{Fe}}{\rho_{Fe}} f^2 d_{Fe}^2 m_{Fe} \sum_{n=1}^{\infty} n^2 [B_{mxn}^2 + B_{mzn}^2]$$

$$= \frac{\pi^2}{6}\frac{\sigma_{Fe}}{\rho_{Fe}} f^2 d_{Fe}^2 m_{Fe} [B_{mx1}^2 + B_{mz1}^2]\eta_d^2 \quad (2.51)$$

where σ_{Fe}, d_{Fe}, ρ_{Fe} and m_{Fe} are the electric conductivity, thickness, specific density and mass of laminations respectively, n are the odd time harmonics, B_{mxn} and B_{mzn} are the harmonic components of the magnetic flux density in the x (tangential) and z (normal) directions and

$$\eta_d = \sqrt{1 + \frac{(3B_{mx3})^2 + (3B_{mz3})^2}{B_{mx1}^2 + B_{mz1}^2} + \frac{(5B_{mx5})^2 + (5B_{mz5})^2}{B_{mx1}^2 + B_{mz1}^2} + \ldots} \quad (2.52)$$

is the *coefficient of distortion of the magnetic flux density*. For $\eta_d = 1$, eqn (2.51) expresses the eddy current losses under sinusoidal magnetic flux density.

In a similar way, the *hysteresis losses* can be expressed with the aid of Richter's formula, i.e.

$$\Delta P_{hFe} = \epsilon \frac{f}{100} m_{Fe} \sum_{n=1}^{\infty} n^2 [B_{mxn}^2 + B_{mzn}^2]$$

$$= \epsilon \frac{f}{100} m_{Fe} [B_{mx1}^2 + B_{mz1}^2]\eta_d^2 \quad (2.53)$$

where $\epsilon = 1.2$ to 2.0 m^4/(H kg) for anisotropic laminations with 4% Si, $\epsilon = 3.8$ m^4/(H kg) for isotropic laminations with 2% Si and $\epsilon = 4.4$ to 4.8 m^4/(H kg) for isotropic siliconless laminations.

Eqns (2.51) and (2.53) exclude the *excess losses* (due to magnetic anomaly) and losses due to metallurgical and manufacturing processes. There is a poor correlation between measured core losses and those calculated using classical methods. The losses calculated according to eqns (2.51) and (2.53) are lower than those obtained from measurements. The coefficient of additional core losses $k_{ad} > 1$ can help to obtain a better agreement of predicted and measured core losses

$$\Delta P_{1Fe} = k_{ad}(\Delta P_{eFe} + \Delta P_{hFe}) \quad (2.54)$$

If the *specific core losses* are known, the stator core losses ΔP_{1Fe} can be calculated on the basis of the specific core losses and masses of teeth and yoke, i.e.

$$\Delta P_{1Fe} = \Delta p_{1/50} \left(\frac{f}{50}\right)^{4/3} [k_{adt} B_{1t}^2 m_{1t} + k_{ady} B_{1y}^2 m_{1y}] \tag{2.55}$$

where $k_{adt} > 1$ and $k_{ady} > 1$ are the factors accounting for the increase in losses due to metallurgical and manufacturing processes, $\Delta p_{1/50}$ is the specific core loss in W/kg at 1 T and 50 Hz, B_{1t} is the magnetic flux density in a tooth, B_{1y} is the magnetic flux density in the yoke, m_{1t} is the mass of the teeth, and m_{1y} is the mass of the yoke. For teeth $k_{adt} = 1.7$ to 2.0 and for the yoke $k_{ady} = 2.4$ to 4.0 [159].

2.6.3 Core Loss Finite Element Model

The core losses within the stator and rotor are calculated using a set of finite element method (FEM) models, assuming constant rotor speed and balanced three-phase armature currents. The eddy current and hysteresis losses, within the cores, in the 2D FEM including distorted flux waveforms can be expressed by eqns (2.51) and (2.53).

The field distribution at several time intervals in the fundamental current cycle is needed to create the magnetic flux density waveforms. This is obtained by rotation of the rotor grid and phase advancement of the stator currents. From a field solution, for a particular rotor position, the magnetic flux density at each element centroid is calculated. Three flux density components in an element can be obtained from a single FEM solution.

2.6.4 Losses in Permanent Magnets

The electric conductivity of sintered NdFeB magnets is from 0.6 to 0.85×10^6 S/m. The electric conductivity of SmCo magnets is from 1.1 to 1.4×10^6 S/m. Since the electric conductivity of rare earth PMs is only 4 to 9 times lower than that of a copper conductor, the *losses in conductive PMs due to higher harmonic magnetic fields* produced by the stator cannot be neglected.

The most important losses in PMs are generated by the fundamental frequency magnetic flux due to the stator slot openings. In practice, those losses are only in AFPM machines with slotted stator ferromagnetic cores. The fundamental frequency of the magnetic flux density component due to the stator slot opening is

$$f_{sl} = s_1 p n \tag{2.56}$$

where s_1 is the number of stator slots, p is the number of pole pairs and n is the rotor speed in rev/s.

The magnetic flux density component due to slot opening is [121]

$$B_{sl} = a_{sl} \beta_{sl} k_C B_{avg} \tag{2.57}$$

where B_{avg} is the mean magnetic flux density over the slot pitch - eqn (2.4), k_C is Carter coefficient (1.2) and

$$a_{sl} = \frac{4}{\pi}\left(0.5 + \frac{\Gamma^2}{0.78 - 2\Gamma^2}\right)\sin(1.6\pi\Gamma) \tag{2.58}$$

$$\beta_{sl} = 0.5\left(1 - \frac{1}{\sqrt{1+\kappa^2}}\right) \tag{2.59}$$

$$\Gamma = \frac{b_{14}}{t_1} \qquad \kappa = \frac{b_{14}}{g'} \tag{2.60}$$

In the above equations (2.58), (2.59) and (2.60) b_{14} is the stator slot opening, $g' = g + h_M/\mu_{rrec}$ is the equivalent air gap, t_1 is the slot pitch and h_M is the magnet thickness per pole.

Assuming that the relative recoil magnetic permeability $\mu_{rrec} \approx 1$, the power losses in PMs can be expressed by the following equation obtained from a 2D electromagnetic field distribution, i.e.

$$\Delta P_{PM} = \frac{1}{2} a_{R\nu} k_z \frac{|\alpha|^2}{\beta^2}\left(\frac{B_{sl}}{\mu_0 \mu_{rrec}}\right)^2 \frac{k}{\sigma_{PM}} S_{PM} \tag{2.61}$$

where $a_{R\nu}$ is according to eqn (1.21) for $\nu = 1$, $\alpha = (1+j)k$ is according to eqn (1.16) for $\nu = 1$, k is according to eqn (1.17) for $\nu = 1$, β is according to eqn (1.11) for $\nu = 1$ and $\tau = 0.5t_1$, and σ_{PM} is the electric conductivity of PMs. Eqn (2.61) can also be used to estimate the reactive losses in PMs if $a_{R\nu}$ is replaced with $a_{X\nu}$ according to eqn (1.22).

The coefficient for including the circumferential component of currents induced in PMs can be found as

$$k_z = 1 + \frac{t_1}{D_{out} - D_{in}} \tag{2.62}$$

where $0.5t_1$ is the induced current loop span and $0.5(D_{out} - D_{in})$ is the radial length of the PM — see eqn (1.35). The active surface area of all PMs (single-sided machines) is

$$S_{PM} = \alpha_i \frac{\pi}{4}(D_{out}^2 - D_{in}^2) \tag{2.63}$$

where the average pole pitch τ is according to eqn (1.14).

2.6.5 Rotor Core Losses

The *rotor core losses*, i.e. losses in backing solid steel discs supporting PMs are due to the pulsating flux produced by rapid changes in air gap reluctance as the rotor passes the stator teeth.

The magnetic permeability of a solid steel disc varies with the z axis (normal axis). To take into account the variable magnetic permeability and hysteresis losses in solid ferromagnetic discs, coefficients $a_{R\nu}$, $a_{X\nu}$ according to eqns (1.21) and (1.22) must be replaced by the following coefficients:

$$a_{RFe} = \frac{1}{\sqrt{2}} \left[\sqrt{4a_R^2 a_X^2 + \left(a_R^2 - a_X^2 + \frac{\beta^2}{k^2} \right)^2} + a_R^2 - a_X^2 + \frac{\beta^2}{k^2} \right]^{\frac{1}{2}} \quad (2.64)$$

$$a_{XFe} = \frac{1}{\sqrt{2}} \left[\sqrt{4a_R^2 a_X^2 + \left(a_R^2 - a_X^2 + \frac{\beta^2}{k^2} \right)^2} - a_R^2 + a_X^2 - \frac{\beta^2}{k^2} \right]^{\frac{1}{2}} \quad (2.65)$$

where $a_R = 1.4$ to 1.5, $a_X = 0.8$ to 0.9 according to Neyman [201], k_ν is according to eqn (1.17) and β is according to eqn (1.11) for $\nu = 1$.

The power losses in solid ferromagnetic discs can be expressed by a similar equation to eqn (2.61), i.e.

$$\Delta P_{2Fe} = \frac{1}{2} a_{RFe} k_z \frac{|\alpha|^2}{\beta^2} \left(\frac{B_{sl}}{\mu_0 \mu_r} \right)^2 \frac{k}{\sigma_{Fe}} S_{Fe} \quad (2.66)$$

where a_{RFe} is according to eqn (2.64) for $\nu = 1$, α is according to eqn (1.16) for $\nu = 1$, k is according to eqn (1.17) for $\nu = 1$, β is according to eqn (1.11) for $\nu = 1$ and $\tau = 0.5t_1$, B_{sl} is according to eqn (2.57), μ_r is the relative magnetic permeability and σ_{Fe} is the electric conductivity of the solid ferromagnetic disc. The frequency in the attenuation coefficient k is according to eqn (2.56). Eqn (2.66) can also be used to estimate the reactive losses in the solid ferromagnetic disc if a_{RFe} is replaced with a_{XFe} according to eqn (2.65). The flux density B_{sl} is according to eqns (2.57) to (2.60). The coefficient k_z is according to eqn (2.62) and the surface of the disc is

$$S_{Fe} = \frac{\pi}{4}(D_{out}^2 - D_{in}^2) \quad (2.67)$$

2.6.6 Eddy Current Losses in Stator Conductors

For slotted AFPM machines the *eddy current losses* in the stator winding are generally ignored as the magnetic flux penetrates through the teeth and yoke and only small leakage flux penetrates through the slot space with conductors.

In slotless and coreless machines, the stator winding is exposed to the air gap magnetic field. The motion of PMs relative to the stator winding produces an alternating field through each conductor and induces eddy currents. In the case of a coreless AFPM machine with a solid ferromagnetic rotor discs, there is also a tangential field component B_{mx} in addition to the axial field component B_{mz}. This can lead to serious additional eddy current loss especially at high frequency. Neglecting the proximity effect, the eddy current loss in the stator winding may be calculated by using a classical equation similar to eqn (2.51) for calculation of the eddy current losses in laminated cores, i.e.

- for round conductors [47]

$$\Delta P_e = \frac{\pi^2}{4}\frac{\sigma}{\rho} f^2 d^2 m_{con} \sum_{n=1}^{\infty} n^2 [B_{mxn}^2 + B_{mzn}^2]$$

$$= \frac{\pi^2}{4}\frac{\sigma}{\rho} f^2 d^2 m_{con} [B_{mx1}^2 + B_{mz1}^2]\eta_d^2 \tag{2.68}$$

- for rectangular conductors [47]

$$\Delta P_e = \frac{\pi^2}{3}\frac{\sigma}{\rho} f^2 a^2 m_{con} \sum_{n=1}^{\infty} n^2 [B_{mxn}^2 + B_{mzn}^2]$$

$$= \frac{\pi^2}{3}\frac{\sigma}{\rho} f^2 a^2 m_{con} [B_{mx1}^2 + B_{mz1}^2]\eta_d^2 \tag{2.69}$$

where d is the diameter of the conductor, a is the width of the conductor (parallel to the stator plane), σ is the electric conductivity, ρ is specific mass density of the conductor, m_{con} is the mass of the stator conductors without end connections and insulation, f is the stator current frequency, B_{mx} and B_{mz} are the peak values of tangential and axial components of the magnetic flux density, respectively, and η_d is the coefficient of distortion according to eqn (2.52).

2.6.7 Rotational Losses

The *rotational* or *mechanical losses* ΔP_{rot} consist of friction losses ΔP_{fr} in bearings, windage losses ΔP_{wind} and ventilation losses ΔP_{vent} (if there is a forced cooling system), i.e.

$$\Delta P_{rot} = \Delta P_{fr} + \Delta P_{wind} + \Delta P_{vent} \tag{2.70}$$

There are many semi-empirical equations for calculating the rotational losses giving various degrees of accuracy. The *friction losses in bearings* of small machines can be evaluated using the following formula

$$\Delta P_{fr} = 0.06 k_{fb}(m_r + m_{sh})n \qquad \text{W} \tag{2.71}$$

where $k_{fb} = 1$ to 3 m^2/s^2, m_r is the mass of the rotor in kg, m_{sh} is the mass of the shaft in kg and n is the speed in rev/s.

The *Reynolds number* for a rotating disc with its outer radius R_{out} is

$$Re = \rho\frac{R_{out}v}{\mu} = \frac{2\pi n\rho R_{out}^2}{\mu} \tag{2.72}$$

where ρ is the specific density of the cooling medium, $v = v_x = 2\pi R_{out} n$ is the linear velocity at the outer radius R_{out}, n is the rotational speed and μ is the dynamic viscosity of the fluid. Most AFPM machines are air cooled. The air density at 1 atm and 20°C is 1.2 kg/m^3. The dynamic viscosity of the air at 1 atm and 20°C is $\mu = 1.8 \times 10^{-5}$ Pa s.

The *coefficient of drag* for turbulent flow can be found as

$$c_f = \frac{3.87}{\sqrt{Re}} \tag{2.73}$$

The *windage losses* for a rotating disc are

$$\Delta P_{wind} = \frac{1}{2} c_f \rho (2\pi n)^3 (R_{out}^5 - R_{sh}^5) \tag{2.74}$$

where R_{sh} is the shaft diameter.

AFPM machines are usually designed without a cooling fan so that the ventilation losses $\Delta P_{vent} = 0$.

2.6.8 Losses for Nonsinusoidal Current

Higher time harmonics generated by static converters produce additional losses. The higher time harmonic frequency in the stator winding is nf where $n = 5, 7, 11, \ldots$. The armature winding losses, the core losses, and the stray losses are frequency-dependent. The mechanical losses do not depend on the shape of the input waveform.

If the stator winding and stator core losses have been calculated for fundamental frequency, the *frequency-dependent losses* of an inverter-fed motor or generator loaded with a rectifier can be found as

- stator (armature) winding losses:

$$\Delta P_{1w} = \sum_{n=1}^{\infty} \Delta P_{1wn} = m_1 \sum_{n=1}^{\infty} I_{an}^2 R_{1n} \approx m_1 R_{1dc} \sum_{n=1}^{\infty} I_{an}^2 k_{1Rn} \tag{2.75}$$

- stator (armature) core losses

$$\Delta P_{1Fe} = \sum_{n=1}^{\infty} \Delta P_{1Fen} = [\Delta P_{1Fen}]_{n=1} \sum_{n=1}^{\infty} \left(\frac{V_{1n}}{V_{1r}} \right)^2 n^{-0.7} \tag{2.76}$$

where k_{1Rn} is the a.c. armature resistance skin effect coefficient for nf, I_{an} is the higher harmonic *rms* armature current, V_{1n} is the higher harmonic inverter output voltage, V_{1r} is the rated voltage, $[\Delta P_{1Fen}]_{n=1}$ are the stator core losses for $n = 1$ and rated voltage.

2.6.9 Efficiency

The total power losses of an AFPM machine are

$$\Delta P = \Delta P_{1w} + \Delta P_{1Fe} + \Delta P_{2Fe} + \Delta P_{PM} + \Delta P_e + \Delta P_{rot} \tag{2.77}$$

The *efficiency* is

$$\eta = \frac{P_{out}}{P_{out} + \Delta P} \tag{2.78}$$

where P_{out} is the mechanical output power for a motor and the electrical output power for a generator.

2.7 Phasor Diagrams

The *synchronous reactance* of a synchronous machine (sine-wave machine) is defined as the sum of the armature reaction (mutual) reactance X_{ad}, X_{aq} and stator (armature) leakage reactance X_1, i.e.

- in the d-axis

$$X_{sd} = X_{ad} + X_1 \tag{2.79}$$

- in the q-axis

$$X_{sq} = X_{aq} + X_1 \tag{2.80}$$

When drawing *phasor diagrams* of synchronous machines, two arrow systems are used:

(a) generator arrow system, i.e.

$$\mathbf{E}_f = \mathbf{V}_1 + \mathbf{I}_a R_1 + j\mathbf{I}_{ad} X_{sd} + j\mathbf{I}_{aq} X_{sq}$$

$$= \mathbf{V}_1 + \mathbf{I}_{ad}(R_1 + jX_{sd}) + \mathbf{I}_{aq}(R_1 + jX_{sq}) \tag{2.81}$$

(b) consumer (motor) arrow system, i.e.

$$\mathbf{V}_1 = \mathbf{E}_f + \mathbf{I}_a R_1 + j\mathbf{I}_{ad} X_{sd} + j\mathbf{I}_{aq} X_{sq}$$

$$= \mathbf{E}_f + \mathbf{I}_{ad}(R_1 + jX_{sd}) + \mathbf{I}_{aq}(R_1 + jX_{sq}) \tag{2.82}$$

where

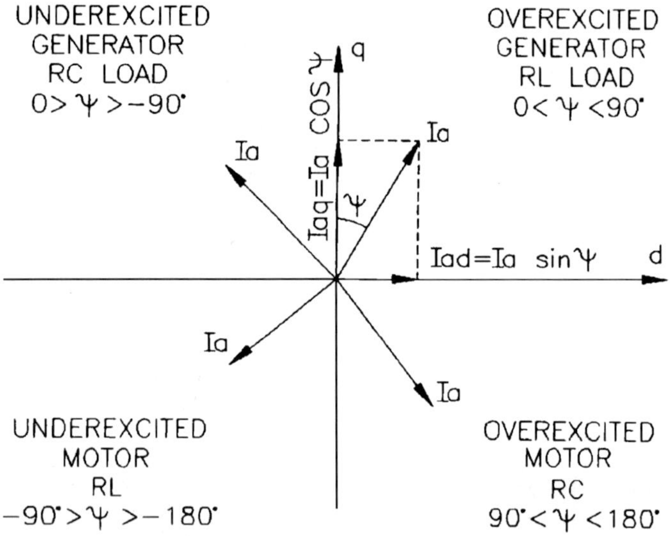

Fig. 2.13. Location of the armature current $\mathbf{I}_a$ in d—q coordinate system.

$$\mathbf{I}_a = \mathbf{I}_{ad} + \mathbf{I}_{aq} \tag{2.83}$$

and

$$I_{ad} = I_a \sin\Psi \qquad\qquad I_{aq} = I_a \cos\Psi \tag{2.84}$$

The angle Ψ is between the q-axis and armature current I_a. When the current arrows are in the opposite direction, the phasors $\mathbf{I}_a$, $\mathbf{I}_{ad}$, and $\mathbf{I}_{aq}$ are reversed by 180°. The same applies to the voltage drops. The location of the armature current $\mathbf{I}_a$ with respect to the d- and q-axis for generator and motor mode is shown in Fig. 2.13.

Phasor diagrams for synchronous generators are constructed using the generator arrow system. The same system can be used for motors, however, the consumer arrow system is more convenient. An overexcited generator (Fig. 2.14a) delivers both active and reactive power to the load or utility grid. An underexcited motor (Fig. 2.14b) draws both active and reactive power from the line. For example, the load current $\mathbf{I}_a$ (Fig. 2.14b) lags the voltage phasor $\mathbf{V}_1$ by the angle ϕ. An overexcited motor, consequently, draws a leading current from the circuit and delivers reactive power to it.

In the phasor diagrams according to Fig. 2.14 [126] the stator core losses have been neglected. This assumption is justified only for power frequency machines with unsaturated armature cores.

For an *underexcited synchronous motor* (Fig. 2.14b) the input voltage V_1 projections on the d and q axes are

$$V_1 \sin\delta = I_{aq} X_{sq} - I_{ad} R_1$$

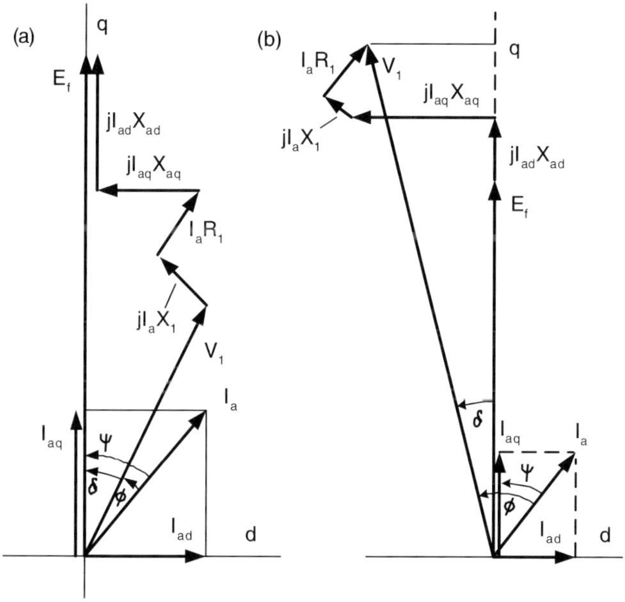

Fig. 2.14. Phasor diagrams of salient-pole synchronous machine: (a) overexcited generator (generator arrow system); (b) underexcited motor (consumer arrow system).

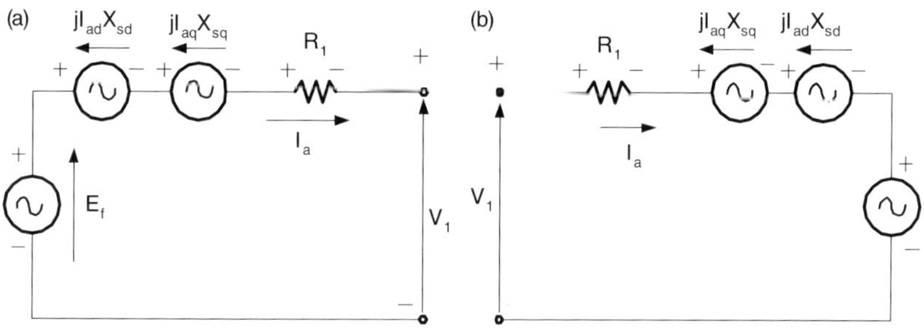

Fig. 2.15. Equivalent circuit per phase of an AFPM synchronous machine: (a) for motoring mode; (b) for generating mode. Stator core losses have been neglected.

$$V_1 \cos\delta = E_f + I_{ad}X_{sd} + I_{aq}R_1 \tag{2.85}$$

where δ is the *load angle* between the voltage V_1 and EMF E_f (q-axis). For an *overexcited motor*

$$V_1 \sin\delta = I_{aq}X_{sq} + I_{ad}R_1$$

$$V_1 \cos\delta = E_f - I_{ad}X_{sd} + I_{aq}R_1 \tag{2.86}$$

The currents of an overexcited motor

$$I_{ad} = \frac{V_1(X_{sq}\cos\delta - R_1\sin\delta) - E_f X_{sq}}{X_{sd}X_{sq} + R_1^2} \tag{2.87}$$

$$I_{aq} = \frac{V_1(R_1\cos\delta + X_{sd}\sin\delta) - E_f R_1}{X_{sd}X_{sq} + R_1^2} \tag{2.88}$$

are obtained by solving the set of eqns (2.85). The *rms* armature current as a function of V_1, E_f, X_{sd}, X_{sq}, δ, and R_1 is

$$I_a = \sqrt{I_{ad}^2 + I_{aq}^2} = \frac{V_1}{X_{sd}X_{sq} + R_1^2}$$

$$\times\sqrt{[(X_{sq}\cos\delta - R_1\sin\delta) - E_f X_{sq}]^2 + [(R_1\cos\delta + X_{sd}\sin\delta) - E_f R_1]^2} \tag{2.89}$$

The phasor diagram can also be used to find the input electric power. For a motor

$$P_{in} = m_1 V_1 I_a \cos\phi = m_1 V_1 (I_{aq}\cos\delta - I_{ad}\sin\delta) \tag{2.90}$$

Thus, the electromagnetic power for a motor mode is

$$\begin{aligned} P_{elm} &= P_{in} - \Delta P_{1w} - \Delta P_{1Fe} \\ &= m_1[I_{aq}E_f + I_{ad}I_{aq}(X_{sd} - X_{sq})] - \Delta P_{1Fe} \end{aligned} \tag{2.91}$$

On the basis of phasor diagrams (Fig. 2.14), equivalent circuits of a AFPM synchronous machine can be drawn (Fig. 2.15).

2.8 Sizing Equations

The *main dimensions* of a double-sided PM brushless motor with internal disc rotor can be determined using the following assumptions: (a) the electric and magnetic loadings are known; (b) the number of turns per phase per one stator is N_1; (c) the phase armature current in one stator winding is I_a; (d) the back EMF per phase per one stator winding is E_f.

The peak line current density at the average radius per one stator is expressed by eqn (2.22) in which the radius may be replaced by an average diameter

$$D = 0.5(D_{out} + D_{in}) = 0.5D_{out}(1 + k_d) \tag{2.92}$$

where D_{out} is the outer diameter, D_{in} is the inner diameter of the stator core and $k_d = D_{in}/D_{out}$ (according to eqn (2.27)). Thus,

$$A_m = \frac{4\sqrt{2} m_1 I_a N_1}{\pi D_{out}(1 + k_d)} \tag{2.93}$$

The EMF induced in the stator winding by the rotor excitation system, according to eqns (2.36) and (2.28) has the following form

$$E_f = \pi\sqrt{2} n_s p N_1 k_{w1} \Phi_f = \frac{\pi}{4}\sqrt{2} n_s N_1 k_{w1} B_{mg} D_{out}^2 (1 - k_d^2) \tag{2.94}$$

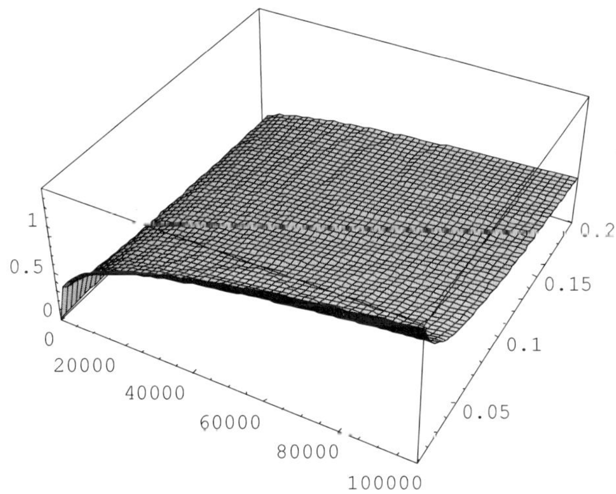

Fig. 2.16. Outer diameter D_{out} as a function of the output power P_{out} and parameter k_D for $\epsilon = 0.9$, $k_{w1}\eta\cos\phi = 0.84$, $n_s = 1000$ rpm $= 16.67$ rev/s and $B_{mg}A_m = 26,000$ TA/m.

The apparent electromagnetic power in two stators is

$$\begin{aligned} S_{elm} &= m_1(2E_f)I_a = m_1 E_f (2I_a) \\ &= \frac{\pi^2}{8} k_{w1} n_s B_{mg} A_m D_{out}^3 (1 + k_d)(1 - k_d^2) \end{aligned} \tag{2.95}$$

For series connection the EMF is equal to $2E_f$ and for parallel connection the current is equal to $2I_a$. For a multidisc motor the number "2" should be replaced by the number of stators. Putting

$$k_D = \frac{1}{8}(1 + k_d)(1 - k_d^2) \tag{2.96}$$

the apparent electromagnetic power is

$$S_{elm} = \pi^2 k_D k_{w1} n_s B_{mg} A_m D_{out}^3 \tag{2.97}$$

The apparent electromagnetic power expressed in terms of active output power is thus

$$S_{elm} = \epsilon \frac{P_{out}}{\eta \cos\phi} \tag{2.98}$$

where the *phase EMF–to–phase voltage* ratio is

$$\epsilon = \frac{E_f}{V_1} \tag{2.99}$$

For motors $\epsilon < 1$ and for generators $\epsilon > 1$.

In connection with eqns (2.97) and (2.98), the PM outer diameter (equal to the outer diameter of the stator core) is

$$D_{out} = \sqrt[3]{\frac{\epsilon P_{out}}{\pi^2 k_D k_{w1} n_s B_{mg} A_m \eta \cos\phi}} \tag{2.100}$$

The outer diameter of PMs is the most important dimension of disc rotor PM motors. Since $D_{out} \propto \sqrt[3]{P_{out}}$ the outer diameter increases rather slowly with the increase of the output power (Fig. 2.16). This is why small power disc motors have relatively large diameters. The disc-type construction is preferred for medium and large power motors. Motors with output power over 10 kW have reasonable diameters. Also, disc construction is recommended for a.c. servo motors fed with high frequency voltage.

The electromagnetic torque is proportional to D_{out}^3, i.e.

$$T_d = \frac{P_{elm}}{2\pi n_s} = \frac{S_{elm}\cos\Psi}{2\pi n_s} = \frac{\pi}{2} k_D k_{w1} D_{out}^3 B_{mg} A_m \cos\Psi \tag{2.101}$$

where P_{elm} is the active electromagnetic power and Ψ is the angle between the stator current I_a and EMF E_f.

2.9 Armature Reaction

The *magnetic fluxes produced by the stator* (armature) can be expressed in a similar way as the field excitation flux — eqn (2.26), i.e.

- in the d-axis

$$\Phi_{ad} = \frac{2}{\pi} B_{mad1} \frac{\pi}{p} \frac{R_{out}^2 - R_{in}^2}{2} \tag{2.102}$$

- in the q-axis

$$\Phi_{aq} = \frac{2}{\pi} B_{maq1} \frac{\pi}{p} \frac{R_{out}^2 - R_{in}^2}{2} \tag{2.103}$$

where B_{mad1} is the peak value of the first harmonic of the stator (armature reaction) magnetic flux density in the d-axis and B_{maq1} is the peak value of the first harmonic of the stator magnetic flux density in the q-axis. The *stator linkage* fluxes are

- in the d-axis

$$\Psi_d = \frac{1}{\sqrt{2}} N_1 k_{w1} \Phi_{ad} = \frac{1}{\sqrt{2}} N_1 k_{w1} \frac{2}{\pi} B_{mad1} \frac{\pi}{p} \frac{R_{out}^2 - R_{in}^2}{2} \tag{2.104}$$

- in the q-axis

$$\Psi_q = \frac{1}{\sqrt{2}} N_1 k_{w1} \Phi_{aq} = \frac{1}{\sqrt{2}} N_1 k_{w1} \frac{2}{\pi} B_{maq1} \frac{\pi}{p} \frac{R_{out}^2 - R_{in}^2}{2} \tag{2.105}$$

where N_1 is the number of stator turns per phase and k_{w1} is the winding factor for the fundamental space harmonic.

Neglecting the magnetic saturation, the first harmonic of the stator magnetic flux density normal components are

- in the d-axis

$$B_{mad1} = k_{fd} B_{mad} = k_{fd} \lambda_d F_{ad} = k_{fd} \frac{\mu_0}{g'} \frac{m_1 \sqrt{2}}{\pi} \frac{N_1 k_{w1}}{p} I_{ad} \tag{2.106}$$

- in the q-axis

$$B_{maq1} = k_{fd} B_{mad} = k_{fd} \lambda_q F_{aq} = k_{fd} \frac{\mu_0}{g_q'} \frac{m_1 \sqrt{2}}{\pi} \frac{N_1 k_{w1}}{p} I_{aq} \tag{2.107}$$

where m_1 is the number of stator phases and the permeance per unit surface in the d- and q-axis are

$$\lambda_d = \lambda_g = \frac{\mu_0}{g'} \qquad \lambda_q = \frac{\mu_0}{g_q'} \tag{2.108}$$

In the above equations (2.106) and (2.107) the *form factors of armature reaction* are defined as the ratios of the first harmonic amplitude–to–maximum value of normal components of armature reaction magnetic flux density in the d- and q-axis respectively, i.e.

$$k_{fd} = \frac{B_{mad1}}{B_{mad}} \qquad k_{fq} = \frac{B_{maq1}}{B_{maq}} \tag{2.109}$$

The equivalent air gaps in the d- and q-axis for surface configuration of PMs are

- for a stator with ferromagnetic core

$$g' = gk_C k_{sat} + \frac{h_M}{\mu_{rrec}} \tag{2.110}$$

$$g'_q = gk_C k_{satq} + h_M \tag{2.111}$$

- for a coreless stator

$$g' = 2\left[(g + 0.5t_w) + \frac{h_M}{\mu_{rrec}}\right] \tag{2.112}$$

$$g'_q = 2[(g + 0.5t_w) + h_M] \tag{2.113}$$

where t_w is the stator winding axial thickness, h_M is the axial height of the PM and μ_{rrec} is the relative recoil permeability of the PM. To take into account the effect of slots, the air gap (mechanical clearance) for a slotted ferromagnetic core is increased by Carter coefficient $k_C > 1$ according to eqn (1.2). The saturation of the magnetic circuit can be included with the aid of saturation factors $k_{sat} \geq 1$ in the d-axis and $k_{satq} \geq 1$ in the q-axis. For a coreless stator the effect of magnetic saturation of the rotor ferromagnetic discs (core) is negligible.

The MMFs $\mathcal{F}_{ad}$ and $\mathcal{F}_{aq}$ in the d-axis and q-axis are

$$\mathcal{F}_{ad} = \frac{m_1\sqrt{2}}{\pi}\frac{N_1 k_{w1}}{p} I_{ad} \tag{2.114}$$

$$\mathcal{F}_{aq} = \frac{m_1\sqrt{2}}{\pi}\frac{N_1 k_{w1}}{p} I_{aq} \tag{2.115}$$

where I_{ad} and I_{aq} are the d- and q-axis stator (armature) currents respectively.

The armature reaction (mutual) inductance is calculated as

- in the d-axis

$$L_{ad} = \frac{\Psi_d}{I_{ad}} = m_1\mu_0\frac{1}{\pi}\left(\frac{N_1 k_{w1}}{p}\right)^2 \frac{(R_{out}^2 - R_{in}^2)}{g'} k_{fd} \tag{2.116}$$

- in the q-axis

$$L_{aq} = \frac{\Psi_q}{I_{aq}} = m_1\mu_0\frac{1}{\pi}\left(\frac{N_1 k_{w1}}{p}\right)^2 \frac{(R_{out}^2 - R_{in}^2)}{g'_q} k_{fq} \tag{2.117}$$

For $\mu_{rrec} \approx 1$ and surface configuration of PMs ($k_{fd} = k_{fq} = 1$), the d- and q-axis armature reaction inductances are equal, i.e.

$$L_a = L_{ad} = L_{aq} = m_1\mu_0\frac{1}{\pi}\left(\frac{N_1 k_{w1}}{p}\right)^2 \frac{(R_{out}^2 - R_{in}^2)}{g'} \tag{2.118}$$

Table 2.1. Armature reaction equations for cylindrical and disc-type machines

Quantity	Cylindrical machine	Disc-type machine
Armature reaction magnetic flux in the d-axis	$\Phi_{ad} = \frac{2}{\pi} B_{mad1} \tau L_i$	$\Phi_{ad} = \frac{2}{\pi} B_{mad1} \frac{\pi}{2p}(R_{out}^2 - R_{in}^2)$
Armature reaction magnetic flux in the q-axis	$\Phi_{aq} = \frac{2}{\pi} B_{maq1} \tau L_i$	$\Phi_{aq} = \frac{2}{\pi} B_{maq1} \frac{\pi}{2p}(R_{out}^2 - R_{in}^2)$
Permeance of the air gap in the d-axis	$\Lambda_d = \frac{\mu_0}{g'} \frac{2}{\pi} \tau L_i$	$\Lambda_d = \frac{\mu_0}{g'} \frac{2}{\pi} \frac{\pi}{2p}(R_{out}^2 - R_{in}^2)$
Permeance of the air gap in the q-axis	$\Lambda_q = \frac{\mu_0}{g'_q} \frac{2}{\pi} \tau L_i$	$\Lambda_q = \frac{\mu_0}{g'_q} \frac{2}{\pi} \frac{\pi}{2p}(R_{out}^2 - R_{in}^2)$
Permeance of the air gap per surface of one pole in the d-axis	$\lambda_d = \frac{\mu_0}{g'}$	
Permeance of the air gap per surface area of one pole in the q-axis	$\lambda_q = \frac{\mu_0}{g'_q}$	
Armature reaction reactance in the d-axis	$X_{ad} = \frac{E_{ad}}{I_{ad}}$ $= 4m_1\mu_0 f \frac{(N_1 k_{w1})^2}{\pi p} \frac{\tau L_i}{g'} k_{fd}$	$X_{ad} = \frac{E_{ad}}{I_{ad}}$ $= 2m_1\mu_0 f \left(\frac{N_1 k_{w1}}{p}\right)^2 \frac{R_{out}^2 - R_{in}^2}{g'} k_{fd}$
Armature reaction reactance in the q-axis	$X_{aq} = \frac{E_{aq}}{I_{aq}}$ $= 4m_1\mu_0 f \frac{(N_1 k_{w1})^2}{\pi p} \frac{\tau L_i}{g'_q} k_{fq}$	$X_{aq} = \frac{E_{aq}}{I_{aq}}$ $= 2m_1\mu_0 f \left(\frac{N_1 k_{w1}}{p}\right)^2 \frac{R_{out}^2 - R_{in}^2}{g'_q} k_{fq}$
Armature reaction inductance in the d-axis	$L_{ad} = \frac{\Psi_{ad}}{I_{ad}}$ $= 2m_1\mu_0 \frac{(N_1 k_{w1})^2}{\pi^2 p} \frac{\tau L_i}{g'} k_{fd}$	$L_{ad} = \frac{\Psi_{ad}}{I_{ad}}$ $= m_1\mu_0 \frac{1}{\pi} \left(\frac{N_1 k_{w1}}{p}\right)^2 \frac{R_{out}^2 - R_{in}^2}{g'} k_{fd}$
Armature reaction inductance in the q-axis	$L_{aq} = \frac{\Psi_{aq}}{I_{aq}}$ $= 2m_1\mu_0 \frac{(N_1 k_{w1})^2}{\pi^2 p} \frac{\tau L_i}{g'_q} k_{fq}$	$L_{aq} = \frac{\Psi_{aq}}{I_{aq}}$ $= m_1\mu_0 \frac{1}{\pi} \left(\frac{N_1 k_{w1}}{p}\right)^2 \frac{R_{out}^2 - R_{in}^2}{g'_q} k_{fq}$

For surface configuration of PMs $k_{fd} = k_{fq} = 1$ [106]. For other configurations, equations for MMFs in the d- and q-axis will contain reaction factors $k_{fd} \neq k_{fq}$ [106].

The armature reaction EMFs in the d- and q-axis are

$$E_{ad} = \pi\sqrt{2} f N_1 k_{w1} \Phi_{ad} \tag{2.119}$$

$$E_{aq} = \pi\sqrt{2} f N_1 k_{w1} \Phi_{aq} \tag{2.120}$$

where the armature fluxes Φ_{ad} and Φ_{aq} are according to eqns (2.102) and (2.103).

The armature reaction reactances can be calculated by dividing EMFs E_{ad} and E_{aq} according to eqns (2.119) and (2.120) by currents I_{ad} and I_{aq}, i.e.

- in the d-axis

$$X_{ad} = 2\pi f L_{ad} = \frac{E_{ad}}{I_{ad}}$$

$$= 2m_1 \mu_0 f \left(\frac{N_1 k_{w1}}{p}\right)^2 \frac{(R_{out}^2 - R_{in}^2)}{g'} k_{fd} \tag{2.121}$$

- in the q-axis

$$X_{aq} = 2\pi f L_{aq} = \frac{E_{aq}}{I_{aq}}$$

$$= 2m_1 \mu_0 f \left(\frac{N_1 k_{w1}}{p}\right)^2 \frac{(R_{out}^2 - R_{in}^2)}{g'_q} k_{fq} \tag{2.122}$$

Table 2.1 compares the armature reaction equations between conventional cylindrical machines and disc-type machines.

2.10 AFPM Motor

2.10.1 Sine-Wave Motor

The three-phase stator winding with distributed parameters produces sinusoidal or quasi-sinusoidal distribution of the MMF. In the case of inverter operation all three solid state switches conduct current at any instant of time. The shape of the stator winding waveforms is shown in Fig. 1.3b. The *sine-wave motor* works as a PM synchronous motor.

For a sinusoidal excitation (synchronous machine) the excitation flux can be found on the basis of eqn (2.26) or (2.28), the EMF per phase induced by PM rotor can be found on the basis of eqn (2.36) and the electromagnetic torque on the basis of eqn (2.33). The EMF constant k_E and torque constant k_T are expressed by eqns (2.37) and (2.34), respectively.

2.10.2 Square-Wave Motor

PM d.c. brushless motors with *square-wave stator current* (Fig. 1.3a) are predominantly designed with large effective pole–arc coefficients, $\alpha_i^{(sq)}$, where $\alpha_i^{(sq)} = b_p(r)/\tau(r)$. Sometimes concentrated stator windings and salient stator poles are used. For the Y-connected windings, as in Fig. 2.17, only two of the three motor phase windings conduct at the same time, i.e. i_{aAB} (T1T4), i_{aAC} (T1T6), i_{aBC} (T3T6), i_{aBA} (T3T2), i_{aCA} (T5T2), i_{aCB} (T5T4), etc. At the on-time interval (120°) for the phase windings A and B, the solid state switches T1 and T4 conduct (Fig. 2.17a). When T1 switches off the current freewheels through diode D2. For the off-time interval both switches T1 and T4 are turned-off and the diodes D2 and D3 conduct the armature current which charges the capacitor C. If the solid state devices are switched at relatively high frequency, the winding inductance keeps the on-off rectangular current waveforms smooth.

For d.c. current excitation $\omega \to 0$, eqn (2.82) is similar to that describing a steady state condition of a d.c. commutator motor, i.e.

$$V_{dc} = E_{fL-L} + 2R_1 I_a^{(sq)} \tag{2.123}$$

where $2R_1$ is the sum of two-phase resistances in series (for Y-connected phase windings), and E_{fL-L} is the sum of two phase EMFs in series, V_{dc} is the d.c. input voltage supplying the inverter and $I_a^{(sq)}$ is the flat-topped value of the square-wave current equal to the inverter input current. The solid switch voltage drops have been neglected in eqn (2.123). The phasor analysis does not apply to this type of operation since the armature current is nonsinusoidal.

For a rectangular distribution of $B_{mg} = const$ with the pole shoe width $b_p < \tau$ being included, the excitation flux is

$$\Phi_f^{(sq)} = \alpha_i^{(sq)} B_{mg} \frac{\pi}{2p} (R_{out}^2 - R_{in}^2) \tag{2.124}$$

For a square-wave excitation the EMF induced in a single turn (two conductors) is $2B_{mg}L_i v = 4pnB_{mg}L_i\tau$. Including b_p and fringing flux, the EMF for $N_1 k_{w1}$ turns $e_f = 4pnN_1k_{w1}\alpha_i^{(sq)}B_{mg}\frac{\pi}{2p}(R_{out}^2 - R_{in}^2) = 4pnN_1k_{w1}\Phi_f^{(sq)}$. For the Y-connection of the armature windings, as in Fig. 2.17, two phases are conducting at the same time. The line–to–line EMF of a Y-connected square-wave motor is

$$E_{fL-L} = 2e_f = 8pN_1k_{w1}\alpha_i^{(sq)}B_{mg}(\pi/2p)(R_{out}^2 - R_{in}^2)n$$

$$= 8pN_1k_{w1}\Phi_f^{(sq)}n = k_{Edc}n \tag{2.125}$$

where the *EMF constant* or armature constant k_{Edc} is

$$k_{Edc} = 8pN_1k_{w1}\Phi_f^{(sq)} \tag{2.126}$$

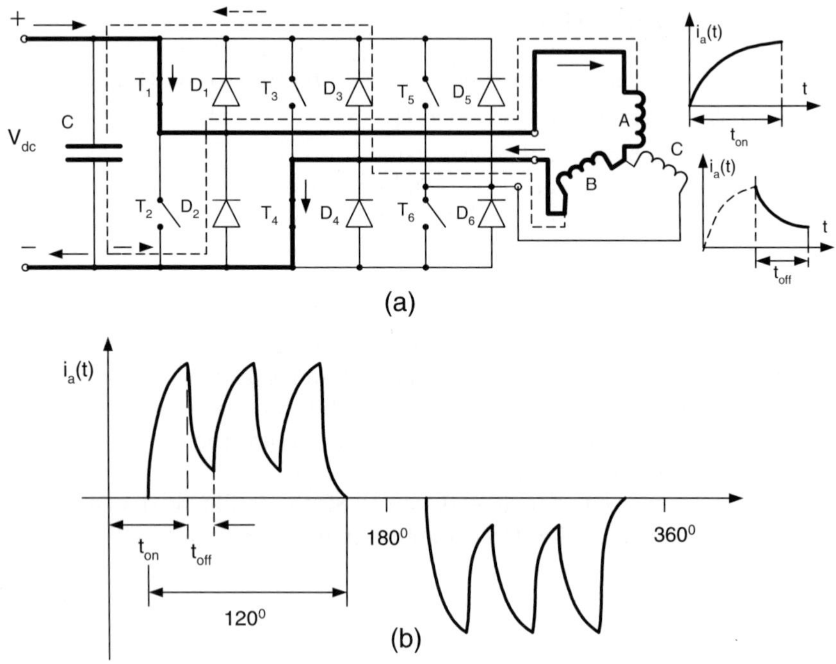

Fig. 2.17. Inverter currents in the stator winding of a PM brushless motor: (a) currents in phases A and B for on–time and off–time intervals, (b) on-off rectangular current waveform.

The *electromagnetic torque* developed by the motor is

$$T_d = \frac{P_{elm}}{2\pi n} = \frac{E_{fL-L} I_a^{(sq)}}{2\pi n}$$

$$= \frac{4}{\pi} p N_1 k_{w1} \Phi_f^{(sq)} I_a^{(sq)} = k_{Tdc} I_a^{(sq)} \tag{2.127}$$

where the *torque constant* of a square-wave motor is

$$k_{Tdc} = \frac{k_{Edc}}{2\pi} = \frac{4}{\pi} p N_1 k_{w1} \Phi_f^{(sq)} \tag{2.128}$$

and $I_a^{(sq)}$ is the flat-top value of the phase current.

The ratio T_d *of a square-wave motor–to–*T_d *of a sinewave motor* is

$$\frac{T_d^{(sq)}}{T_d} = \frac{4}{\pi} \frac{\sqrt{2}}{m_1} \frac{\Phi_f^{(sq)}}{\Phi_f} \frac{I_a^{(sq)}}{I_a} \approx 0.6 \frac{\Phi_f^{(sq)}}{\Phi_f} \frac{I_a^{(sq)}}{I_a} \tag{2.129}$$

Assuming the same motor ($\alpha_i^{(sq)} = \alpha_i$) and the same values of air gap magnetic flux densities, the ratio of the square-wave motor flux to sinewave motor flux is

$$\frac{\Phi_f^{(sq)}}{\Phi_f} = \frac{\alpha_i^{(sq)} B_{mg}[\pi/(2p)](R_{out}^2 - R_{in}^2)}{\alpha_i k_f B_{mg}[\pi/(2p)](R_{out}^2 - R_{in}^2)} = \frac{1}{k_f} \tag{2.130}$$

where [106]

$$k_f = \frac{4}{\pi}\sin\left(\frac{\alpha_i \pi}{2}\right) \tag{2.131}$$

and

$$\Phi_f = \Phi_{f1} = \alpha_i k_f B_{mg} \frac{\pi}{2p}(R_{out}^2 - R_{in}^2) \tag{2.132}$$

On the basis of eqns (2.125) and (2.127) the torque–speed characteristic can be expressed in the following simplified form

$$\frac{n}{n_0} = 1 - \frac{I_a}{I_{ash}} = 1 - \frac{T_d}{T_{dst}} \tag{2.133}$$

where the no-load speed, locked rotor armature current and stall torque are respectively

$$n_0 = \frac{V_{dc}}{k_E} \qquad I_{ash} = \frac{V_{dc}}{R} \qquad T_{dst} = k_{Tdc} I_{ash} \tag{2.134}$$

For half-wave operation $R = R_1$ while for full-wave operation $R = 2R_1$. Note that eqn (2.133) neglects the armature reaction, rotational and switching losses.

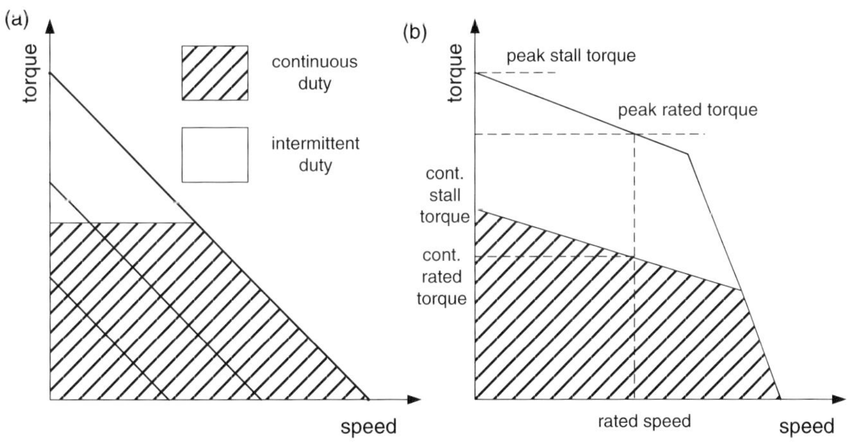

Fig. 2.18. Torque–speed characteristics of a PM brushless motor: (a) theoretical, (b) practical.

The torque-speed characteristics are shown in Fig. 2.18. Eqns (2.133) and (2.134) are very approximate and cannot be used in calculation of performance characteristics of practical PM d.c. brushless motors. Theoretical torque-speed characteristics (Fig. 2.18a) differ from practical characteristics (Fig. 2.18b). The *continuous torque* line is set by the maximum rated temperature of the motor. The intermittent duty operation zone is bounded by the *peak torque* line and the maximum input voltage.

The *rms* stator current of a d.c. brushless motor for a 120° square wave ($T = 2\pi/\omega$) is

$$I_a = \sqrt{\frac{2}{T}\int_0^{T/2} i_a^2(t)dt} = \sqrt{\frac{\omega}{\pi}\int_{\pi/(6\omega)}^{5\pi/(6\omega)} [I_a^{(sq)}]^2 dt}$$

$$= I_a^{(sq)}\sqrt{\frac{\omega}{\pi}\left(\frac{5}{6}\frac{\pi}{\omega} - \frac{1}{6}\frac{\pi}{\omega}\right)} = I_a^{(sq)}\sqrt{\frac{2}{3}} \qquad (2.135)$$

2.11 AFPM Synchronous Generator

2.11.1 Performance Characteristics of a Stand Alone Generator

An AFPM machine driven by a prime mover and connected to an electric load operates as a *stand alone synchronous generator*. For the same synchronous reactances in the *d*- and *q*-axis $X_{sd} = X_{sq} = X_s = \omega L_s$, and load impedance per phase

$$\mathbf{Z}_L = R_L + j\omega L_L - j\frac{1}{\omega C} \qquad (2.136)$$

the input current in the stator (armature) winding is

$$I_a = \frac{E_f}{\sqrt{(R_1 + R_L)^2 + [\omega L_s + \omega L_L - 1/(\omega C)]^2}} \qquad (2.137)$$

The voltage across the output terminals is

$$V_1 = I_a\sqrt{R_L^2 + \left(\omega L_L - \frac{1}{\omega C}\right)^2} \qquad (2.138)$$

Characteristics of EMF per phase E_f, phase voltage V_1, stator (load) current I_a, output power P_{out}, input power P_{in}, efficiency η and power factor $pf = \cos\phi$ versus speed n of a stand alone AFPM synchronous generator for inductive load $\mathbf{Z}_L = R_L + j\omega L_L$ are plotted in Fig. 2.19.

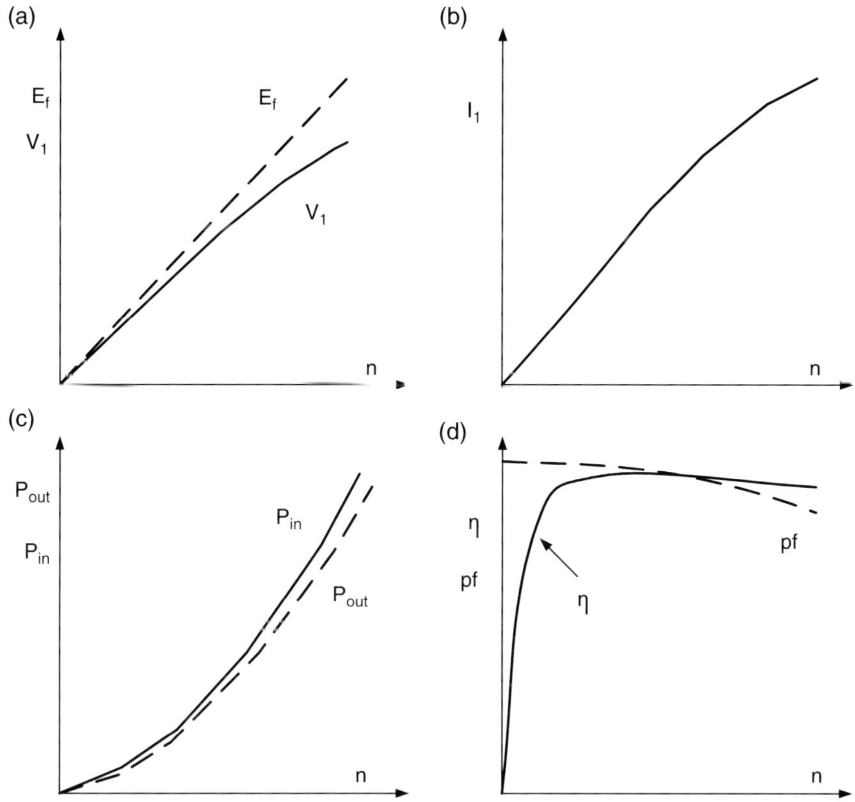

Fig. 2.19. Characteristics of a stand alone AFPM synchronous generator for inductive load $\mathbf{Z}_L = R_L + j\omega L_L$: (a) EMF E_f per phase and phase voltage V_1 versus speed n, (b) load current I_a versus speed n, (c) output power P_{out} and input power P_{in} versus speed n, (d) efficiency η and power factor pf$= \cos\phi$ versus speed n.

2.11.2 Synchronization With Utility Grid

An AFPM synchronous generator can be *synchronized* directly (connected in parallel) with the utility grid. The number of pole pairs is usually minimum $2p = 6$, so that to synchronize a 6-pole generator with a 50 Hz power system (infinite bus), the speed of the prime mover must be $n = 60(f/p) = 60(50/3) = 1000$ rpm. For $2p = 12$ and the same frequency, the speed will drop to $n = 500$ rpm. In general, direct paralleling of AFPM generators requires low speed prime movers. Before connecting the generator to the infinite bus, the incoming generator and the infinite bus must have the same:

- voltage
- frequency

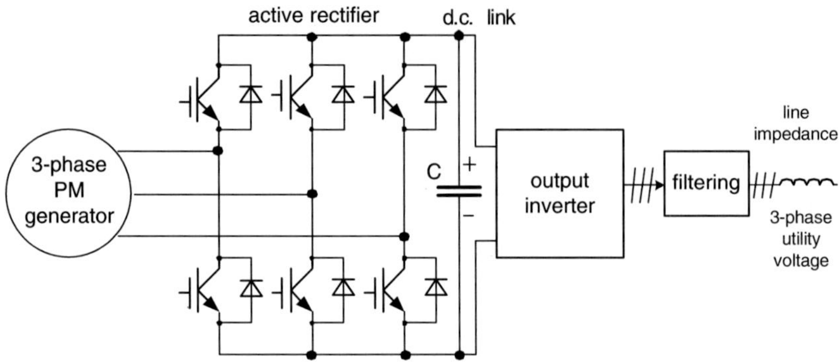

Fig. 2.20. Power circuit of a microturbine driven AFPM generator.

- phase sequence
- phase

In power plants, those conditions are checked using a *synchroscope.*

AFPM synchronous generators have recently been used in distributed generator systems as microturbine-driven high speed self-excited generators. A *microturbine* is a small, single shaft gas turbine of which the rotor is integrated with a high speed electric generator (up to 200 000 rpm), typically rated from 30 to 200 kW. If an AFPM generator is driven by a microturbine, in most cases the speed exceeds 20 000 rpm. The high frequency current of the generator must first be rectified and then inverted to obtain the same frequency as that of the power system. To minimize the higher harmonic contents in the stator windings, an active rather than passive rectifier is used (Fig. 2.20).

Numerical Example 2.1

Find the armature current, torque, electromagnetic power and winding losses of the S802F AFPM d.c. brushless servo motor (Table 1.1) at $n = 3000$ rpm and input voltage 230 V (line-to-line).

Solution

The line to line EMF is

$$E_{fL-L} = \sqrt{3}k_{Edc}n = \sqrt{3} \times \frac{42}{1000} \times 3000 = 218.24 \text{ V}$$

where the EMF constant $k_{Edc} = 42$ V/1000 rpm is for the phase EMF.

Assuming that the d.c. bus voltage is approximately equal to the input voltage, the armature current at 3000 rpm is

$$I_a^{(sq)} \approx \frac{230 - 218.24}{2 \times 0.76} = 7.74 \text{ A}$$

where 0.76 Ω is the resistance per phase.

The shaft torque at 3000 rpm is

$$T = k_{Tdc} I_a^{(sq)} = 0.64 \times 7.74 = 4.95 \text{ Nm}$$

where the torque constant is $k_{Tdc} = 0.64$ Nm/A.

The electromagnetic power (only two phases conduct current at the same time) is

$$P_{elm} = E_{fL-L} I_a^{(sq)} = 218.24 \times 7.74 \approx 1689 \text{ W}$$

The winding losses for 2 × 0.76 Ω line-to-line resistance are

$$\Delta P_w = (2 \times 0.76) \times 7.74^2 = 91 \text{ W}$$

Numerical Example 2.2

A three phase, Y-connected, $2p = 12$ pole, AFPM brushless motor has the surface PM of inner radius $R_{in} = 0.06$ m, outer radius $R_{out} = 0.11$ m and the pole shoe width–to–pole pitch ratio $\alpha_i = 0.84$ (square wave mode), the number of turns per phase $N_1 = 222$, the winding factor $k_{w1} = 0.96$ and the peak value of the air gap magnetic flux density $B_{mg} = 0.65$ T. Neglecting the armature reaction, find approximate values of the EMF, electromagnetic torque developed by the motor and electromagnetic power at $n = 1200$ rpm and *rms* current $I_a = 13.6$ A for:

(a) sinewave operation ($\Psi = 0^\circ$)
(b) 120° square wave operation

Solution

(a) sinewave operation at $\Psi = 0^\circ$
The speed in rev/s

$$n = \frac{1200}{60} = 20/\text{rev/s}$$

The form factor of the excitation field

$$k_f = \frac{4}{\pi} \sin\frac{\alpha_i \pi}{2} = \frac{4}{\pi} \times \sin\frac{0.84 \times \pi}{2} = 1.233$$

The excitation flux according to eqn (2.26) is

$$\Phi_f = \Phi_{f1} = \frac{2}{\pi} \times 1.233 \times 0.65 \times \frac{\pi}{12} \times (0.11^2 - 0.06^2) = 0.00114 \text{ Wb}$$

where $B_{mg1} = k_f B_{mg}$. The EMF constant according to eqn (2.37) is

$$k_E = \pi p \sqrt{2} N_1 k_{w1} \Phi_f = \pi \times 6 \times \sqrt{2} \times 222 \times 0.96 \times 0.00114 = 6.452 \text{ Vs}$$

The EMF per phase according to eqn (2.36) is

$$E_f = k_E n = 6.452 \times 20 = 129 \text{ V}$$

The line-to-line EMF for Y connection is

$$E_{fL-L} = \sqrt{3} E_f = \sqrt{3} \times 129 = 223.5 \text{ V}$$

The torque constants according to eqn (2.34) is

$$k_T = \frac{m_1}{\sqrt{2}} p N_1 k_{w1} \Phi_f = \frac{3}{\sqrt{2}} \times 6 \times 222 \times 0.96 \times 0.00114 = 3.08 \text{ Nm/A}$$

The electromagnetic torque developed at 13.6 A according to eqn (2.32) is

$$T_d = k_T I_a = 3.08 \times 13.6 = 41.9 \text{ Nm}$$

The electromagnetic power is

$$P_{elm} = m_1 E_f I_a \cos \Psi = 3 \times 129 \times 13.6 \times 1 = 5264.5 \text{ W}$$

where Ψ is the angle between the EMF E_f and stator current I_a.

(b) 120° **square-wave operation**

The flat-topped value of the phase current according to eqn (2.135) is

$$I_a^{(sq)} = \sqrt{\frac{3}{2}} I_a = \sqrt{\frac{3}{2}} \times 13.6 = 16.66 \text{ A}$$

The excitation flux according to eqn (2.124) is

$$\Phi_f^{(sq)} = 0.84 \times 0.65 \times \frac{\pi}{12} \times (0.11^2 - 0.06^2) = 0.00122 \text{ Wb}$$

The ratio of the square-wave flux to sinewave flux

$$\frac{\Phi_f^{(sq)}}{\Phi_f} = \frac{0.00122}{0.00114} = 1.07$$

Note that it has been assumed that $\alpha_i = 2/\pi = 0.6366$ for sinewave mode and $\alpha_i^{(sq)} = 0.84$ for square wave mode. For the same α_i the flux ratio is equal to $1/k_f$.

The EMF constant according to eqn (2.126) is

$$k_{Edc} = 8pN_1k_{w1}\Phi_f^{(sq)} = 8 \times 6 \times 222 \times 0.96 \times 0.00122 = 12.43 \text{ Vs}$$

The line–to–line EMF (two phases in series) according to eqn (2.125) is

$$E_{fL-L} = 12.43 \times \frac{1200}{60} = 248.6 \text{ V}$$

The torque constant according to eqn (2.128) is

$$k_{Tdc} = \frac{k_{Edc}}{2\pi} = \frac{12.43}{2\pi} = 1.978 \text{ Nm/A}$$

The electromagnetic torque at $I_a^{(sq)} = 16.66$ A according to eqn (2.127) is

$$T_d = k_{Tdc}I_a^{(sq)} = 1.978 \times 16.66 = 32.95 \text{ Nm}$$

The electromagnetic power is

$$P_{elm} = E_{fL-L}I_a^{(sq)} = 248.6 \times 16.66 = 4140.6 \text{ W}$$

Numerical Example 2.3

Find the main dimensions, approximate number of turns per phase and approximate cross section of the stator slot for a three-phase, double-sided, double-stator disc rotor PM brushless motor with a laminated stator core rated at: $P_{out} = 75$ kW, $V_{1L} = 460$ V (Y connection), $f = 100$ Hz, $n_s = 1500$ rpm. The stator windings are connected in series.

Solution

For $f = 100$ Hz and $n_s = 1500$ rpm $= 25$ rev/s the number of poles is $2p = 8$. Assuming $k_d = D_{in}/D_{out} = 1/\sqrt{3}$, the parameter k_D according to eqn (2.96) is

$$k_D = \frac{1}{8}\left(1 + \frac{1}{\sqrt{3}}\right)\left[1 - \left(\frac{1}{\sqrt{3}}\right)^2\right] = 0.131$$

For a 75 kW motor the product $\eta \cos\phi \approx 0.9$. The phase current for series connected stator windings is

$$I_a = \frac{P_{out}}{m_1(2V_{1)}\eta\cos\phi} = \frac{75\,000}{3 \times 265.6 \times 0.9} = 104.6 \quad \text{A}$$

where $2V_1 = 460/\sqrt{3} = 265.6$ V. The electromagnetic loading can be assumed as $B_{mg} = 0.65$ T and $A_m = 40,000$ A/m. The ratio $\epsilon = E_f/V_1 \approx 0.9$ and the stator winding factor has been assumed $k_{w1} = 0.96$. Thus, the stator outer diameter according to eqn (2.100) is

$$D_{out} = \sqrt[3]{\frac{0.9 \times 75\ 000}{\pi^2 \times 0.131 \times 0.96 \times 25 \times 0.65 \times 40\ 000 \times 0.9}} = 0.452 \text{ m}$$

The inner diameter according to eqns (2.27) is

$$D_{in} = \frac{D_{out}}{\sqrt{3}} = \frac{0.452}{\sqrt{3}} = 0.261 \text{ m},$$

The magnetic flux according to eqn (2.28) is

$$\Phi_f = \frac{2}{\pi} \times 0.65 \times \frac{\pi}{8 \times 4} \times 0.452^2 \left(1 - \frac{1}{(\sqrt{3})^2}\right) = 0.00555 \text{ Wb}$$

The number of stator turns per phase per stator calculated on the basis of line current density according to eqn (2.93) is

$$N_1 = \frac{\pi D_{out}(1 + k_d)A_m}{4m_1\sqrt{2}I_a} = \frac{\pi \times 0.452 \times (1 + 1/\sqrt{3}) \times 40\ 000}{4 \times 3\sqrt{2} \times 104.6} \approx 51$$

The number of stator turns per phase per stator calculated on the basis of eqn (2.94) and (2.99) is

$$N_1 = \frac{\epsilon V_1}{\pi\sqrt{2} f k_{w1} \Phi_f}$$

$$= \frac{0.9 \times 265.6/2}{\pi\sqrt{2} \times 100 \times 0.96 \times 0.00555} \approx 50$$

A double layer winding can be located, say, in 16 slots per phase, i.e. $s_1 = 48$ slots for a three phase machine. The number of turns should be rounded to 48. This is an approximate number of turns which can be calculated exactly only after performing detailed electromagnetic and thermal calculations of the machine.

The number of slots per pole per phase according to eqn (2.2) is

$$q_1 = \frac{s_1}{2pm_1} = \frac{48}{8 \times 3} = 2$$

The number of stator coils (double-layer winding) is the same as the number of slots, i.e. $2pq_1m_1 = 8 \times 2 \times 3 = 48$. If the stator winding is made of four parallel conductors $a_w = 4$, the number of conductors in a single coil for one parallel current path $a_p = 1$ according to eqn (2.4) is

$$N_c = \frac{a_w N_1}{(s_1/m_1)} = \frac{4 \times 48}{(48/3)} = 12$$

The current density in the stator conductor can be assumed $J_a \approx 4.5 \times 10^6$ A/m^2 (totally enclosed a.c. machines rated up to 100 kW). The cross section area of the stator conductor is

$$s_a = \frac{I_a}{a_w J_a} = \frac{104.6}{4 \times 4.5} = 5.811\text{mm}^2$$

The stator winding of a 75 kW machine is made of a copper conductor of rectangular cross section. The slot fill factor for rectangular conductors and low voltage machines can be assumed to be 0.6. The cross section of the stator slot should, approximately, be

$$\frac{5.811 \times 12 \times 2}{0.6} \approx 233 \text{ mm}^2$$

where the number of conductors in a single slot is $12 \times 2 = 24$. The minimum stator slot pitch is

$$t_{1min} = \frac{\pi D_{in}}{z_1} = \frac{\pi \times 0.262}{48} = 0.0171 \text{ m} = 17.1 \text{ mm}$$

The stator slot width can be chosen to be 11.9 mm; this means that the stator slot depth is $233/11.9 \approx 20$ mm, and the stator narrowest tooth width is $c_{1min} = 17.1 - 11.9 = 5.2$ mm. Magnetic flux density in the narrowest part of the stator tooth is

$$B_{1tmax} \approx \frac{B_{mg} t_{1min}}{c_{1min}} = \frac{0.65 \times 17.1}{5.2} = 2.14 \text{ T}$$

This is a permissible value for the narrowest part of the tooth and silicon electrical steel with saturation magnetic flux density of 2.2 T. The maximum stator slot pitch is

$$t_{1max} = \frac{\pi D_{ext}}{z_1} = \frac{\pi \times 0.452}{48} = 0.0296 \text{ m} = 29.6 \text{ mm}$$

The magnetic flux density in the widest part of the stator tooth is

$$B_{1tmin} \approx \frac{B_{mg} t_{1max}}{c_{1max}} = \frac{0.65 \times 29.6}{29.6 - 11.9} = 1.09 \text{ T}$$

Numerical Example 2.4

A 7.5-kg cylindrical core is wound of isotropic silicon steel ribbon. It can be assumed that the magnetic field inside the core is uniform and the vector of the magnetic flux density is parallel to steel laminations. The magnitudes of magnetic flux density time harmonics are: $B_{m1} = 1.7$ T, $B_{m3} = 0.25$ T, $B_{m5} = 0.20$ T and $B_{m7} = 0.05$ T. Harmonics $n > 7$ are negligible. The electric conductivity of laminations is $\sigma_{Fe} = 3.3 \times 10^6$ S/m, specific mass density $\rho_{Fe} = 7600$ kg/m3, thickness $d_{Fe} = 0.5$ mm and Richter's coefficient of hysteresis losses $\epsilon = 3.8$. Find the core losses at 50 Hz.

Solution

Eddy-current losses
The coefficient of distortion of the magnetic flux density according to eqn (2.52) is

$$\eta_d = \sqrt{1 + \left(3 \times \frac{0.25}{1.7}\right)^2 + \left(5 \times \frac{0.20}{1.7}\right)^2 \times \left(7 \times \frac{0.05}{1.7}\right)^2} = 1.258$$

The eddy current losses according to eqn (2.51) are

- specific losses at sinusoidal magnetic flux density

$$\Delta p_{esin} = \frac{\pi^2}{6} \times \frac{3.3 \times 10^6}{7600} \times 50^2 \times 0.0005^2 \times 1.7^2 = 1.29 \text{ W/kg}$$

- specific losses at nonsinusoidal magnetic flux density

$$\Delta p_e = \Delta p_{esin} \times \eta_d^2 = 1.29 \times 1.258^2 = 2.04 \text{ W/kg}$$

- losses at sinusoidal magnetic flux density

$$\Delta P_{esin} = \Delta p_{esin} m_{Fe} = 1.29 \times 7.5 = 9.68 \text{ W}$$

- losses at nonsinusoidal magnetic flux density

$$\Delta P_e = \Delta p_e m_{Fe} = 2.04 \times 7.5 = 15.32 \text{ W}$$

Hysteresis losses
The hysteresis losses may be calculated according to eqn (2.53), in which Richter's coefficient of hysteresis losses $\epsilon = 3.8$

- specific losses at sinusoidal magnetic flux density

$$\Delta p_{hsin} = 3.8 \times \frac{50}{100} \times 1.7^2 = 5.49 \text{ W/kg}$$

- specific losses at nonsinusoidal magnetic flux density

$$\Delta p_h = \Delta p_{hsin} \times \eta_d^2 = 5.49 \times 1.258^2 = 8.69 \text{ W/kg}$$

- losses at sinusoidal magnetic flux density

$$\Delta P_{hsin} = \Delta p_{hsin} m_{Fe} = 5.49 \times 7.5 = 41.18 \text{ W}$$

- losses at nonsinusoidal magnetic flux density

$$\Delta P_h = \Delta p_h m_{Fe} = 8.69 \times 7.5 = 65.19 \text{ W}$$

Specific eddy current and hysteresis losses plotted against frequency are shown in Fig. 2.21.

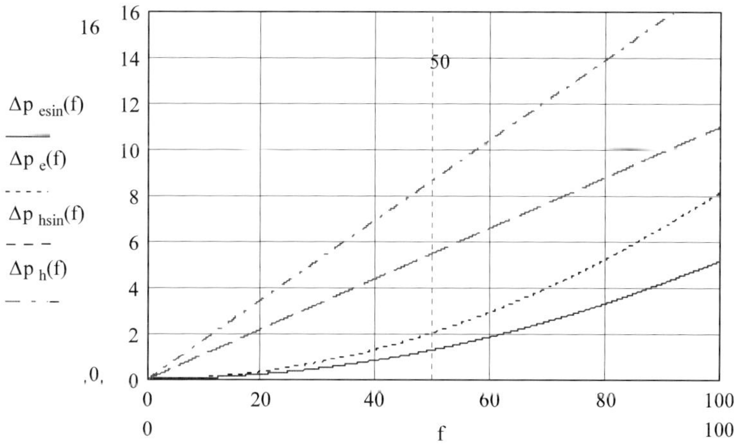

Fig. 2.21. Specific eddy-current and hysteresis losses versus frequency. *Numerical example 2.4.*

Total losses

The eddy current and hysteresis losses at sinusoidal magnetic flux density are calculated as

$$\Delta P_{Fesin} = \Delta P_{esin} + \Delta P_{hsin} = 9.68 + 41.18 = 50.86 \text{ W}$$

The eddy current and hysteresis losses at nonsinusoidal magnetic flux density are

$$\Delta P_{Fe} = \Delta P_e + \Delta P_h = 15.32 + 65.19 = 80.51 \text{ W}$$

The ratio of hysteresis–to–total losses is the same for both sinusoidal and nonsinusoidal magnetic flux density, i.e.

$$\frac{41.18}{50.86} = \frac{65.19}{80.51} = 0.81$$

Numerical Example 2.5

Find the power losses in PMs and solid rotor backing steel disc of a single-sided AFPM with slotted stator at ambient temperature of 20°C. The inner diameter of PMs is $D_{in} = 0.14$ m, the outer diameter of PMs is $D_{out} = 0.242$ m, number of poles $2p = 8$, magnet width–to–pole pitch ratio is $\alpha_i = 0.75$, height of surface permanent magnet is $h_M = 6$ mm, air gap thickness $g = 1.2$ mm, number of slots $s_1 = 36$ and speed $n = 3000$ rpm. Assume the peak value of the air gap magnetic flux density $B_{mg} = 0.7$ T, conductivity of NdFeB PMs $\sigma_{Pm} = 0.8 \times 10^6$ S/m at 20°C, its relative recoil magnetic permeability $\mu_{rrec} = 1.05$, conductivity of solid disc $\sigma_{Fe} = 4.5 \times 10^6$ S/m at 20°C and its relative recoil magnetic permeability $\mu_r = 300$.

Solution

The average magnetic flux density is $B_{avg} = (2/\pi)B_{mg} = (2/\pi) \times 0.7 = 0.446$ T, the average diameter $D = 0.5(D_{out} + D_{in}) = 0.5(0.242 + 0.14) = 0.191$ m, the average pole pitch is $\tau = \pi D/(2p) = \pi 0.191/8 = 0.075$ m, the average slot pitch $t_{s1} = \pi D/s_1 = \pi 0.191/36 = 0.017$ m. The frequency of the fundamental harmonic of the slot component of the magnetic flux density according to eqn (2.56) is

$$f_{sl} = s_1 pn = 36 \times 4 \times \frac{3000}{60} = 7200 \text{ Hz}$$

The corresponding angular frequency is $\omega_{sl} = 2\pi 7200 = 45\ 238.9$ rad/s.

The equivalent air gap in the d-axis for $k_C = 1$ according to eqn (2.110) is

$$g' = 0.0012 + \frac{0.006}{1.05} = 0.006914\text{m}$$

The parameters $\kappa = 0.434$, $\beta_{sl} = 0.041$, $\Gamma = 0.18$ and $a_{sl} = 0.546$ have been found on the basis of eqns (2.60), (2.59) and (2.58) respectively. The Carter coefficient is $k_C = 1.012$ - eqns (1.2) and (1.3). Thus, eqn (2.57) gives the magnetic flux density component due to slot opening as

$$B_{sl} = 0.546 \times 0.041 \times 1.012 \times 0.446 = 0.01018 \text{ T}$$

The parameters $k = 139.283$ 1/m for $\mu_{rrec} = 1.05$, $\Delta = 1/k = 0.0072$ m, $\alpha = 139.283 + j139.283$ 1/m, $\beta = \pi/(0.5t_{s1}) = \pi/(0.5 \times 0.017) = 377.241$ 1/m, $\kappa = 386.668 + j50.962$ 1/m, $a_{R\nu} = 2.733$ and $a_{X\nu} = 0.366$ for $\nu = 1$ have been found on the basis of eqns (1.17), (1.16), (1.11), (1.17), (1.21) and (1.22) respectively. The transverse edge effect coefficient k_z according to eqn (2.62) is

$$k_z = 1 + \frac{0.01666}{0.242 - 0.14} = 1.163$$

The active surface area of PMs according to eqn (2.63) is

$$S_{PM} = 0.75\pi \times [(0.5 \times 0.242)^2 - (0.5 \times 0.14)^2] = 0.023 \text{ m}^2$$

The active losses in PMs according to eqn (2.61) are

$$\Delta P_{PM} = \frac{1}{2} \times 2.733 \times 1.081 \times \left(\frac{|139.283 + j139.283|^2}{377.241^2} \right)$$

$$\times \left(\frac{0.01018}{0.4\pi \times 10^{-6} \times 1.05} \right)^2 \times \frac{139.283}{0.65 \times 10^6} \times 0.023 = 127.2 \text{ W}$$

The reactive power losses in PMs can be estimated as

$$\Delta Q_{PM} = \frac{a_{X\nu}}{a_{R\nu}} \Delta P_{PM} = \frac{0.366}{2.733} \times 127.2 = 17.03 \text{ VAr}$$

For calculation of power losses in the solid steel disc of the rotor, the magnetic flux density component due to slot openings $B_{sl} = 0.01018$ T is the same. The relative magnetic permeability $\mu_r = 300$ and conductivity of the disc $\sigma_{Fe} = 4.5 \times 10^6$ S/m are different, so that the attenuation factor (1.17) calculated for these values will be $k = 6194.6$ 1/m. The coefficients $a_{RFe} = 1.451$ and $a_{XFe} = 0.849$ are calculated using eqns (2.64) and (2.65) for $a_R = 1.45$ and $a_X = 0.85$. The remaining field parameters, i.e. $\alpha = 8982 + j5265$ 1/m, $\beta = \pi/(0.5t_{s1}) = \pi/(0.5 \times 0.017) = 377.241$ 1/m and $\kappa = 8988 + j5362$ 1/m have been found from eqns (1.16), (1.11) and (1.17) respectively.

The surface area of the rotor disc according to eqn (2.67) is

$$S_{Fe} = \frac{\pi}{4} \times (0.242^2 - 0.14^2) = 0.031 \text{ m}^2$$

The active power losses in the rotor disc according to eqn (2.66) is

$$\Delta P_{Fe2} = \frac{1}{2} \times 1.451 \times 1.081 \times \left(\frac{|8982 + j5265|^2}{377.241^2} \right)$$

$$\times \left(\frac{0.01018}{0.4\pi \times 10^{-6} \times 300} \right)^2 \times \frac{6194.6}{4.5 \times 10^6} \times 0.031 = 19.79 \text{ W}$$

The coefficients $k_z = 1.081$ is the same as for calculation of losses in PMs. The reactive power losses in the rotor disc can be estimated as

$$\Delta Q_{Fe2} = \frac{a_{XFe\nu}}{a_{RFe\nu}} \Delta P_{Fe2} = \frac{0.849}{1.451} \times 118.2 = 11.6 \text{ VAr}$$

Numerical Example 2.6

Find the rotational loss of a 6000 rpm disc type machine with outer radius of the rotor $R_{out} = 0.14$ m, radius of shaft behind the rotor core $R_{sh} = 0.025$ m, mass of the rotor $m_r = 1.31$ kg and mass of the shaft $m_{sh} = 1.49$ kg. The machine is naturally cooled with air. The ambient temperature is 20°C. The coefficient of bearing friction is $k_{fb} = 1.5$.

Solution

The bearing friction losses according to eqn (2.71) are

$$\Delta P_{fr} = 0.06 \times 1.5 \times (1.31 + 1.49) \times \frac{6000}{60} = 25.23 \text{ W}$$

The Reynolds number according to eqn (2.72) is

$$Re = 1.2 \times \frac{2\pi}{1.8 \times 10^{-5}} \times \frac{6000}{60} \times 0.2^2 = 8.21 \times 10^5$$

where the air density is $\rho = 1.2$ kg/m^2 and dynamic viscosity of the air is $\mu = 1.8 \times 10^{-5}$ Pa s. The coefficient of drag according to eqn (2.73) is

$$c_f = \frac{3.87}{\sqrt{8.21 \times 10^5}} = 4.271 \times 10^{-3}$$

The windage losses according to eqn (2.74) are

$$\Delta P_{wind} = \frac{1}{2} \times 4.271 \times 10^{-3} \times 1.2 \times \left(2\pi \times \frac{6000}{60}\right) (0.14^5 - 0.025^5) = 34.18 \text{ W}$$

The rotational losses are thus

$$\Delta P_{rot} = 25.23 + 34.18 = 59.41 \text{ W}$$

3

Materials and Fabrication

3.1 Stator Cores

Stator cores of AFPM brushless machines are made of laminated steels or soft magnetic powder materials. Soft magnetic powders simplify the manufacturing process and reduce the cost of AFPM machines.

3.1.1 Nonoriented Electrical Steels

Most laminated cores for stators of AFPM brushless machines are made of *nonoriented (isotropic) silicon steel* ribbons with standard thickness from 0.12 to 0.64 mm. Nonoriented steels are Fe-Si alloys with a random orientation of crystal cubes. Magnetic properties are practically the same in any direction in the plane of the sheet or ribbon. A secondary recrystallization process is not needed and high temperature annealing is not essential. Nonoriented grades contain between 0.5% and 3.25% Si with up to 0.5% Al addition to increase the resistivity and lower the temperature of the primary recrystallization.

Nonoriented electrical steels are available as both *fully processed* and *semi-processed* products. Fully processed nonoriented electrical steels are completely processed by the steel manufacturer, ready for use without any additional processing required to achieve the desired magnetic quality. Semi-processed nonoriented electrotechnical steels are those which have not been given the full annealing treatment by the steel producer. In some cases, users prefer to develop the final magnetic quality and achieve relief of fabricating stresses in laminations or assembled cores for small machines.

The most universally accepted grading of electrical steels by core losses is the American Iron and Steel Industry (AISI) system (Table 3.1), the so called "M-grading". For small and medium power electrical machines (output power less that 75 kW), the following grades can be used: M-27, M-36, M-43, M-45 and M-47.

Laminating a magnetic core is ineffective in keeping excessive eddy currents from circulating within the entire core unless the surfaces of laminations

Table 3.1. The most important silicon steel designations specified by different standards

Europe IEC 404-8-4 (1986)	U.S.A. AISI	Japan JIS 2552 (1986)	Russia GOST 21427 0-75
250-35-A5	M-15	35A250	2413
270-35-A5	M-19	35A270	2412
300-35-A5	M-22	35A300	2411
330-35-A5	M-36	—	—
270-50-A5	—	50A270	—
290-50-A5	M-15	50A290	2413
310-50-A5	M-19	50A310	2412
330-50-A5	M-27	—	—
350-50-A5	M-36	50A350	2411
400-50-A5	M-43	50A400	2312
470-50-A5	—	50A470	2311
530-50-A5	M-45	—	2212
600-50-A5	—	50A600	2112
700-50-A5	M-47	50A700	—
800-50-A5	—	50A800	2111
350-65-A5	M-19	—	—
400-65-A5	M-27	—	—
470-65-A5	M-43	—	—
530-65-A5	—	—	2312
600-65-A5	M-45	—	2212
700-65-A5	—	—	2211
800-65-A5	—	65A800	2112
1000-65-A5	—	65A1000	—

Table 3.2. Specific core losses of *Armco* DI-MAX nonoriented electrical steels M-27, M-36 and M-43 at 60 Hz

Magnetic flux density	Specific core losses W/kg							
	0.36 mm		0.47 mm			0.64 mm		
T	M-27	M-36	M-27	M-36	M-43	M-27	M-36	M-43
0.20	0.09	0.10	0.10	0.11	0.11	0.12	0.12	0.13
0.50	0.47	0.52	0.53	0.56	0.59	0.62	0.64	0.66
0.70	0.81	0.89	0.92	0.97	1.03	1.11	1.14	1.17
1.00	1.46	1.61	1.67	1.75	1.87	2.06	2.12	2.19
1.30	2.39	2.58	2.67	2.80	2.99	3.34	3.46	3.56
1.50	3.37	3.57	3.68	3.86	4.09	4.56	4.70	4.83
1.60	4.00	4.19	4.30	4.52	4.72	5.34	5.48	5.60
1.70	4.55	4.74	4.85	5.08	5.33	5.99	6.15	6.28
1.80	4.95	5.14	5.23	5.48	5.79	6.52	6.68	6.84

Table 3.3. Magnetization curves of fully processed *Armco* DI-MAX nonoriented electrical steels M-27, M-36 and M-43

Magnetic flux density T	Magnetic field intensity A/m M-27	M-36	M-43
0.20	36	41	47
0.40	50	57	64
0.70	74	80	89
1.00	116	119	130
1.20	175	174	187
1.50	859	785	777
1.60	2188	2109	1981
1.70	4759	4727	4592
1.80	8785	8722	8682
2.00	26 977	26 022	26 818
2.10	64 935	65 492	66 925
2.20	137 203	136 977	137 075

are adequately insulated. Types of surface insulation include the natural oxide surface, inorganic insulation, enamel, varnish or chemically treated surface. The thickness of the insulation is expressed with the aid of the *stacking factor*:

$$k_i = \frac{d}{d + 2\Delta} \tag{3.1}$$

where d is the thickness of bare lamination and Δ is the thickness of the insulation layer measured on one side. The stacking factor k_i is typically from 0.94 to 0.97.

Core loss curves of *Armco* DI-MAX nonoriented electrical steels M-27, M-36 and M-43 tested at 60 Hz are given in Table 3.2. Core losses when tested at 50 Hz would be approximately 0.79 times the core loss at 60 Hz. Magnetization curves of the same electrical steels are given in Table 3.3. The specific mass density of DI-MAX M-27, M-36 and M-43 is 7650, 7700 and 7750 kg/m^3 respectively. The stacking factor is $k_i = 0.95$ to 0.96. The prefix DI MAX, e.g. DI-MAX M27, designates a registered trademark with a special strip-annealed process that maximizes *punchability*. DI-MAX grades have superior permeability at high magnetic flux density, low core losses and good gauge uniformity. A smooth surface, excellent flatness and high stacking factor is obtained as a result of cold finishing and strip annealing.

For frequencies exceeding the power frequency of 50 or 60 Hz nonoriented laminations thinner than 0.2 mm must be used. Table 3.4 shows magnetization and specific core losses of nonoriented grades NO 12, NO 18 and NO 20 capable of operating up to 2.5 kHz (*Cogent Power Ltd.*, Newport, UK). Typical chemical composition is 3.0 % Si, 0.4 % Al, 96.6 % Fe. The standard thickness of an inorganic phosphate based insulation is 1 μm (one side). The maximum continuous operating temperature in the air is 230°C, maximum

Table 3.4. Specific core losses and d.c. magnetization curve of nonoriented thin electrical steels manufactured by *Cogent Power Ltd.*, Newport, U.K.

Magn. flux dens. T	Specific core losses W/kg NO 12 0.12 mm			NO 18 0.18 mm			NO 20 0.2 mm			Magn. field stren. kA/m
	50 Hz	400 Hz	2.5 kHz	50 Hz	400 Hz	2.5 kHz	50 Hz	400 Hz	2.5 kHz	
0.10	0.02	0.16	1.65	0.02	0.18	2.18	0.02	0.17	2.79	0.025
0.20	0.08	0.71	6.83	0.08	0.73	10.6	0.07	0.72	10.6	0.032
0.30	0.16	1.55	15.2	0.16	1.50	19.1	0.14	1.49	24.4	0.039
0.40	0.26	2.57	2.54	0.26	2.54	31.7	0.23	2.50	40.4	0.044
0.50	0.37	3.75	37.7	0.36	3.86	45.9	0.32	3.80	58.4	0.051
0.60	0.48	5.05	52.0	0.47	5.22	61.5	042	5.17	78.4	0.057
0.70	0.62	6.49	66.1	0.61	6.77	81.1	0.54	6.70	103.0	0.064
0.80	0.76	8.09	83.1	0.75	8.47	104.0	0.66	8.36	133.0	0.073
0.90	0.32	9.84	103.0	0.90	10.4	161.0	0.80	10.3	205.0	0.084
1.00	1.09	11.8	156	1.07	12.3	198.0	0.95	12.2	253.0	0.099
1.10	1.31	14.1		1.28	14.9		1.14	14.8		0.124
1.20	1.56	16.7		1.52	18.1		1.36	17.9		0.160
1.30	1.89	19.9		1.84	21.6		1.65	21.4		0.248
1.40	2.29	24.0		2.23	25.6		2.00	25.3		0.470
1.50	2.74	28.5		2.67	30.0		2.40	29.7		1.290
1.60	3.14			3.06			2.75			3.550
1.70	3.49			3.40			3.06			7.070
1.80	3.78			3.69			3.32			13

intermittent operating temperature in an inert gas is 850°C, hardness 180 HV and density 7650 kg/m^3.

3.1.2 Amorphous Ferromagnetic Alloys

To minimize core losses at high frequencies, nonoriented electrotechnical steels should be replaced with *amorphous magnetic alloys* (Tables 3.5 and 3.6). Amorphous ferromagnetic alloys, in comparison with electrical steels with crystal structure, do not have arranged in order, regular inner crystal structure (lattice).

Amorphous alloy ribbons based on alloys of iron, nickel and cobalt are produced by rapid solidification of molten metals at cooling rates of about 10^{6}°C/s. The alloys solidify before the atoms have a chance to segregate or crystallize. The result is a metal alloy with a glass-like structure, i.e. a non-crystalline frozen liquid.

Application of amorphous alloy ribbons to the mass production of electrical machines is limited by hardness, up to 1100 in Vicker's scale. Standard cutting methods like a guillotine or blank die are not suitable. The mechanically

Table 3.5. Physical properties of iron based METGLAS amorphous alloy ribbons (*Honeywell*, Morristown, NJ, U.S.A.)

Quantity	2605CO	2605SA1
Saturation magnetic flux density, T	1.8	1.59 annealed 1.57 cast
Specific core losses at 50 Hz and 1 T, W/kg	less than 0.28	about 0.125
Specific density, kg/m^3	7560	7200 annealed 7190 cast
Electric conductivity, S/m	0.813×10^6 S/m	0.769×10^6 S/m
Hardness in Vicker's scale	810	900
Elasticity modulus, GN/m^2	100...110	100...110
Stacking factor	less than 0.75	less than 0.79
Crystallization temperature, °C	430	507
Curie temperature, °C	415	392
Maximum service temperature, °C	125	150

Table 3.6. Specific core losses of iron based METGLAS amorphous alloy ribbons (*Honeywell*, Morristown, NJ, U.S.A.)

Magnetic flux density, B T	Specific core losses, Δp, W/kg			
	2605CO		2605SA1	
	50 Hz	60 Hz	50 Hz	60 Hz
0.05	0.0024	0.003	0.0009	0.0012
0.10	0.0071	0.009	0.0027	0.0035
0.20	0.024	0.030	0.0063	0.008
0.40	0.063	0.080	0.016	0.02
0.60	0.125	0.16	0.032	0.04
0.80	0.196	0.25	0.063	0.08
1.00	0.274	0.35	0.125	0.16

stressed amorphous material cracks. Laser and electro-discharge machining (EDM) cutting methods melt the amorphous material and cause undesirable crystallization. In addition, these methods make electrical contacts between laminations which contribute to the increased eddy-current and additional losses. In the early 1980s chemical methods were used in *General Electric* to cut amorphous materials but these methods were very slow and expensive [191]. The problem of cutting hard amorphous ribbons can be overcome by using a liquid jet [246]. This method makes it possible to cut amorphous materials in ambient temperature without cracking, melting, crystallization and electric contacts between isolated ribbons.

3.1.3 Soft Magnetic Powder Composites

Powder metallurgy is used in the production of ferromagnetic cores of small electrical machines or ferromagnetic cores with complicated shapes. The components of *soft magnetic powder composites* are iron powder, dielectric (epoxy resin) and filler (glass or carbon fibers) for mechanical strengthening. Powder composites for ferromagnetic cores of electrical machines and apparatus can be divided into [265]:

- dielectromagnetics and magnetodielectrics,
- magnetic sinters.

Dielectromagnetics and magnetodielectrics are names referring to materials consisting of the same basic components: ferromagnetic (mostly iron powder) and dielectric (mostly epoxy resin) material [265]. The main tasks of the dielectric material is insulation and binding of ferromagnetic particles. In practice, composites containing up to 2% (of their mass) of dielectric materials are considered as *dielectromagnetics*. Those with a higher content of dielectric material are considered as *magnetodielectrics* [265].

TSC International, Wadsworth, IL, U.S.A., has developed a new soft powder material, *Accucore*, which is competitive to traditional steel laminations [4]. The magnetization curve and specific core loss curves of the non-sintered *Accucore* are given in Table 3.7. When sintered, *Accucore* has higher saturation magnetic flux density than the non-sintered material. The specific density is 7550 to 7700 kg/m^3.

Höganäs, Höganäs, Sweden, manufactures soft magnetic composite (SMC) powders that are surface-coated metal powders with excellent compressibility [242]. *Somaloy^{TM} 500* (Tables 3.8, 3.9) has been developed for 3D magnetic circuits of electrical machines, transformers, ignition systems and sensors.

3.1.4 Fabrication of Stator Cores

Fabrication of Laminated Stator Cores

Normally, the stator cores are wound from electrotechnical steel strips and the slots are machined by shaping or planing. An alternative method is first

Table 3.7. Magnetization and specific core loss characteristics of non-sintered *Accucore* (*TSC Ferrite International*, Wadsworth, IL, U.S.A.)

Magnetization curve		Specific core loss curves		
Magnetic flux density, B T	Magnetic field intensity, H A/m	60 Hz W/kg	100 Hz W/kg	400 Hz W/kg
0.10	152	0.132	0.242	1.058
0.20	233	0.419	0.683	3.263
0.30	312	0.772	1.323	6.217
0.40	400	1.212	2.072	9.811
0.50	498	1.742	2.976	14.088
0.60	613	2.315	3.968	18.850
0.70	749	2.954	5.071	24.295
0.80	909	3.660	6.305	30.490
0.90	1107	4.431	7.650	37.346
1.00	1357	5.247	9.039	44.489
1.10	1677	6.129	10.582	52.911
1.20	2101	7.033	12.214	61.377
1.30	2687	7.981	13.845	70.151
1.40	3525	8.929	15.565	79.168
1.50	4763	9.965	17.394	90.302
1.60	6563	10.869	19.048	99.671
1.70	9035	11.707	20.635	109.880
1.75	10,746	12.125	21.407	

Table 3.8. Specific core losses of Somaloy^{TM}500 +0.5% Kenolube, 800 MPa, treated at 500°C for 30 min in the air, *Höganäs*, Höganäs, Sweden

Magnetic flux density, A/m	Specific losses W/kg					
	50 Hz	100 Hz	300 Hz	500 Hz	700 Hz	1000 Hz
0.4	1.5	3	12	18	27	45
0.5	1.9	3.6	17	27	40	60
0.6	2.7	6	21	34	55	90
0.8	4.6	10	32	52	92	120
1.0	6.8	16	48	80	140	180
2.0	30	50	170	270	400	570

to punch the slots with variable distances between them and then to wind the steel strip into the form of the slotted toroidal core (*R & D Institute of Electrical Machines VÜES* in Brno, Republic of Czech). In addition, this manufacturing process allows for making skewed slots to minimize the cogging torque and the effect of slot harmonics. Each stator core has skewed slots in opposite directions. It is recommended that a wave stator winding should be made to obtain shorter end connections and more space for the shaft. An odd

Table 3.9. Magnetization curves of Somaloy^{TM}500 +0.5% Kenolube, treated at 500°C for 30 min in the air, *Höganäs*, Höganäs, Sweden

Magnetic field intensity H A/m	B at density 6690 kg/m^3 T	B at density 7100 kg/m^3 T	B at density 7180 kg/m^3 T
1 500	0.7	0.83	0.87
3 200	0.8	1.13	1.19
4 000	0.91	1.22	1.28
6 000	1.01	1.32	1.38
10 000	1.12	1.42	1.51
15 000	1.24	1.52	1.61
20 000	1.32	1.59	1.69
40 000	1.52	1.78	1.87
60 000	1.65	1.89	1.97
80 000	1.75	1.97	2.05
100 000	1.82	2.02	2.10

number of slots, e.g. 25 instead of 24 can help to reduce the cogging torque (*VÚES* Brno).

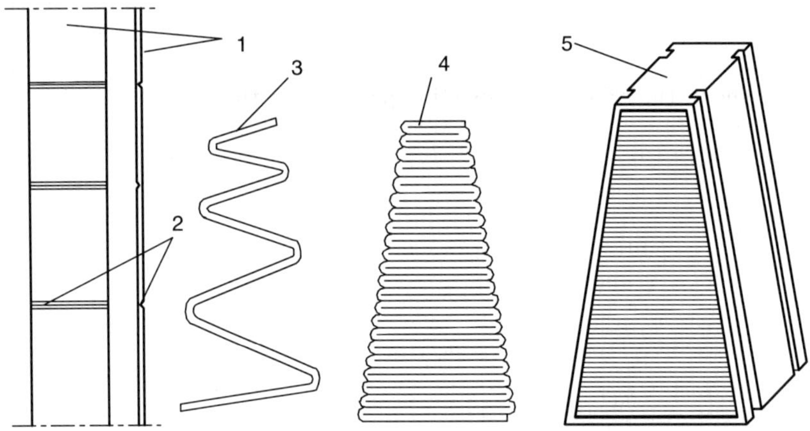

Fig. 3.1. Stator core segment formed from lamination strip: 1 — lamination strip, 2 — groove, 3 — folding, 4 — compressed segment, 5 — finished segment.

Another technique is to form the stator core using trapezoidal segments [257]. Each segment corresponds to one slot pitch (Fig. 3.1). The lamination strip of constant width is folded at distances proportional to the radius. To make folding easy, the strip has transverse grooves on opposite sides of the alternative steps. The zigzag laminated segment is finally compressed and fixed using a tape or thermosetting, as shown in Fig. 3.1 [257].

Fabrication of Soft Magnetic Powder Stator Cores

The laminated cores of axial flux machines are much more difficult to fabricate than those of radial flux machines. SMC powders simplify the manufacturing process of stator cores with complicated shapes, in general, 3D cores. Mass production of AFPM machines is much more cost effective if soft magnetic powder composites are used as materials for stator cores.

Using SMC powders the stator core of an AFPM machine can be made as a slotted core, slotless cylindrical core and salient pole core with one coil per pole.

The slotted and slotless cylindrical cores for AFPM machines can be made in a *powder metallurgy* process using a ferromagnetic powder with a small amount of lubricants or binders. The powder metallurgy process generally consists of four basic steps, namely: (1) powder manufacture, (2) mixing or blending, (3) compacting and (4) sintering. Most compacting is done with mechanical, hydraulic or pneumatic presses and rigid tools. Compacting pressures generally range between 70 to 800 MPa with 150 to 500 MPa being the most common. The outer diameter of the core is limited by the press capability. Frequently, the stator core must be divided into smaller segments. Most powder metallurgy products must have cross sections of less than 2000 mm^2. If the press capacity is sufficient, sections up to 6500 mm^2 can be pressed. Figure 3.2 shows the effect of compacting pressure on the density of *Höganäs* SMC powders.

For *Somaloy*^{TM}500 the heat treatment temperature (sintering) is typically 500°C for 30 min. After heat treatment the compacted powder has much less mechanical strength than solid steel.

The thermal expansion of conductors within the stator slots creates thermal expansion stresses in the stator teeth. The magnitude of these stresses depends upon the difference in the temperature of the winding and core, difference in coefficients of thermal expansion of both materials and slot fill factor. This problem is more important in powder cores than in laminated cores since the tensile stress of powder cores is at least 25 times lower and their modulus of elasticity is less than 100 GPa (versus 200 GPa for steel laminations).

Slotted stators for small disc-type motors fabricated from SMC powders are shown in Fig. 3.3 [160, 265]. SMC powder salient-pole stators for small single-sided AFPM motors manufactured by *Mii Technologies, LLC*, Lebanon, NH, U.S.A. are shown in Fig. 3.4. The three-phase stator has 9 poles. A single SMC powder salient pole manufactured by *Höganäs* is shown in Fig. 3.5.

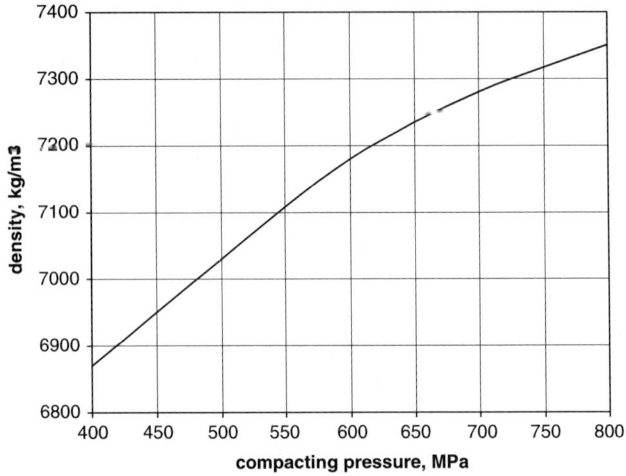

Fig. 3.2. The effect of compacting pressure on the specific mass density of Höganäs soft magnetic composite powders.

Fig. 3.3. Slotted stators for small single-sided disc-type motors [265].

3.2 Rotor Magnetic Circuits

Magnetic circuits of rotors consist of PMs and mild steel backing rings or discs. Since the air gap is somewhat larger that that in similar RFPM counterparts, high energy density PMs should be used.

Normally, surface magnets are glued to smooth backing rings or rings with cavities of the same shape as magnets without any additional mechanical protection against normal attractive forces. Epoxy, acrylic or silicon based adhesives are used for gluing between magnets and backing rings or between magnets. The minimum required shearing strength of adhesives is 20×10^6 Pa. There were attempts to develop interior PM rotor for AFPM machines. According to [221], rotor poles can only be fabricated by using soft magnetic powders [221]. The main advantage of this configuration is the improved flux

Fig. 3.4. Powder salient pole stators for small single-sided AFPM motors. Courtesy of *Mii Technologies, LLC*, Lebanon, NH, U.S.A.

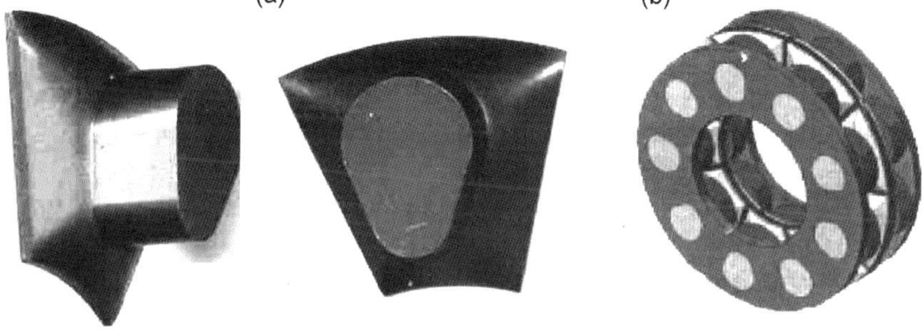

Fig. 3.5. SMC powder salient pole for small single-sided AFPM motors: (a) single SMC pole; (b) double-sided AFPM motor. Courtesy of *Höganäs*, Höganäs, Sweden.

weakening performance. However, the complexity and high cost of the rotor structure discourage further commercializing development.

3.2.1 PM Materials

A PM can produce magnetic flux in an air gap with no exciting winding and no dissipation of electric power. As any other ferromagnetic material, a PM can be described by its B—H hysteresis loop. PMs are also called *hard magnetic materials*, which means ferromagnetic materials with a wide hysteresis loop.

The basis for the evaluation of a PM is the portion of its hysteresis loop located in the upper left-hand quadrant, called the *demagnetization curve* (Fig. 3.6). If reverse magnetic field intensity is applied to a previously magnetized, say, toroidal specimen, the magnetic flux density drops down to the magnitude determined by the point K. When the reversal magnetic flux density is removed, the flux density returns to the point L according to a minor hysteresis loop. Thus, the application of a reverse field has reduced the *remanence*, or *remanent magnetism*. Reapplying magnetic field intensity will again reduce the flux density, completing the minor hysteresis loop by returning the core to approximately the same value of flux density at the point K as before. The minor hysteresis loop may usually be replaced with little error by a straight line called the *recoil line*. This line has a slope called the *recoil permeability* μ_{rec}.

As long as the negative value of applied magnetic field intensity does not exceed the maximum value corresponding to the point K, the PM may be regarded as being reasonably permanent. If, however, greater negative field intensity H is applied, the magnetic flux density will be reduced to a value lower than that at point K. On the removal of H, a new and lower recoil line will be established.

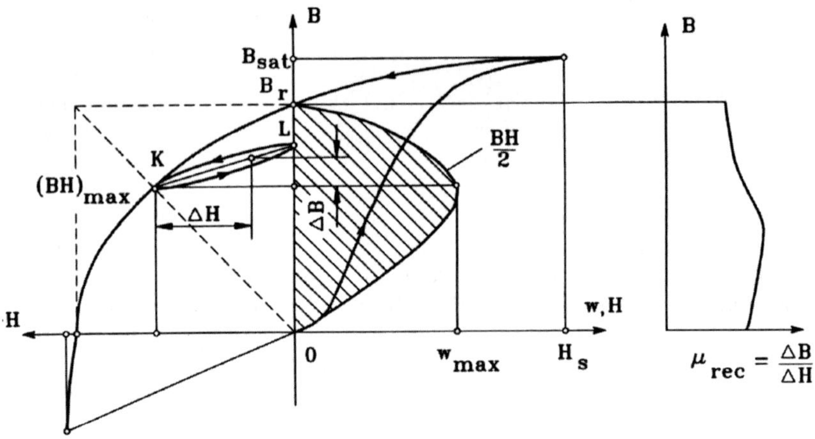

Fig. 3.6. Demagnetization curve, recoil loop, energy of a PM, and recoil magnetic permeability.

Remanent magnetic flux density B_r, or *remanence*, is the magnetic flux density corresponding to zero magnetic field intensity.

Coercive field strength H_c, or *coercivity*, is the value of demagnetizing field intensity necessary to bring the magnetic flux density to zero in a material previously magnetized.

Both B_r and H_c decrease as the magnet temperature increases, i.e.

$$B_r = B_{r20}[1 + \frac{\alpha_B}{100}(\vartheta_{PM} - 20)] \tag{3.2}$$

$$H_c = H_{c20}[1 + \frac{\alpha_H}{100}(\vartheta_{PM} - 20)] \tag{3.3}$$

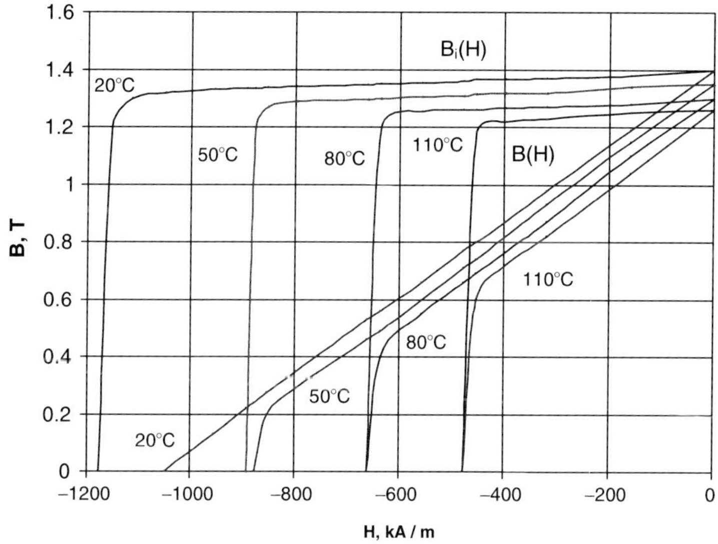

Fig. 3.7. Comparison of B—H and B_i—H demagnetization curves and their variations with the temperature for sintered N48M NdFeB PMs. Courtesy of *ShinEtsu*, Takefu-shi, Fukui Prefecture, Japan.

where ϑ_{PM} is the temperature of PM, B_{r20} and H_{c20} are the remanent magnetic flux density and coercive force at 20°C and $\alpha_B < 0$ and $\alpha_H < 0$ are *temperature coefficients* for B_r and H_c in %/°C respectively. Thus, demagnetization curves are sensitive to the temperature (Fig. 3.7).

Intrinsic demagnetization curve (Fig. 3.7) is the portion of the $B_i = f(H)$ hysteresis loop located in the upper left-hand quadrant, where $B_i = B - \mu_0 H$. For $H = 0$ the intrinsic magnetic flux density $B_i = B_r$.

Intrinsic coercivity ${}_iH_c$ is the magnetic field strength required to bring to zero the intrinsic magnetic flux density B_i of a magnetic material described by the $B_i = f(H)$ curve. For PM materials ${}_iH_c > H_c$.

Saturation magnetic flux density B_{sat} corresponds to high values of the magnetic field intensity, when an increase in the applied magnetic field produces no further effect on the magnetic flux density. In the *saturation region* the alignment of all the *magnetic moments of domains* is in the direction of the external applied magnetic field.

Recoil magnetic permeability μ_{rec} is the ratio of the magnetic flux density–to–magnetic field intensity at any point on the demagnetization curve, i.e.

$$\mu_{rec} = \mu_o \mu_{rrec} = \frac{\Delta B}{\Delta H} \tag{3.4}$$

where the *relative recoil permeability* $\mu_{rrec} = 1 \ldots 4.5$.

Maximum magnetic energy per unit produced by a PM in the external space is equal to the maximum magnetic energy density per volume, i.e.

$$w_{max} = \frac{(BH)_{max}}{2} \qquad \text{J/m}^3 \tag{3.5}$$

where the product $(BH)_{max}$ corresponds to the maximum energy density point on the demagnetization curve with coordinates B_{max} and H_{max} (Fig. 3.6).

Form factor of the demagnetization curve characterizes the concave shape of the demagnetization curve, i.e.

$$\gamma = \frac{(BH)_{max}}{B_r H_c} = \frac{B_{max} H_{max}}{B_r H_c} \tag{3.6}$$

for a square demagnetization curve $\gamma = 1$ and for a straight line (rare-earth PMs) $\gamma = 0.25$.

The leakage flux causes the magnetic flux to be distributed nonuniformly along the height $2h_M$ of a PM, where h_M is the height per pole. As a result, the MMF produced by the PM is not constant. The magnetic flux is higher in the neutral cross section and lower at the ends, but the behavior of the MMF distribution is the opposite [106].

The PM surface is not equipotential. The magnetic potential at each point on the surface is a function of the distance to the neutral zone. To simplify the calculation, the magnetic flux which is a function of the MMF distribution along the height h_M per pole is replaced by an equivalent flux. This equivalent flux goes through the whole height h_M and exits from the surface of the poles. To find the equivalent leakage flux and the whole flux of a PM, the equivalent magnetic field intensity needs to be found, i.e.

$$H = \frac{1}{h_M} \int_0^{h_M} H_x dx = \frac{\mathcal{F}_M}{h_M} \tag{3.7}$$

where H_x is the magnetic field intensity at a distance x from the neutral cross section and $\mathcal{F}_M$ is the MMF of the PM per pole (MMF $= 2\mathcal{F}_M$ per pole pair).

The equivalent magnetic field intensity (3.7) allows the equivalent leakage flux of the PM to be found, i.e.

$$\Phi_{lM} = \Phi_M - \Phi_g \tag{3.8}$$

where Φ_M is the full equivalent flux of the PM and Φ_g is the air gap magnetic flux. The *coefficient of leakage flux* of the PM,

$$\sigma_{lM} = \frac{\Phi_M}{\Phi_g} = 1 + \frac{\Phi_{lM}}{\Phi_g} > 1 \tag{3.9}$$

simply allows the air gap magnetic flux to be expressed as $\Phi_g = \Phi_M/\sigma_{lM}$.

The following leakage permeance expressed in the flux Φ–MMF coordinate system corresponds to the equivalent leakage flux of the PM:

$$G_{lM} = \frac{\Phi_{lM}}{\mathcal{F}_M} \tag{3.10}$$

An accurate estimation of the leakage permeance G_{lM} is the most difficult task in analytical calculation of magnetic circuits with PMs. Using the field approach, e.g. the FEM, the leakage permeance can be found fairly accurately.

The average equivalent magnetic flux and equivalent MMF mean that the magnetic flux density and magnetic field intensity are assumed to be the same in the whole volume of a PM. The full energy produced by the magnet in the outer space is

$$W = \frac{BH}{2} V_M \qquad \text{J} \tag{3.11}$$

where V_M is the volume of the PM or a system of PMs.

For a PM circuit with a rectangular cross section, single PM and two mild steel pole shoes, the magnetic flux density B_g in a given air gap volume $V_g = g w_M l_M$ is directly proportional to the square root of the magnetic energy product $(B_M H_M)$ [106], i.e.

$$B_g = \mu_0 H_g = \sqrt{\frac{\mu_0}{\sigma_{lM}} \left(1 + \frac{2H_{Fe} l_{Fe}}{H_g q}\right)^{-1} \frac{V_M}{V_g} B_M H_M}$$

$$\approx \sqrt{\mu_o \frac{V_M}{V_g} B_M H_M} \tag{3.12}$$

where H_{Fe} is the magnetic field intensity in the mild steel yoke, H_g is the magnetic field intensity in the air gap, $V_M = 2h_M w_M l_M$ is the magnet volume, w_M is the PM width, l_M is the PM length and $2l_{Fe}$ is the length of the magnetic flux path in two mild steel pole shoes. Following the trend to smaller packaging, smaller mass and higher efficiency, the material research in the field of PMs has focused on finding materials with high values of the maximum energy product $(BH)_{max}$.

The air gap magnetic flux density B_g can be estimated analytically on the basis of the demagnetization curve, air gap and leakage permeance lines and recoil lines [106]. Approximately, it can be found on the basis of the balance of the magnetic voltage drops, i.e.

$$\frac{B_r}{\mu_0 \mu_{rrec}} h_M = \frac{B_g}{\mu_0 \mu_{rrec}} h_M + \frac{B_g}{\mu_0} g$$

where μ_{rrec} is the relative permeability of the PM (relative recoil permeability). Hence,

$$B_g \approx \frac{B_r h_M}{h_M + \mu_{rrec} g} = \frac{B_r}{1 + \mu_{rrec} g / h_M} \tag{3.13}$$

The air gap magnetic flux density is proportional to the remanent magnetic flux density B_r and decreases as the air gap g increases. Eqn (3.13) can only be used for preliminary calculations.

For rare-earth PMs the approximation of the demagnetization curve is simple due to practically linear demagnetization curve, i.e.

$$B(H) = B_r \left(1 - \frac{H}{H_c}\right) \tag{3.14}$$

The approximation of more complicated demagnetizations curves (Alnico or ferrites) is given e.g. in [106].

The intersection point of the above demagnetization curve (3.14) and the following line representing the permeance of the air gap

$$B(H) = \mu_0 \frac{h_M}{g} H \tag{3.15}$$

gives a point called the *operating point*. This point corresponds to the air gap magnetic flux density B_g multiplied by the leakage coefficient σ_{lM} according to eqn (3.9).

3.2.2 Characteristics of PM Materials

There are three classes of PMs currently used for electric motors:

- Alnicos (Al, Ni, Co, Fe);
- Ceramics (ferrites), e.g. barium ferrite $BaO{\times}6Fe_2O_3$ and strontium ferrite $SrO{\times}6Fe_2O_3$;
- Rare-earth materials, i.e. samarium-cobalt SmCo and neodymium-iron-boron NdFeB.

The demagnetization curves of the above three types of permanent magnet materials are given in Fig. 3.8.

Alnico

Alnico magnets dominated the PM motor market in the range from a few watts to 150 kW between the mid 1940s and the late 1960s. The main advantages of Alnico are its high magnetic remanent flux density and low temperature coefficients (Table 3.10). The temperature coefficient of B_r is $-0.02\%/^\circ$C and maximum service temperature is 520°C. Unfortunately, the coercive force is very low and the demagnetization curve is extremely non-linear. Therefore, it is very easy not only to magnetize but also to demagnetize Alnico. Alnico has been used in PM d.c. commutator motors of the disc type with relatively large air gaps. This results in a negligible armature reaction magnetic flux acting on the PMs. Sometimes, Alnico PMs are protected from the armature flux, and consequently from demagnetization, using additional mild steel pole shoes.

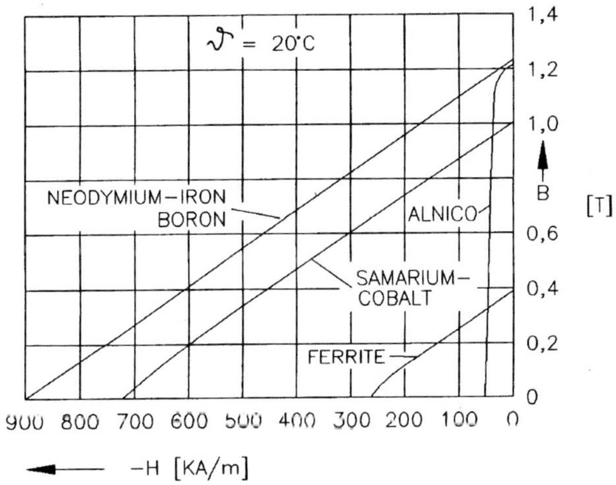

Fig. 3.8. Demagnetization curves for different PM materials.

Table 3.10. Physical properties of a representative selection of PM materials for small motors offered by *Magnaquench GmBH*, Essen, Germany

Property	Alnico sintered Koerzit 500	Hard ferrite bonded Koerox 12/22p	Hard ferrite sintered Koerox 350
Remanent flux density, B_r, T	1.24	0.26	0.39
Coercivity, H_c, kA/m	51	180	270
Intrinsic coercivity, iH_c, kA/m	51	225	310
$(BH)_{max}$, kJ/m^3	41.4	13	30
Relative recoil magnetic permeability	3 to 4.5	1.1	1.1
Temperature coefficient α_B of B_r at 20 to 100°C, %/°C	−0.02	−0.2	−0.2
Temperature coefficient α_{iH} of iH_c at 20 to 100°C, %/°C	+0.03 to −0.07	+0.4	+0.3
Curie temperature, °C	850	450	450
Maximum continuous service temperature, °C	500	100 to 200	200
Thermal conductivity, W/(m °C)	10 to 100	–	4
Specific mass density, ρ_{PM}, kg/m^3	7300	3400	4800
Electric conductivity, S/m	1.4 to 2.5 $\times 10^6$	< 0.0001	< 0.0001
Coefficient of thermal expansion at 20 to 100°C, $\times 10^{-6}$/°C	11 to 13	20 to 50	12 parallel 8 vertical
Specific heat, J /(°C kg)	350 to 500	–	800

Ferrites

Barium and strontium *ferrites* produced by powder metallurgy were invented in the 1950s. Their chemical formulation may be expressed as $MO\times 6(Fe_2O_3)$, where M is Ba, Sr, or Pb. Ferrite magnets are available in *isotropic* and *anisotropic* grades.

A ferrite has a higher coercive force than Alnico, but at the same time has a lower remanent magnetic flux density (Table 3.10). Temperature coefficients are relatively high, i.e. the coefficient of B_r is $-0.20\%/^{\circ}$C and the coefficient of H_c is -0.27 to $-0.4\%/^{\circ}$C. The maximum service temperature is 450°C. The main advantages of ferrites are their low cost and very high electric resistance, which means practically no eddy-current losses in the PM volume. Ferrite magnets are most economical in fractional horsepower motors. Barium ferrite PMs are commonly used in small d.c. commutator motors for automobiles (blowers, fans, windscreen wipers, pumps, etc.) and electric toys.

Rare-Earth Permanent Magnets

The first generation of *rare-earth permanent magnets*, i.e. alloys based on the composition of $SmCo_5$ has been commercially produced since the early 1970s (invented in the 1960s). $SmCo_5$ has the advantage of a high remanent flux density, high coercive force, high energy product, a linear demagnetization curve and a low temperature coefficient (Table 3.11). The temperature coefficient of B_r is -0.02 to $-0.045\%/^{\circ}$C and the temperature coefficient of H_c is -0.14 to $-0.40\%/^{\circ}$C. Maximum service temperature is 300 to 350°C. It is suitable for motors with low volumes and motors operating at increased temperatures, e.g. brushless generators for microturbines. Both Sm and Co are relatively expensive due to their supply restrictions.

With the discovery in the recent years of a second generation of rare-earth magnets on the basis of inexpensive neodymium (Nd), remarkable progress with regard to lowering raw material costs has been achieved. The new generation of rare-earth PMs based on inexpensive neodymium (Nd) was announced by *Sumitomo Special Metals*, Japan, in 1983 at the 29th Annual Conference of Magnetism and Magnetic Materials held in Pittsburgh, PA, U.S.A. The Nd is a much more abundant rare-earth element than Sm. NdFeB magnets, which are now produced in increasing quantities have better magnetic properties (Table 3.12) than those of SmCo, but unfortunately only at room temperature. The demagnetization curves, especially the coercive force, are strongly temperature dependent. The temperature coefficient of B_r is -0.09 to $-0.15\%/^{\circ}$C and the temperature coefficient of H_c is -0.40 to $-0.80\%/^{\circ}$C. The maximum service temperature is 250°C and Curie temperature is 350°C). The NdFeB is also susceptible to corrosion. NdFeB magnets have great potential for considerably improving the *performance–to–cost* ratio for many applications. For this reason they will have a major impact on the development and application of PM machines in the future.

Table 3.11. Physical properties of Vacomax sintered Sm_2Co_{17} PM materials at room temperature 20°C manufactured by *Vacuumschmelze GmbH*, Hanau, Germany

Property	Vacomax 240 HR	Vacomax 225 HR	Vacomax 240
Remanent flux density, B_r, T	1.05 to 1.12	1.03 to 1.10	0.98 to 1.05
Coercivity, H_c, kA/m	600 to 730	720 to 820	580 to 720
Intrinsic coercivity, iH_c, kA/m	640 to 800	1590 to 2070	640 to 800
$(BH)_{max}$, kJ/m^3	200 to 240	190 to 225	180 to 210
Relative recoil magnetic permeability	1.22 to 1.39	1.06 to 1.34	1.16 to 1.34
Temperature coefficient α_B of B_r at 20 to 100°C, %/°C		−0.030	
Temperature coefficient α_{iH} of iH_c at 20 to 100°C, %/°C	−0.15	−0.18	−0.15
Temperature coefficient α_B of B_r at 20 to 150°C, %/°C		−0.035	
Temperature coefficient α_{iH} of iH_c at 20 to 150°C, %/°C	−0.16	−0.19	−0.16
Curie temperature, °C		approximately 800	
Maximum continuous service temperature, °C	300	350	300
Thermal conductivity, W/(m °C)		approximately 12	
Specific mass density, ρ_{PM}, kg/m^3		8400	
Electric conductivity, $\times 10^6$ S/m		1.18 to 1.33	
Coefficient of thermal expansion at 20 to 100°C, $\times 10^{-6}$/°C		10	
Young's modulus, $\times 10^6$ MPa		0.150	
Bending stress, MPa		90 to 150	
Vicker's hardness		approximately 640	

Chemical reactivity of rare-earth magnets is similar to that of alkaline earth metals, e.g. magnesium. The reaction is accelerated at increased temperature and humidity. The NdFeB alloy if exposed to hydrogen gas, usually at a slightly elevated temperature and/or elevated pressure, becomes brittle and with very little effort, it can be crushed. Diffusion of hydrogen into the alloy causes it literally to fall apart.

Corrosion protective coatings can be divided into metallic and organic. For metallic coatings, e.g. nickel and tin, galvanic processes are used as a rule. Organic coatings include powder coatings applied electrostatically, varnishes and resins.

Nowadays, for the industrial production of rare-earth PMs the powder metallurgical route is mainly used [212]. Apart from some material specific parameters, this processing technology is, in general, the same for all rare-earth magnet materials. The alloys are produced by vacuum induction melting or by a calciothermic reduction of the oxides. The material is then size-reduced

Table 3.12. Physical properties of Hicorex-Super sintered NdFeB PM materials at room temperature 20°C manufactured by *Hitachi Metals, Ltd.*, Tokyo, Japan

Property	Hicorex-Super HS-38AV	Hicorex-Super HS-25EV	Hicorex-Super HS-47AH
Remanent flux density, B_r, T	1.20 to 1.30	0.98 to 1.08	1.35 to 1.43
Coercivity, H_c, kA/m	875 to 1035	716 to 844	1018 to 1123
Intrinsic coercivity, iH_c, kA/m	min. 1114	min. 1989	min. 1114
$(BH)_{max}$, kJ/m^3	278 to 319	183 to 223	342 to 390
Relative recoil magnetic permeability		1.03 to 1.06	
Temperature coefficient α_B of B_r at 20 to 100°C, %/°C		−0.11 to −0.13	
Temperature coefficient α_{iH} of iH_c at 20 to 100°C, %/°C		−0.65 to −0.72	
Curie temperature, °C		$\approx$ 310	
Maximum continuous service temperature, °C	160	180	140
Thermal conductivity, W/(m°C)		$\approx$ 7.7	
Specific mass dens., ρ_{PM}, kg/m^3		7500	
Electric conductivity, $\times 10^6$ S/m		$\approx$ 0.67	
Coefficient of thermal expansion at 20 to 100°C, $\times 10^{-6}$/°C		−1.5	
Young's modulus, $\times 10^6$ MPa		0.150	
Bending stress, MPa		260	
Vicker's hardness		$\approx$ 600	
Features	High energy product	High temperature	Super high performance

by crushing and milling to a single crystalline powder with particle sizes less than 10 μm. In order to obtain anisotropic PMs with the highest possible $(BH)_{max}$ value, the powders are then aligned in an external magnetic field, pressed and densified to nearly theoretical density by sintering. The most economical method for mass production of simply shaped parts like blocks, rings or arc segments is die pressing of the powders in approximately the final shape.

Researchers at *General Motors*, U.S.A., developed a fabrication method based on the melt-spinning casting system originally invented for the production of amorphous metal alloys. In this technology a molten stream of NdFeCoB material is first formed into ribbons 30 to 50-μm thick by rapid quenching, then cold pressed, extruded and hot pressed into bulk. Hot pressing and hot working are carried out while maintaining the fine grain to provide a high density close to 100% which eliminates the possibility of internal corrosion. The standard electro-deposited epoxy resin coating provides excellent corrosion resistance.

The prices of NdFeB magnets ordered in large quantities are now below US$20 per kg. Owing to a large supply of NdFeB magnets from China it is expected that the prices will fall further.

3.2.3 Operating Diagram

The energy of a PM in the external space only exists if the reluctance of the external magnetic circuit is higher than zero. If a previously magnetized PM is placed inside a closed ideal ferromagnetic circuit, i.e. toroid, this PM does not show any magnetic properties in the external space, in spite of the fact that there is the magnetic flux

$$\Phi_r = B_r S_M = B_r w_M l_M \tag{3.16}$$

corresponding to the remanent flux density B_r inside the PM.

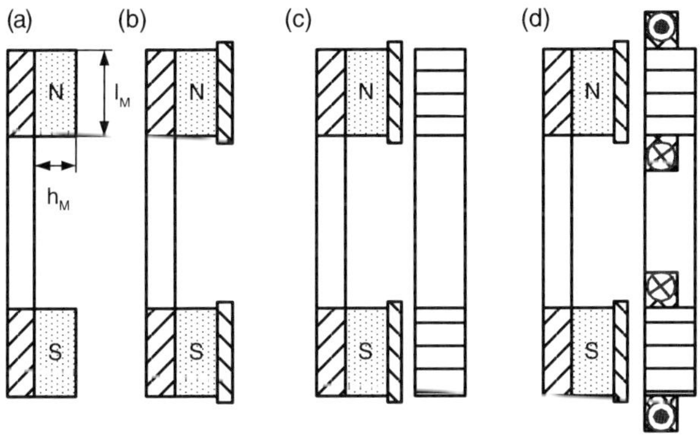

Fig. 3.9. Stabilization of a PM: (a) PM alone, (b) PM with pole shoes, (c) PM inside an external magnetic circuit, (d) PM with a complete external armature system.

A PM previously magnetized and placed alone in an open space, as in Fig. 3.9a, generates a magnetic field. To sustain a magnetic flux in the external open space, an MMF developed by the magnet is necessary. The state of the PM is characterized by the point K on the demagnetization curve (Fig. 3.10). The location of the point K is at the intersection of the demagnetization curve with a straight line representing the permeance of the external magnetic circuit (open space):

$$G_{ext} = \frac{\Phi_K}{\mathcal{F}_K}, \qquad \tan\alpha_{ext} = \frac{\Phi_K/\Phi_r}{\mathcal{F}_K/\mathcal{F}_c} = G_{ext}\frac{\mathcal{F}_c}{\Phi_r} \tag{3.17}$$

The permeance G_{ext} corresponds to the flux Φ—MMF coordinate system and is referred to as MMF at the ends of the PM. In the Φ—MMF coordinate system the remanent flux Φ_r is according to eqn (3.16) and the MMF corresponding to the coercivity H_c is

$$\mathcal{F}_c = H_c h_M \tag{3.18}$$

The magnetic energy per unit produced by the PM in the external space is $w_K = B_K H_K/2$. This energy is proportional to the rectangle limited by the coordinate system and lines perpendicular to the Φ and $\mathcal{F}$ coordinates projected from the point K. It is obvious that the maximum magnetic energy is for $B_K = B_{max}$ and for $H_K = H_{max}$.

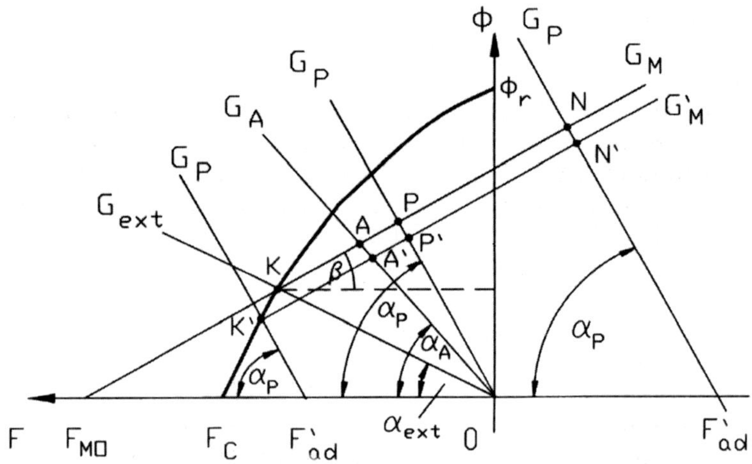

Fig. 3.10. Diagram of a PM for finding the origin of the recoil line and operating point.

If the poles are furnished with pole shoes (Fig. 3.9b) the permeance of the external space increases. The point which characterizes a new state of the PM in Fig. 3.10 moves along the recoil line from the point K to the point A. The recoil line KG_M is the same as the internal permeance of the PM, i.e.

$$G_M = \mu_{rec}\frac{w_M l_M}{h_M} = \mu_{rec}\frac{S_M}{h_M} \tag{3.19}$$

The point A is the intersection of the recoil line KG_M and the straight line OG_A representing the leakage permeance of the PM with pole shoes, i.e.

$$G_A = \frac{\Phi_A}{\mathcal{F}_A}, \qquad \tan\alpha_A = G_A\frac{\mathcal{F}_c}{\Phi_r} \tag{3.20}$$

The energy produced by the PM in the external space decreases as compared with the previous case, i.e. $w_A = B_A H_A/2$.

The next stage is to place the PM in an external ferromagnetic circuit as shown in Fig. 3.9c. The resultant permeance of this system is

$$G_P = \frac{\Phi_P}{\mathcal{F}_P}, \qquad \tan \alpha_P = G_P \frac{\mathcal{F}_c}{\Phi_r} \tag{3.21}$$

which meets the condition $G_P > G_A > G_{ext}$. For an external magnetic circuit without any electric circuit carrying the armature current, the magnetic state of the PM is characterized by the point P (Fig. 3.10), i.e. the intersection of the recoil line KG_M and the permeance line OG_P.

When the external magnetic circuit is furnished with an armature winding and when this winding is fed with a current which produces an MMF magnetizing the PM (Fig. 3.9d), the magnetic flux in the PM increases to the value Φ_N. The d-axis MMF $\mathcal{F}'_{ad}$ of the external (armature) field acting directly on the PM corresponds to Φ_N. The magnetic state of the PM is described by the point N located on the recoil line on the right-hand side of the origin of the coordinate system. To obtain this point it is necessary to lay off the distance $O\mathcal{F}'_{ad}$ and to draw a line G_P from the point $\mathcal{F}'_{ad}$ inclined by the angle α_P to the F-axis. The intersection of the recoil line and the permeance line G_P gives the point N. If the exciting current in the external armature winding is increased further, the point N will move further along the recoil line to the right, up to the saturation of the PM.

When the excitation current is reversed, the external armature magnetic field will demagnetize the PM. In this case it is necessary to lay off the distance $O\mathcal{F}'_{ad}$ from the origin of the coordinate system to the left (Fig. 3.10). The line G_P drawn from the point $\mathcal{F}'_{ad}$ with the slope α_P intersects the demagnetization curve at the point K'. This point can be above or below the point K (for the PM alone in the open space). The point K' is the origin of a new recoil line $K'G'_M$. Now if the armature exciting current decreases, the operating point will move along the new recoil line $K'G'_M$ to the right. If the armature current drops down to zero, the operating point takes the position P' (intersection of the new recoil line $K'G'_M$ with the permeance line G_P drawn from the origin of the coordinate system).

On the basis of Fig. 3.10 the energies $w_{P'} = B_{P'}H_{P'}/2$, $w_P = B_P H_P/2$, and $w_{P'} < w_P$. The location of the origin of the recoil line, as well as the location of the operating point, determine the level of utilization of the energy produced by the PM. A *PM behaves differently than a d.c. electromagnet*: the energy of a PM is not constant if the permeance and exciting current of the external armature changes.

The location of the origin of the recoil line is determined by the minimum value of the permeance of the external magnetic circuit or the demagnetization action of the external field.

To obtain the properties of PMs more independent of the external magnetic fields, PMs need to be stabilized. *Stabilization* means the PM is demagnetized up to a value which is slightly higher than the most dangerous demagnetization field during the operation of a system where the PM is installed. In magnetic circuits with stabilized PMs the operating point describing the state of the PM is located on the recoil line.

More details about how to find the operating point of a PM graphically and analytically can be found in [106].

3.2.4 Permeances for Main and Leakage Fluxes

Permeances of air gaps and permeances for leakage fluxes can be found analytically by dividing the magnetic field into simple solids. Permeances of simple solids shown in Fig. 3.11 can be found using the following formulae:

(a) Rectangular prism (Fig. 3.11a)

$$G = \mu_0 \frac{w_M l_M}{g} \quad \text{H} \tag{3.22}$$

(b) Cylinder (Fig. 3.11b)

$$G = \mu_0 \frac{\pi d_M^2}{4g} \tag{3.23}$$

(c) Half-cylinder (Fig. 3.11c)

$$G = 0.26 \mu_0 l_M \tag{3.24}$$

where the average air gap $g_{av} = 1.22g$ and the average surface $S_{av} = 0.322 g l_M$ [15]

(d) One-quarter of a cylinder (Fig. 3.11d)

$$G = 0.52 \mu_0 l_M \tag{3.25}$$

(e) Half-ring (Fig. 3.11e)

$$G = \mu_0 \frac{2 l_M}{\pi (g/w_M + 1)} \tag{3.26}$$

For $g < 3w_M$,

$$G = \mu_0 \frac{l_M}{\pi} \ln \left(1 + \frac{2 w_M}{g}\right) \tag{3.27}$$

(f) One-quarter of a ring (Fig. 3.11f)

$$G = \mu_0 \frac{2 l_M}{\pi (g/c + 0.5)} \tag{3.28}$$

For $g < 3c$,

$$G = \mu_0 \frac{2 l_M}{\pi} \ln \left(1 + \frac{c}{g}\right) \tag{3.29}$$

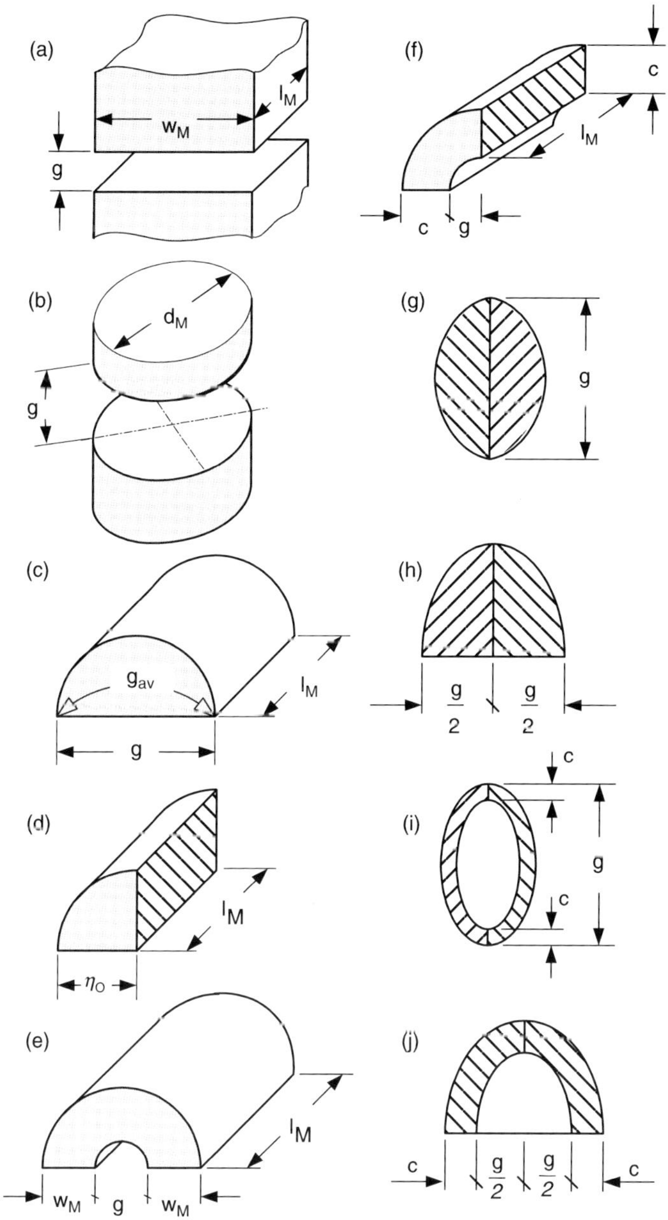

Fig. 3.11. Simple solids: (a) rectangular prism, (b) cylinder, (c) half-cylinder, (d) one-quarter of a cylinder, (e) half-ring, (f) one-quarter of a ring, (g) one quarter of a sphere, (h) one-eighth of a sphere, (i) one-quarter of a shell, (j) one-eighth of a shell.

(g) One-quarter of a sphere (Fig. 3.11g)

$$G = 0.077\mu_0 g \tag{3.30}$$

(h) One-eighth of a sphere (Fig. 3.11h)

$$G = 0.308\mu_0 g \tag{3.31}$$

(i) One-quarter of a shell (Fig. 3.11i)

$$G = \mu_0 \frac{c}{4} \tag{3.32}$$

(j) One-eighth of a shell (Fig. 3.11j)

$$G = \mu_0 \frac{c}{2} \tag{3.33}$$

Fig. 3.12 shows a model of a flat electrical machine with smooth armature core (without slots) and surface PM excitation system. The armature is of steel laminations. The PMs are fixed to the mild steel yoke.

The pole pitch is τ, the width of each PM is w_M, and its length is l_M. In an AFPM machine

$$l_M = 0.5(D_{out} - D_{in}) \tag{3.34}$$

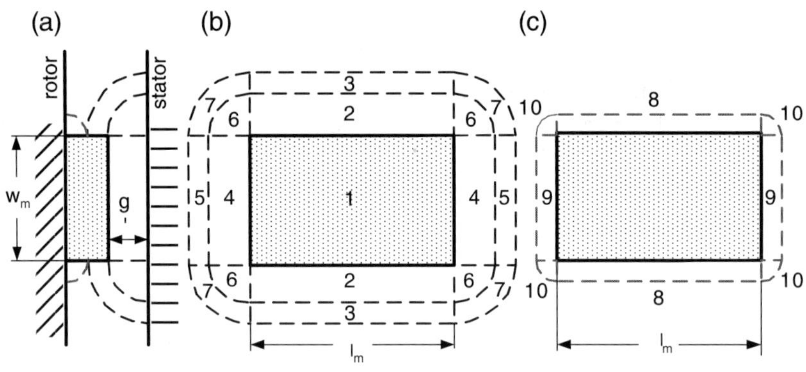

Fig. 3.12. Electrical machine with flat slotless armature and flat PM excitation system — division of the space occupied by the magnetic field into simple solids: (a) longitudinal section, (b) air gap field, (c) leakage field (between the PM and steel yoke).

The space between the pole face and the armature core is divided into a prism (1), four quarters of cylinders (2 and 4), four quarters of rings (3 and 5), four pieces of 1/8 of a sphere (6) and four pieces of 1/8 of a shell (7). Formulae for the permeance calculations have been found on the assumption that the permeance of a solid is equal to its average cross section area to the average length of the flux line. If we ignore the fringing flux, the permeance of a rectangular air gap per pole (prism 1 in Fig. 3.12) is

$$G_{g1} = \mu_0 \frac{w_M l_M}{g'} \tag{3.35}$$

The equivalent air gap g' is only equal to the nonmagnetic gap (mechanical clearance) g for a slotless and unsaturated armature. To take into account slots (if they exist) and magnetic saturation, the air gap g is increased to $g' = gk_C k_{sat}$, where $k_C > 1$ is Carter's coefficient taking into account slots (1.2), and $k_{sat} > 1$ is the saturation factor of the magnetic circuit defined as the ratio of the MMF per pole pair to the air gap magnetic voltage drop (MVD) taken twice [106].

To take into account the fringing flux it is necessary to include all paths for the magnetic flux coming from the excitation system through the air gap to the armature system (Fig. 3.12), i.e.

$$G_g = G_{g1} + 2(G_{g2} + G_{g3} + G_{g4} + G_{g5}) + 4(G_{g6} + G_{g7}) \tag{3.36}$$

where G_{g1} is the air gap permeance according to eqn (3.35) and G_{g2} to G_{g7} are the air gap permeances for fringing fluxes. The permeances G_{g2} to G_{g5} can be found using eqns (3.25), (3.28), (3.31) and (3.32).

In a similar way the resultant permeance for the leakage flux of the PM can be found, i.e.

$$G_{lM} = 2(G_{l8} + G_{l9}) + 4G_{l10} \tag{3.37}$$

where G_{l8}, G_{l9} (one-quarter of a cylinder) and G_{l10} (one-eight of a sphere) are the permeances for leakage fluxes between the PM and rotor yoke according to Fig. 3.12c — eqns (3.25) and (3.31).

3.2.5 Calculation of Magnetic Circuits With PMs

The *equivalent magnetic circuit of a PM system* with armature is shown in Fig. 3.13. The reluctances of pole shoes (mild steel) and armature stack (electrotechnical laminated steel) are much smaller than those of the air gap and PM and have been neglected. The "open circuit" MMF acting along the internal magnet permeance $G_M = 1/\Re_{\mu M}$ is $\mathcal{F}_{M0} = H_{M0} h_M$, the d-axis armature reaction MMF is $\mathcal{F}_{ad}$, the total magnetic flux of the PM is Φ_M, the leakage flux of the PM is Φ_{lM}, the useful air gap magnetic flux is Φ_g, the leakage flux of the external armature system is Φ_{la}, the flux in the d-axis produced by the armature is Φ_{ad} (demagnetizing or magnetizing), the reluctance for the PM leakage flux is $\Re_{\mu lM} = 1/G_{lM}$, the air gap reluctance is $\Re_{\mu g} = 1/G_g$, and the external armature leakage reactance is $\Re_{\mu la} = 1/G_{gla}$. The following Kirchhoff's equations can be written on the basis of the equivalent circuit shown in Fig. 3.13

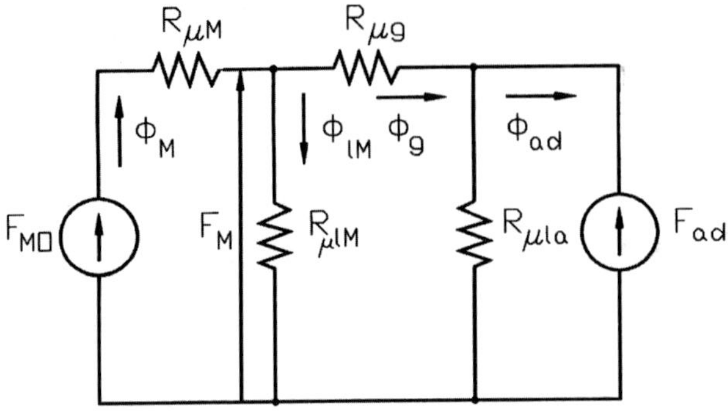

Fig. 3.13. Equivalent circuit (in the d-axis) of a PM system with armature.

$$\Phi_M = \Phi_{lM} + \Phi_g$$

$$\Phi_{la} = \frac{\pm \mathcal{F}_{ad}}{\Re_{\mu la}}$$

$$\mathcal{F}_{M0} - \Phi_M \Re_{\mu M} - \Phi_{lM} \Re_{\mu lM} = 0$$

$$\Phi_{lM} \Re_{lM} - \Phi_g \Re_{\mu g} \mp \mathcal{F}_{ad} = 0$$

The solution to the above equation system yields the air gap magnetic flux:

$$\Phi_g = \left[\mathcal{F}_{Mo} \mp \mathcal{F}_{ad} \frac{G_g}{G_g + G_{lM}} \frac{(G_g + G_{lM})(G_M + G_{lM})}{G_g G_M} \right] \frac{G_g G_M}{G_g + G_{lM} + G_M}$$

or

$$\Phi_g = \left[\mathcal{F}_{M0} \mp \mathcal{F}'_{ad} \frac{G_t (G_M + G_{lM})}{G_g G_M} \right] \frac{G_g G_M}{G_t + G_M} \tag{3.38}$$

where the total resultant permeance G_t for the flux of the PM is

$$G_t = G_g + G_{lM} = \sigma_{lM} G_g \tag{3.39}$$

and the direct-axis armature MMF acting directly on the PM is

$$\mathcal{F}'_{ad} = \mathcal{F}_{ad} \frac{G_g}{G_g + G_{lM}} = \mathcal{F}_{ad} \left(1 + \frac{G_{lM}}{G_g} \right)^{-1} = \frac{\mathcal{F}_{ad}}{\sigma_{lM}} \tag{3.40}$$

The upper sign in eqn (3.38) is for the demagnetizing armature flux and the lower sign is for the magnetizing armature flux.

The coefficient of the PM leakage flux (3.9) can also be expressed in terms of permeances, i.e.

$$\sigma_{lM} = 1 + \frac{\Phi_{lM}}{\Phi_g} = 1 + \frac{G_{lM}}{G_g} \tag{3.41}$$

3.2.6 Fabrication of Rotor Magnetic Circuits

Magnetic circuits of rotors of AFPM brushless machines provide the excitation flux and are designed as:

- PMs glued to a ferromagnetic ring or disc which serves as a backing magnetic circuit (yoke);
- PMs arranged into Halbach array without any ferromagnetic core.

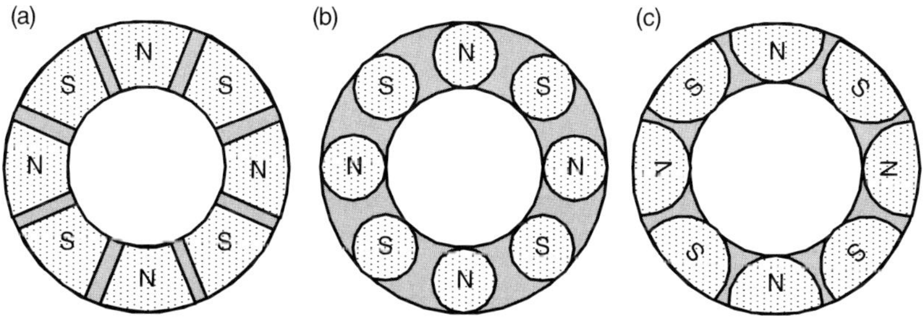

Fig. 3.14. Shapes of PM rotors of disc-type machines: (a) trapezoidal, (b) circular, (c) semicircular.

Shapes of PMs are usually trapezoidal, circular or semicircular (Fig. 3.14). The shape of PMs affects the distribution of the air gap magnetic field and contents of higher space harmonics. The output voltage quality (harmonics of the EMF) of AFPM generators depends on the PM geometry (circular, semicircular, trapezoidal) and distance between adjacent magnets [83].

Since the magnetic flux in the rotor magnetic circuit is stationary, mild steel (carbon steel) backing rings can be used. Rings can be cut from 4 to 6 mm mild steel sheets. Table 3.13 shows magnetization characteristics B–H of a mild carbon steel and cast iron. Electrical conductivities of carbon steels are from 4.5×10^6 to 7.0×10^6 S/m at 20°C.

Halbach Array

Twin rotors of double-sided coreless AFPM machines (Fig. 1.4d) may use PMs arranged in *Halbach array* [114, 115, 116]. The key concept of Halbach array is that the magnetization vector of PMs should rotate as a function of

Table 3.13. Magnetization curves of solid ferromagnetic materials: 1 — carbon steel (0.27%C), 2 — cast iron

Magnetic flux density, B	Magnetic field intensity, H	
	Mild carbon steel 0.27% C	Cast iron
T	A/m	A/m
0.2	190	900
0.4	280	1600
0.6	320	3000
0.8	450	5150
1.0	900	9500
1.2	1500	18,000
1.4	3000	28,000
1.5	1500	
1.6	6600	
1.7	11,000	

distance along the array (Fig. 3.15) [114, 115, 116]. Halbach array has the following advantages:

- the fundamental field is stronger by a factor of 1.4 than in a conventional PM array, and thus the power efficiency of the machine is doubled;
- the array of PMs does not require any backing steel magnetic circuit and PMs can be bonded directly to a non-ferromagnetic supporting structure (aluminum, plastics);
- the magnetic field is more sinusoidal than that of a conventional PM array;
- Halbach array has very low back-side fields.

The peak value of the magnetic flux density at the active surface of Halbach array is

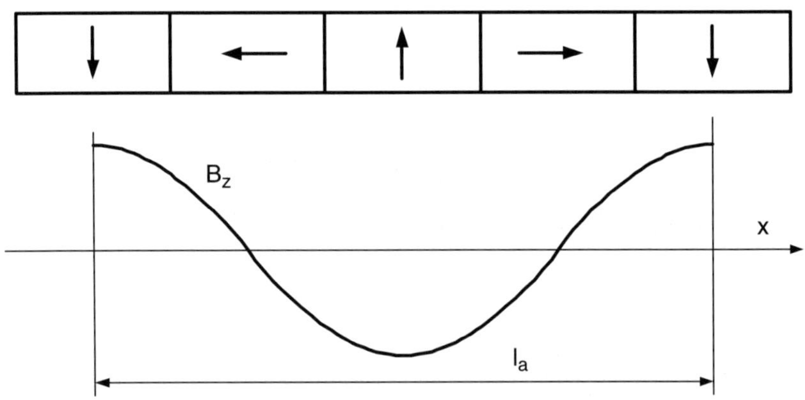

Fig. 3.15. Cartesian Halbach array.

$$B_{m0} = B_r[1 - \exp(-\beta h_M)]\frac{\sin(\pi/n_M)}{\pi/n_M} \tag{3.42}$$

where B_r is the remanent magnetic flux density of the magnet, $\beta = 2\pi/l_a$ - see also eqn (1.11), l_a is the spatial period (wavelength) of the array and n_M is the number of PM pieces per wavelength. For the array shown in Fig. 3.15 $n_M = 4$. For example, assuming $B_r = 1.25$ T, $h_M = 6$ mm, $l_a = 24$ mm, $n_m = 4$ (rectangular PMs), the peak magnetic flux density at the surface of Halbach array $B_{M0} = 0.891$ T.

The tangential B_x and normal B_z components of Halbach array at the distance z from the surface of PMs are

$$B_x(x, z) = B_{m0}cos(\beta x)exp(-\beta z) \tag{3.43}$$

$$B_z(x, z) = B_{m0}sin(\beta x)exp(-\beta z) \tag{3.44}$$

For a double-sided configuration of Halbach arrays, i.e. twin disc external rotors, the tangential and normal component of the magnetic flux density distribution in the space between discs are

$$B_x(x, z) = B_{m0}\frac{1}{\beta}\cos(\beta x)\frac{1}{\cosh(\beta t/2)}\sinh(\beta z) \tag{3.45}$$

$$B_z(x, z) = B_{m0}sin(\beta x)\frac{1}{\cosh(\beta t/2)}\cosh(\beta z) \tag{3.46}$$

where B_{m0} is according to eqn (3.42) and t is magnet–to–magnet distance between two halves of the disc. The origin of the $0xyz$ coordinate system is as in Fig. 1.9.

3.3 Windings

3.3.1 Conductors

Stator (armature) windings of electric motors are made of solid *copper conductor wires* with round or rectangular cross sections.

The *electric conductivity* at 20°C of copper wires is $57 \times 10^6 \geq \sigma_{20} \geq 56 \times 10^6$ S/m. For aluminum wires $\sigma_{20} \approx 33 \times 10^6$ S/m. The electric conductivity is temperature dependent and for $\theta - 20^{\circ} \leq 150^{\circ}$C can be expressed as

$$\sigma = \frac{\sigma_{20}}{1 + \alpha(\theta - 20^{\circ})} \tag{3.47}$$

where α is the *temperature coefficient of electric resistance*. For copper wires $\alpha = 0.00393$ 1/°C and for aluminum wires $\alpha = 0.00403$ 1/°C. For $\vartheta - 20^{\circ} >$

150°C eqn (3.47) contains two temperature coefficients α and β of the electric resistance, i.e.

$$\sigma = \frac{\sigma_{20}}{1 + \alpha(\vartheta - 20^{\circ}) + \beta(\vartheta - 20^{\circ})^2} \tag{3.48}$$

Table 3.14. Maximum temperature rise $\Delta\vartheta$ for armature windings of electrical machines according to IEC and NEMA (based on 40°C ambient temperature)

Rated power of machines, length of core and voltage	Insulation class				
	A °C	E °C	B °C	F °C	H °C
IEC a.c. machines < 5000 kVA (resistance method)	60	75	80	100	125
IEC a.c. machines ≥ 5000 kVA or length of core ≥ 1 m (embedded detector method)	60	70	80	100	125
NEMA a.c. machines ≤ 1500 hp (embedded detector method)	70	–	90	115	140
NEMA a.c. machines > 1500 hp and ≤ 7 kV (embedded detector method)	65	–	85	110	135

The *maximum temperature rise* for the windings of electrical machines is determined by the temperature limits of insulating materials. The maximum temperature rise in Table 3.14 assumes that the temperature of the cooling medium $\vartheta_c \leq 40$°C. The maximum temperature of windings is actually

$$\vartheta_{max} = \vartheta_c + \Delta\vartheta \tag{3.49}$$

where $\Delta\vartheta$ is the maximum allowable temperature rise according to Table 3.14. A polyester-imide and polyamide-imide coat can provide an operating temperature of 200°C. The highest operating temperatures (over 600°C) can be achieved using *nickel clad copper* or *palladium-silver* conductor wires and ceramic insulation.

3.3.2 Fabrication of Slotted Windings

Stator windings are usually made of insulated copper conductors. The cross section of conductors can be circular or rectangular. For large AFPM machines a direct water cooling system and consequently hollow conductors can be considered.

It is difficult to make and form stator coils if the round conductor is thicker than 1.5 mm. If the current density is too high, parallel conductor wires of smaller diameter are recommended rather than one thicker wire. Stator windings can also have parallel current paths.

The armature windings can be either *single-layer* or *double layer* (Section 2.2).

After coils are wound, they must be secured in place, somehow, so as to avoid conductor movement. Two standard methods are used to secure the conductors of electrical machines in place:

- *dipping* the whole component into a varnish-like material, and then baking off its solvent,
- *trickle impregnation* method, which uses heat to cure a catalyzed resin which is dripped onto the component.

Polyester, epoxy or silicon resins are used most often as impregnating materials for treatment of stator windings. Silicon resins of high thermal endurance are able to withstand $\vartheta_{max} > 225^{\circ}$C.

Recently, a new method of conductor securing that does not require any additional material, and uses very low energy input, has emerged [178]. The solid conductor wire (usually copper) is coated with a *heat and/or solvent activated adhesive*. The adhesive which is usually a polyvinyl butyral, utilizes a low temperature thermoplastic resin [178]. This means that the bonded adhesive can come apart after a certain minimum temperature is reached, or it again comes in contact with the solvent. Normally this temperature is much lower than the thermal rating of the base insulation layer. The adhesive is activated by either passing the wire through a solvent while winding or heating the finished coil as a result of passing electric current through it.

The conductor wire with a heat activated adhesive overcoat costs more than the same class of non-bondable conductor. However, a less than two second current pulse is required to bond the heat activated adhesive layer and bonding machinery costs about half as much as trickle impregnation machinery [178].

3.3.3 Fabrication of Coreless Windings

Stator coreless windings of AFPM machines are fabricated as uniformly distributed coils on a disc-type cylindrical supporting structure (hub) made of nonmagnetic and nonconductive material. There are two types of windings:

(a) winding comprised of multi-turn coils wound with turns of insulated conductor of round or rectangular cross section;
(b) printed winding also called *film coil winding*.

Coils are connected in groups to form the phase windings typically connected in star or delta. Coils or groups of coils of the same phase can be connected in parallel to form parallel paths.

To assemble the winding of the same coils and obtain high density packing, coils should be formed with *offsetting bends*, as shown in Fig. 3.16. The space between two sides of the same coil is filled with coil sides from each of the adjacent coils.

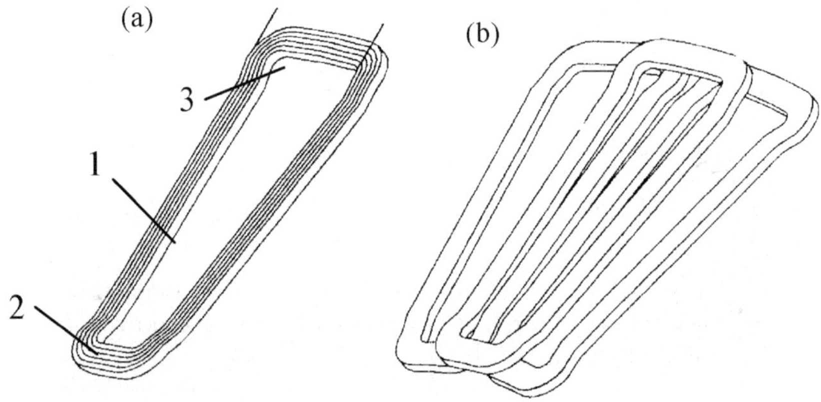

Fig. 3.16. Disc-type coreless winding assembled of coils of the same shape according to U.S. Patent No. 5 744 896 [P83]: (a) single coil; (b) three adjacent coils. 1 — coil side, 2 — inner offsetting bend, 3 — outer offsetting bend.

Coils can be placed in a slotted structure of the mould (Fig. 3.17). With all the coils in position, the winding (often with a supporting structure or hub) is moulded into a mixture of epoxy resin and hardener and then cured in a heated oven. Because of the difficulty of releasing the cured stator from the slotted structure of the mould (Fig. 3.17a), each spacing block that forms a guide slot consists of several removable pins of different size (Fig. 3.17b).

For very small AFPM machines and micromachines printed circuit coreless windings allow for automation of production. Printed circuit windings for AFPM brushless machines fabricated in a similar way as printed circuit boards have not been commercialized due to poor performance. A better performance has been achieved using film coil windings made through the same process as flexible printed circuits [93]. The coil pattern is formed by etching two copper films that are then attached to both sides of a board made of insulating materials (Fig. 3.18). Compact coil patterns are made possible by connecting both sides of coil patterns through holes [93].

Numerical Example 3.1

A simple stationary magnetic circuit is shown in Fig. 3.19. There are two Vacomax 240 HR SmCo PMs (Table 3.11) with $B_r = 1.10$ T, $H_c = 680$ kA/m, temperature coefficients $\alpha_B = -0.03\%/^{\circ}\text{C}$ and $\alpha_H = -0.15\%/^{\circ}\text{C}$ at

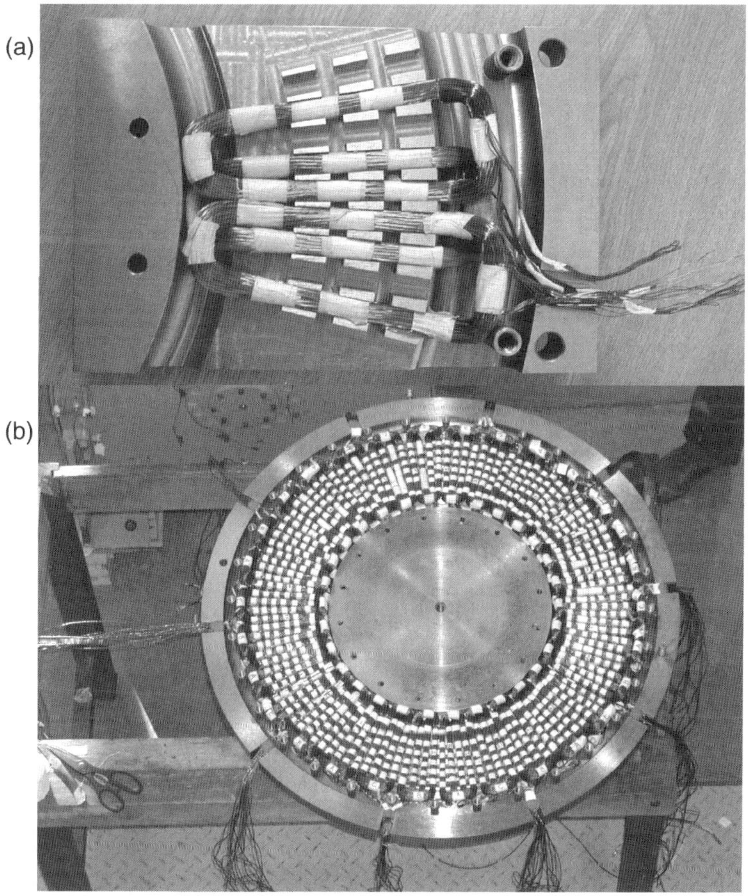

Fig. 3.17. Moulds for positioning the coils: (a) mould with guide slots; (b) mould with guide pins.

Fig. 3.18. Film coils for AFPM micromotors. Courtesy of *EMbest*, Soeul, Korea.

$20 \leq \vartheta_{PM} \leq 100°$C. The height of the PM per pole is $h_M = 6$ mm and the air gap thickness $g = 1$ mm. The U-shaped and I-shaped (top) ferromagnetic cores are made of a laminated electrotechnical steel. The width of the magnets and cores is 17 mm. Calculate the air gap magnetic flux density, air gap magnetic field strength, the useful energy of PMs and normal attractive force per two poles at: (a) $\vartheta_{PM} = 20°$C and (b) $\vartheta_{PM} = 100°$C. The MVD in the laminated core, leakage and fringing magnetic flux can be neglected.

Solution:

(a) Magnet temperature $\vartheta_{PM} = 20°$C

The relative recoil magnetic permeability according to eqn (3.4) for a straight line demagnetization curve is

$$\mu_{rrec} = \frac{1}{\mu_0}\frac{\Delta B}{\Delta H} = \frac{1}{0.4\pi \times 10^{-6}}\frac{1.10 - 0}{680,000 - 0} \approx 1.29$$

The air gap magnetic flux density according to eqn (3.13) is

$$B_g \approx \frac{1.10}{1 + 1.29 \times 1.0/6.0} = 0.906 \text{ T}$$

The air gap magnetic field strength according to eqn (3.14) in which $H = H_g$ and $B = B_g$ is

$$H_g = H_c\left(1 - \frac{B_g}{B_r}\right) = 680 \times 10^3\left(1 - \frac{1.064}{1.10}\right) = 120.12 \times 10^3 \text{ A/m}$$

The useful energy per magnet volume according to eqn (3.5) is

$$w_g = \frac{B_g H_g}{2} = \frac{1.064 \times 120,120}{2} = 54\ 395.8 \text{ J/m}^3$$

The useful energy per pole pair is

$$W_g = w_g V_M = 54,395.8(2 \times 6 \times 15 \times 17 \times 10^{-9}) = 0.166 \text{ J}$$

The normal attractive force per 2 poles is

$$F = \frac{B_g^2}{2\mu_0}(2S_M) = \frac{1.064^2}{0.4\pi \times 10^{-6}}(15 \times 17 \times 10^{-6}) = 166.5 \text{ N}$$

(b) Magnet temperature $\vartheta_{PM} = 100°$C

The remanent magnetic flux density and coercivity at 100°C according to eqns (3.2) and (3.3) is

$$B_r = 1.10\left[1 + \frac{-0.03}{100}(100 - 20)\right] = 1.074 \text{ T}$$

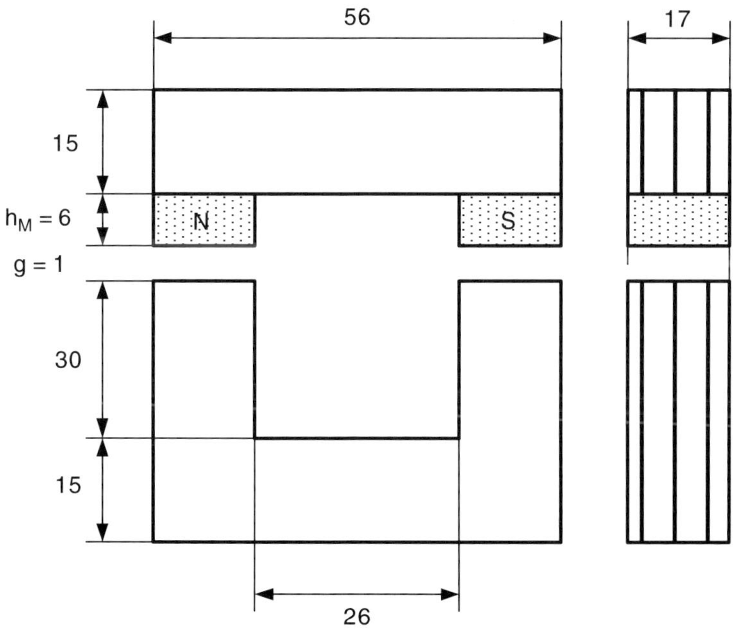

Fig. 3.19. A simple stationary magnetic circuit with PMs and air gap. *Numerical example 3.1.*

$$H_c = 680 \times 10^3 \left[1 + \frac{-0.15}{100}(100 - 20)\right] = 598.4 \times 10^3 \text{ A/m}$$

At $\vartheta_{PM} = 100°$C the demagnetization curve is nonlinear. Its linear part is only between 0.5 T and B_r parallel to the demagnetization curve at 20°C. Thus, the relative recoil magnetic permeability μ_{rrec} at 100°C is approximately the same as that at room temperature.

The air gap magnetic flux density according to (3.13) is

$$B_g \approx \frac{1.074}{1 + 1.29 \times 1.0/6.0} = 0.884 \text{ T}$$

The air gap magnetic field strength at 100°C according to eqn (3.14)

$$H_g = 598.4 \times 10^3 \left(1 - \frac{0.884}{1.074}\right) = 105.7 \times 10^3 \text{ A/m}$$

The useful energy per magnet volume is

$$w_g \approx \frac{0.884 \times 105.7 \times 10^3}{2} = 46\ 719.5 \text{ J/m}^3$$

The useful energy per pole pair is

$$W_g = 46\ 719.5(2 \times 6 \times 15 \times 17 \times 10^{-9}) = 0.143 \text{ J}$$

The normal attractive force per 2 poles is

$$F = \frac{0.884^2}{0.4\pi \times 10^{-6}}(15 \times 17 \times 10^{-6}) = 158.6 \text{ N}$$

Numerical Example 3.2

A single-sided, 8-pole AFPM machine with slotted ferromagnetic stator has the PM outer diameter $D_{out} = 0.22$ m and inner diameter $D_{in} = 0.12$ m. The air gap including the effect of slotting (Carter coefficient) is $g = 1.9$ mm. Trapezoidal sintered NdFeB magnets have $B_r = 1.15$ T and $H_c = 900$ kA/m at 20°C. The temperature coefficient for B_r is $\alpha_B = -0.15$ % /°C and the temperature coefficient for H_c is $\alpha_H = -0.64$ % /°C. The coefficient of PM leakage flux is $\sigma_{lM} = 1.15$ and pole width-to-pole pitch coefficient is $\alpha_i = 0.72$.

Find PM dimensions to obtain the air gap magnetic flux density $B_g = 0.64$ T at no-load and temperature $\vartheta_{PM} = 80$°C. Sketch operating diagrams in the B–H and Φ–MMF coordinate system at no load. Assume that the magnetic circuit is unsaturated.

Solution:

The remanence and coercivity at $\vartheta_{PM} = 80$°C according to eqns (3.2) and (3.3) is

$$B_r = 1.15\left[1 + \frac{-0.15}{100}(\vartheta_{PM} - 20)\right] = 1.046 \text{ T}$$

$$H_c = 900\left[1 + \frac{-0.64}{100}(\vartheta_{PM} - 20)\right] = 554.4 \text{ kA/m}$$

Approximation of demagnetization curves according to eqn (3.14) at 20°C and 80°C respectively

$$B_{20}(H) = 1.15\left(1 - \frac{H}{900\ 000}\right) \qquad B(H) = 1.046\left(1 - \frac{H}{554\ 400}\right)$$

The relative recoil magnetic permeability according to eqn (3.4) is

$$\mu_{rrec} = \frac{1.046}{0.4\pi \times 10^{-6} \times 554\ 400} = 1.5$$

The axial height of the PM per pole according to eqn (3.13) is

$$h_M = \mu_{rrec}\frac{\sigma_{lM}B_g}{B_r - \sigma_{lM}B_g}g = 1.5\frac{1.15 \times 0.64}{1.046 - 1.15 \times 0.64}0.0019 = 0.0068 \text{ m}$$

The equivalent air gap is

$$g_{eq} = g + \frac{h_M}{\mu_{rrec}} = 1.9 + \frac{6.8}{1.5} = 6.4 \text{ mm}$$

The average diameter, pole pitch (1.14), length (3.34) and width of the magnet are respectively

$$D = 0.5(0.22 + 0.12) = 0.17 \text{ m} \qquad \tau = \frac{\pi D}{2p} = \frac{\pi 0.17}{8} = 0.0668 \text{ m}$$

$$l_M = 0.5(0.22 - 0.12) = 0.05 \text{ m} \qquad w_M = \alpha_i \tau = 0.72 \times 0.0668 = 0.048 \text{ m}$$

The permeance of the air gap according to eqn (2.30) is

$$G_g = \mu_0 \frac{1}{g} \frac{\alpha_i \pi}{8p} (D_{out}^2 - D_{in}^2)$$

$$= 0.4\pi 10^{-6} \frac{1}{0.0019} \frac{0.72\pi}{8 \times 4} (0.22^2 - 0.12^2) = 1.59 \times 10^{-6} \text{ H}$$

or

$$G_g = \mu_0 \frac{w_M l_M}{g} = 0.4 \times \pi 10^{-6} \frac{0.048 \times 0.05}{0.0019} = 1.59 \times 10^{-6} \text{ H}$$

The total permeance for magnetic flux including leakage permeances is

$$G_t = \sigma_{lm} G_g = 1.15 \times 1.59 \times 10^{-6} = 1.828 \times 10^{-6} \text{ H}$$

Approximation of the total permeance (air gap and leakage) line can be expressed as a linear function of H

$$\lambda_t(H) = G_t \frac{h_M}{w_M l_M} H = 1.828 \times 10^{-6} \frac{0.0068}{0.048 \times 0.05} H = 5.146 \times 10^{-6} H$$

The magnetic field intensity corresponding to the operating point of the magnet is calculated as

$$H_M = \frac{B_r}{G_t h_M/(w_M l_M) + B_r/H_c}$$

$$= \frac{1.046}{5.146 \times 10^{-6} + 1.046/554,400} = 148\ 795 \text{ A/m}$$

The air gap magnetic flux density as obtained from the operating diagram (Fig. 3.20a) is

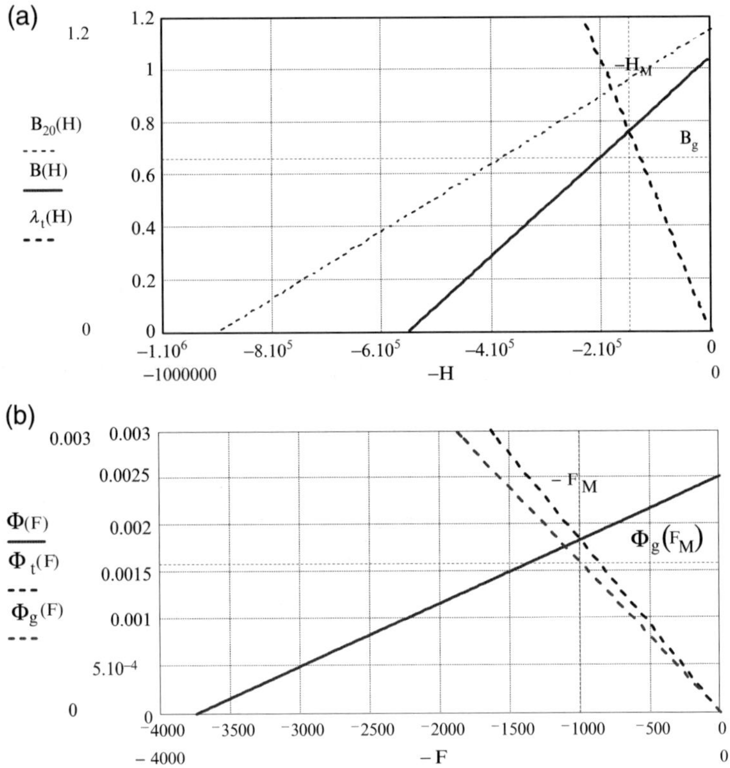

Fig. 3.20. Operating diagram of PM at no load: (a) in B–H coordinate system; (b) in Φ–MMF coordinate system. *Numerical example 3.2.*

$$B_g = G_g \frac{h_M}{w_M l_M} H_M = 1.59 \times 10^{-6} \frac{0.0068}{0.048 \times 0.05} 148\ 795 = 0.666\ \mathrm{T}$$

To plot the operating diagram in the Φ–MMF coordinate system, eqns (3.16) and (3.18) are used to find the magnetic flux Φ_r corresponding to B_r and the MMF $\mathcal{F}_c$ per pole corresponding to H_c, i.e.

$$\Phi_r = 1.046 \times 0.048 \times 0.05 = 0.00252\ \mathrm{Wb}$$

$$\mathcal{F}_c = 554\ 400 \times 0.0068 = 3750.6\ \mathrm{A}$$

Approximation of the magnetic flux line is

$$\Phi(\mathcal{F}) = \Phi_r \left(1 - \frac{\mathcal{F}}{\mathcal{F}_c}\right) = 0.00252 \left(1 - \frac{\mathcal{F}}{3750.6}\right)$$

The MMF F_M corresponding to the operating point of the magnet in the Φ–MMF coordinate system (Fig. 3.20b) is

$$\mathcal{F}_M = \frac{\Phi_r}{G_t + \Phi_r/\mathcal{F}_c} = \frac{0.00252}{1.828 \times 10^{-6} + 0.00252/3750.6} = 1006.6 \text{ A}$$

The air gap flux line (Fig. 3.20b) is

$$\Phi_g(\mathcal{F}) = G_g\mathcal{F} = 1.59 \times 10^{-6}\mathcal{F}$$

The total flux line (Fig. 3.20b) is

$$\Phi_t(\mathcal{F}) = G_t\mathcal{F} = 1.828 \times 10^{-6}\mathcal{F}$$

The air gap magnetic flux corresponding to the operating point is $\Phi_g(F_M) = 0.0016$ Wb. The air gap magnetic flux density is $B_g = 0.0016/(0.048 \times 0.05) = 0.666$ T.

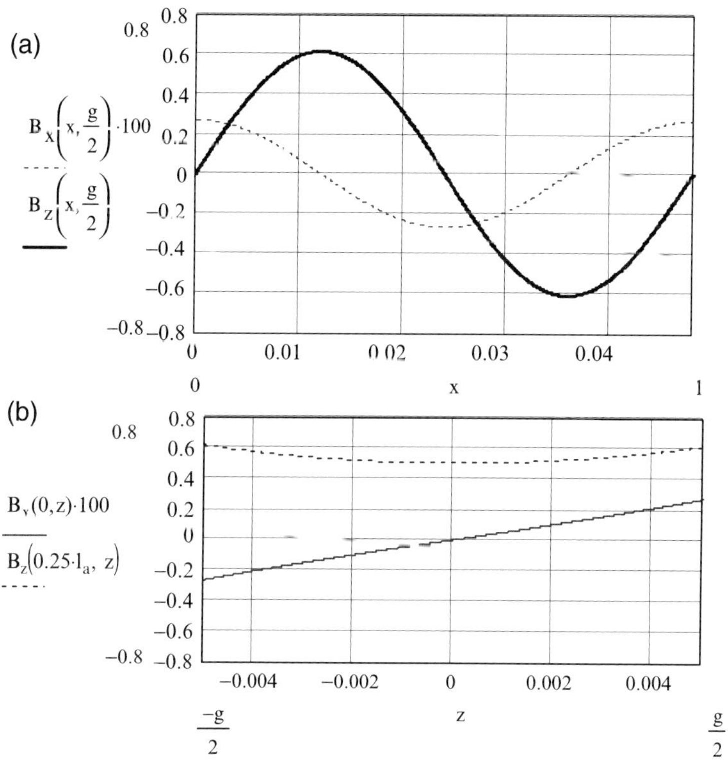

Fig. 3.21. Distribution of B_x and B_z components: (a) in the x direction; (b) in the z direction. *Numerical example 3.3.*

Numerical Example 3.3

The magnetic field in the air gap of an AFPM machine with coreless stator is excited by sintered NdFeB magnets arranged in Halbach array. The external twin rotors do not have any backing steel discs (Fig. 1.4d). The remanent magnetic flux density is $B_r = 1.25$ T, the height of the PM is $h_M = 6$ mm, the wavelength at the average diameter is $l_a = 2\tau = 48$ mm and the magnet-to-magnet distance is $t = 10$ mm.

Find the distribution of the magnetic flux density in the space between magnets for the 90-degree Halbach array, i.e. $n_M = 4$. Estimate, how the number n_M of PMs per wavelength affects the magnetic flux density at the active surface of PMs.

Solution

The peak value of the magnetic flux density at the active surface of Halbach array according to eqn (3.42) is

$$B_{m0} = 1.25[1 - \exp(-130.9 \times 0.006)] \frac{\sin(\pi/4)}{\pi/4} = 0.612 \text{ T}$$

where $\beta = 2 \times \pi/0.048 = 130.9$ 1/m.

The distribution of the tangential component B_x in the space between magnets is described by eqn (3.45) and the distribution of the normal component B_z is described by eqn (3.46). Both components B_x and B_z are plotted in Fig. 3.21.

For the 90-degree Halbach array ($n_M = 4$) the peak value of the magnetic flux density at the active surface of PMs is $B_{m0} = 0.612$ T. Similarly, using eqn (3.42), the peak value B_{m0} can be calculated for other Halbach configurations. For 60-degree Halbach array ($n_M = 6$) $B_{m0} = 0.649$ T and for 45-degree Halbach array ($n_M = 8$) $B_{m0} = 0.663$ T. In general,

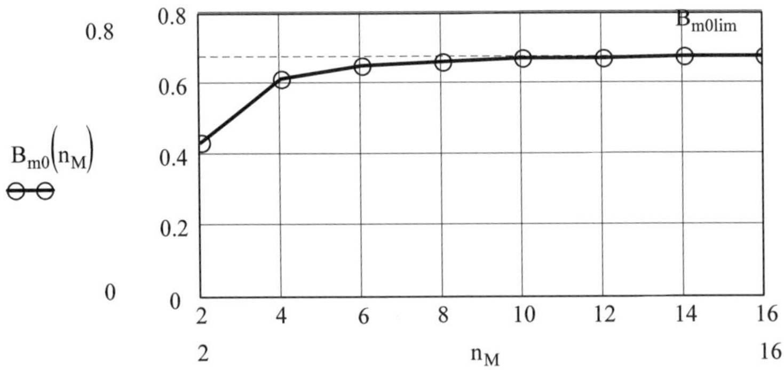

Fig. 3.22. Peak value of magnetic flux density B_m as a function of number n_M of PMs per wavelength. *Numerical example 3.3.*

$$\lim_{n_M \to \infty} B_{m0} = B_r[1 - \exp(1 - \beta h_M)] \frac{\sin(\pi/n_M)}{\pi/n_M}$$

$$= 1.25[1 - \exp(130.9 \times 0.006)] \times 1 = 0.68 \text{ T}$$

since $\lim_{x \to 0} \sin x/x = 1$. The peak value B_{m0} as a function of number n_M of PMs per wavelength is shown in Fig. 3.22.

4

AFPM Machines With Iron Cores

In Chapter 2, principles of operation, topologies and fundamental equations of a broad family of AFPM machines were discussed. In this chapter, the focus is on those types of AFPM machines that make use of the stator and rotor ferromagnetic cores. The AFPM machines with ferromagnetic cores are designed both as single-sided and double-sided. The stator core can be fabricated using either laminated steels or SMC powders. General equations given in Chapter 2 for the performance calculations will be developed further and adjusted to the construction of AFPM machines with ferromagnetic cores. Application of the FEM analysis to performance calculations is also emphasized.

4.1 Geometries

Single-sided AFPM machines with stator ferromagnetic cores have a single PM rotor disc opposite to a single stator unit consisting of a polyphase winding and ferromagnetic core (Fig. 2.1). The stator ferromagnetic cores can be slotted or slotless. The stator winding is always made of flat wound coils (Fig. 2.8). The PMs can be mounted on the surface of the rotor or embedded (buried) in the rotor disc. In the case of a slotless stator the magnets are almost always surface mounted, while in the case of a slotted stator with a small air gap between the rotor and stator core, the magnets can be either surface mounted on the disc (Fig. 2.1) or buried in the rotor disc (Fig. 2.6). Large axial magnetic forces on bearings are the main drawback of single-sided AFPM machines with ferromagnetic stator cores.

In *double-sided AFPM machines* with ideal mechanical and magnetic symmetry, the axial magnetic forces are balanced. Double-sided AFPM machines with stator ferromagnetic cores have either a single PM rotor disc with iron cored stators on both sides of the discs (Figs 1.4c and 2.3) or outer PM rotor discs with iron-cored stator fixed in the middle (Figs 2.4, 2.5 and 2.6). As with single-sided AFPM machines the stator ferromagnetic cores can be slotted or slotless, and the rotor magnets can be surface mounted, embedded or buried

(Fig. 2.6). Again, in the case of a slotless stator with a large air gap between the rotor and stator core the magnets are almost always surface mounted. The stator windings of double-sided AFPM machines can be flat wound (slotted or slotless) as shown in Fig. 2.8 or toroidally wound (normally slotless) as shown in Fig. 2.9.

4.2 Commercial AFPM Machines With Stator Ferromagnetic Cores

An example of a commercial double-sided AFPM servo motor with ferromagnetic core is shown in Figs. 4.1 and 4.2. External stators have slotted ring-shaped cores made of nonoriented electrotechnical steel ribbon. The inner rotor does not have any ferromagnetic material. PMs are mounted on a nonmagnetic rotating disc.

Fig. 4.1. Double-sided AFPM brushless servo motor with the stator ferromagnetic core and built-in brake. Courtesy of *Mavilor*, Barcelona, Spain.

Fig. 4.3 shows a double-sided AFPM synchronous generator with the stator core wound of amorphous alloy ribbon manufactured by *LE Incorporated*, Indianapolis, IN, U.S.A. The volume of *LE* AFPM generators is approximately 60% lower than that of classical synchronous generators of the same rating.

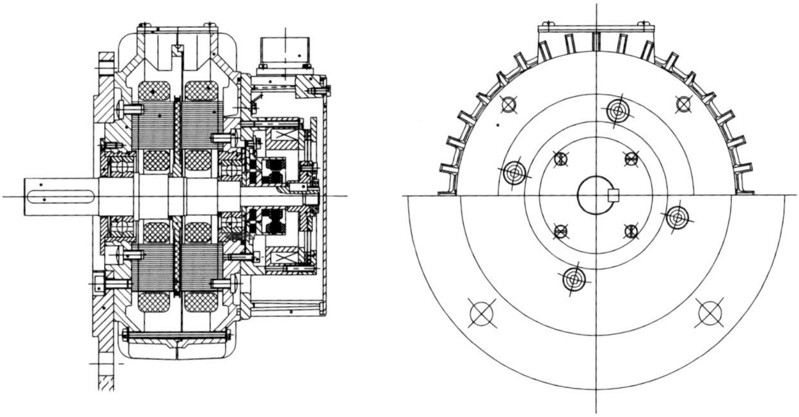

Fig. 4.2. Longitudinal section of the double-sided AFPM brushless servo motor shown in Fig. 4.1. Courtesy of *Mavilor*, Barcelona, Spain.

Fig. 4.3. *GenSmart*^{TM} AFPM synchronous generator with amorphous alloy stator core. Courtesy of *LE Incorporated*, Indianapolis, IN, U.S.A.

4.3 Some Features of Iron-Cored AFPM Machines

Iron-cored AFPM machines are distinguished in two ways from coreless (air-cored) AFPM machines, namely: (i) iron-cored machines have core losses while coreless machines do not and (ii) the per unit values of the synchronous reactances of iron-cored machines are much higher than those of coreless machines.

The core losses are functions of amongst other things the frequency and density of the magnetic flux in the ferromagnetic core – see eqn (2.55). The frequency of the flux variation in the core is in turn determined by the speed and number of pole pairs. As AFPM brushless machines tend to have a high number of poles (minimum $2p = 6$), the speeds of iron-cored AFPM machines are limited to keep the frequency, and hence the core losses of the machine, within limits; unless the machines are designed to have low flux densities in ferromagnetic cores. Flux frequencies are normally kept below, say, 100 Hz for laminated ferromagnetic cores. If higher frequencies are required, then silicon steel laminations with a thickness of less than 0.2 mm, amorphous alloy ribbons or SMC powders must be used.

The higher synchronous inductance of iron-cored AFPM machines affects negatively the voltage regulation of the machine in generator mode, which might be considered as a disadvantage. However, for solid-state converter-fed applications the higher synchronous inductance of iron-cored AFPM motors is an important advantage as it helps reduce the current-ripple due to converter switching. Thus the relatively low inductance of coreless AFPM machines can be a significant disadvantage in switched solid state converter applications.

4.4 Magnetic Flux Density Distribution in the Air Gap

AFPM brushless machines with ferromagnetic stator and rotor cores can produce strong magnetic flux density in the air gap with a minimum volume of PMs. The distribution of the normal component of the magnetic flux density in the x-direction (circumferential) excited by PMs without armature reaction at the radius corresponding to the average pole pitch τ can be described by the following equation

$$B_g(x) = \frac{1}{k_C} B_{zPM}(x) + B_{sl}(x) \tag{4.1}$$

where the PM excitation flux density for smooth stator core is

$$B_{zPM}(x) = \sum_{\nu=1}^{\infty} \frac{1}{\sigma_{lM}} B_g b_\nu \cos\left[\nu\left(\frac{\pi}{\tau}x - \frac{\pi}{2}\right)\right] \qquad \nu = 1, 3, 5, \ldots \tag{4.2}$$

and the magnetic flux density component due to stator slots is

$$B_{sl}(x) = \lambda_{sl}(x) \frac{g'}{\mu_0} B_{zPM}(x) \tag{4.3}$$

The peak values of the higher harmonics of the magnetic flux density distribution of eqn (4.2) are $B_{mg\nu} = (B_g/\sigma_{lM})b_\nu$.

In the above equations (4.1), (4.2) and (4.3) k_C is Carter's coefficient according to eqn (1.2), b_ν is according to eqn (1.13) usually for $\alpha = 0$, τ is the average pole pitch according to eqn (1.14), g' is the equivalent air gap in the d-axis according to eqn (2.110), B_g is the flat-topped value of the magnetic flux density according to eqn (3.13), σ_{lM} is the coefficient of the PM leakage

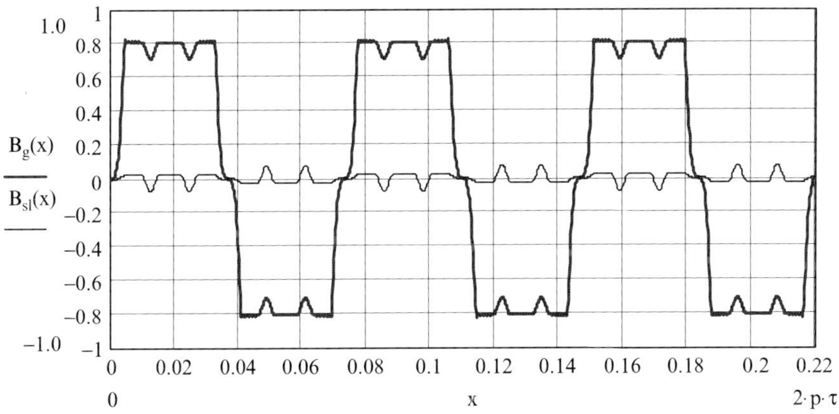

Fig. 4.4. Magnetic flux density distribution in the air gap of an AFPM machine with iron slotted core obtained on the basis of eqn (4.1) for $p = 3$, $s_1 = 18$, $\tau = 36.7$ mm, $\alpha_i = 0.78$, $b_p = \alpha_i \tau$, $B_r = 1.2$ T, $\mu_{rrec} = 1.061$, $g = 1.5$ mm, $b_{14} = 3.5$ mm, $h_M = 5$ mm, $\alpha = 0$ and $\sigma_{lM} = 1.15$.

flux according to eqn (3.9) and λ_{s1} is the relative slot leakage permeance given by

$$\lambda_{sl}(x) = -\sum_{\nu=1}^{\infty} a_\nu \cos\left(\nu \frac{2\pi}{t_1} x\right) \tag{4.4}$$

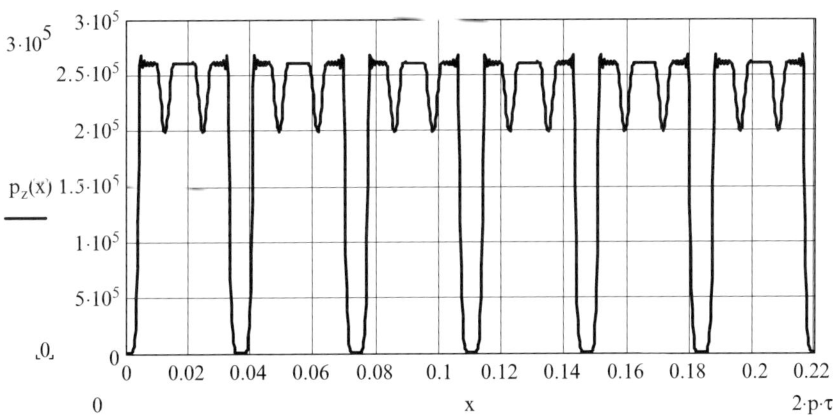

Fig. 4.5. Magnetic pressure distribution in an AFPM machine with iron slotted core obtained on the basis of eqn (4.7) for $p = 3$, $s_1 = 18$, $\tau = 36.7$ mm, $\alpha_i = 0.78$, $b_p = \alpha_i \tau$, $B_r = 1.2$ T, $\mu_{rrec} = 1.061$, $g = 1.5$ mm, $b_{14} = 3.5$ mm, $h_M = 5$ mm, $\alpha = 0$ and $\sigma_{lM} = 1.15$.

in which the average slot pitch $t_1 = 2p\tau/s_1$, s_1 is the number of stator slots and the amplitude of slot harmonics as derived by W. Weber [71, 121] is

$$a_\nu = \mu_0 \frac{\beta(\kappa)}{g'} \frac{1}{\nu} \frac{4}{\pi} \left(0.5 + \frac{\nu^2 \Gamma^2}{0.78 - 2\nu^2 \Gamma^2}\right) \sin(1.6\nu\pi\Gamma) \tag{4.5}$$

The remaining quantities $\beta(\kappa)$, κ and Γ are expressed by the following equations:

$$\beta(\kappa) = \frac{1}{2}\left(1 - \frac{1}{\sqrt{1-\kappa^2}}\right), \qquad \kappa = \frac{b_{14}}{g + h_M/\mu_{rrec}}, \qquad \Gamma = \frac{b_{14}}{t_1} \tag{4.6}$$

The distribution of the magnetic flux density in the air gap obtained on the basis of eqn (4.1) is shown in Fig. 4.4. The magnetic pressure on the stator and rotor active surfaces can be found on the basis of the normal component (4.1) of the magnetic flux density distribution in the air gap, i.e.

$$p_z(x) = \frac{1}{2} \frac{B_g(x)^2}{\mu_0} \tag{4.7}$$

The distribution of the magnetic pressure (4.7) is visualized in Fig. 4.5. The magnetic pressure resolved into Fourier series yields harmonics of magnetic forces acting on the stator. It is necessary to calculate those harmonics, e.g. in the noise analysis of electromagnetic origin radiated by the AFPM machine.

4.5 Calculation of Reactances

4.5.1 Synchronous and Armature Reaction Reactances

Calculations of the stator current, other machine parameters and characteristics require knowledge of the *synchronous reactance* per phase which is expressed by eqns (2.79) and (2.80). The synchronous reactance X_{sd} and X_{sq} is the sum of the *armature reaction* (mutual) reactance X_{ad}, X_{aq} and the stator winding *leakage reactance* X_1. The armature reaction reactances X_{ad}, X_{aq} and armature reaction inductances L_{ad}, L_{aq} can be calculated with the aid of equations given in Section 2.9 and Table 2.1. The analytical approach to calculation of the stator leakage reactance X_1 is discussed in Section 4.5.2.

4.5.2 Stator Leakage Reactance

The armature leakage reactance is the sum of the slot leakage reactance X_{1s}, the end connection leakage reactance X_{1e}, and the differential leakage reactance X_{1d} (for higher space harmonics), i.e.

$$X_1 = X_{1s} + X_{1e} + X_{1d}$$

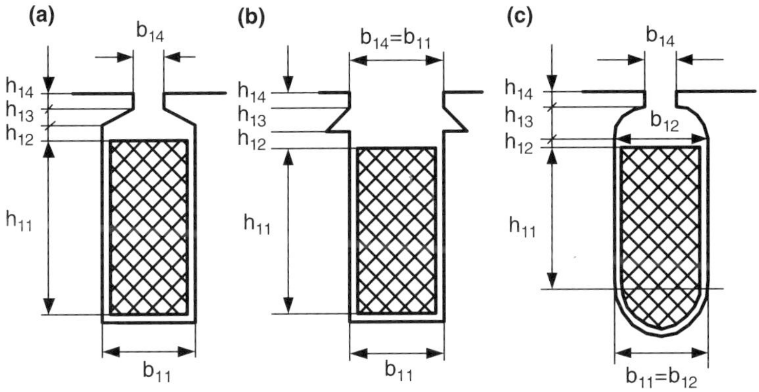

Fig. 4.6. Stator slots of AFPM brushless machines: (a) rectangular semi-open slot; (b) rectangular open slot; (c) oval semi-open slot.

$$= 4\pi f \mu_o \frac{L_i {N_1}^2}{p q_1} (\lambda_{1s} k_{1X} + \frac{l_{1in}}{L_i} \lambda_{1ein} + \frac{l_{1out}}{L_i} \lambda_{1eout} + \lambda_{1d}) \tag{4.8}$$

where N_1 is the number of turns per phase, k_{1X} is the skin-effect coefficient for leakage reactance, p is the number of pole pairs, $q_1 = s_1/(2pm_1)$ is the number of stator slots s_1 per pole per phase (2.2), l_{1in} is the length of the stator winding inner end connection, l_{1out} is the length of the stator winding outer end connection, λ_{1s} is the coefficient of the slot leakage permeance (slot-specific permeance), λ_{1ein} is the coefficient of the leakage permeance of the inner end connection, λ_{1eout} is the coefficient of the leakage permeance of the outer end connection and λ_{1d} is the coefficient of the differential leakage

The coefficients of leakage permeances of the slots shown in Fig. 4.6 are:

- for a rectangular semi-open slot (Fig. 4.6a):

$$\lambda_{1s} = \frac{h_{11}}{3b_{11}} + \frac{h_{12}}{b_{11}} + \frac{2h_{13}}{b_{11} + b_{14}} + \frac{h_{14}}{b_{14}} \tag{4.9}$$

- for a rectangular open slot (Fig. 4.6b):

$$\lambda_{1s} \approx \frac{h_{11}}{3b_{11}} + \frac{h_{12} + h_{13} + h_{14}}{b_{11}} \tag{4.10}$$

- for an oval semi-open slot (Fig. 4.6c):

$$\lambda_{1s} \approx 0.1424 + \frac{h_{11}}{3b_{11}} + \frac{h_{12}}{b_{12}} + 0.5 \arcsin[\sqrt{1 - (b_{14}/b_{12})^2}] + \frac{h_{14}}{b_{14}} \tag{4.11}$$

The coefficients of leakage permeances for other shapes of slots than those shown in Fig. 4.6 are given, e.g. in [159]. The above specific-slot permeances

(4.9), (4.10) and (4.11) are for single-layer windings. To obtain the specific permeances of slots containing double-layer windings, it is necessary to multiply eqns (4.9), (4.10) and (4.11) by the factor

$$\frac{3\beta+1}{4} \tag{4.12}$$

where β is according to eqn (2.7). Such an approach is justified if $2/3 \leq \beta \leq 1.0$.

The specific permeance of the end connection (overhang) is estimated on the basis of experiments. For double-layer, low-voltage, small and medium-power machines the specific permeances of the inner and outer end connections are:

- inner end connection

$$\lambda_{1ein} \approx 0.17 q_1 (1 - \frac{2}{\pi}\frac{w_{cin}}{l_{1in}}) \tag{4.13}$$

- outer end connection

$$\lambda_{1eout} \approx 0.17 q_1 (1 - \frac{2}{\pi}\frac{w_{cout}}{l_{1out}}) \tag{4.14}$$

where l_{1in} is the length of the inner end connection, l_{1out} is the length of the outer end connection, w_{cin} is the inner coil span, and w_{cout} is the outer coil span. The specific permeance of two end connections is a sum of permeances (4.13) and (4.14), i.e.

$$\lambda_{1e} = \lambda_{1ein} + \lambda_{1eout} = 0.34 q_1 \left(1 - \frac{1}{\pi}\frac{w_{cin} l_{1out} + w_{cout} l_{1in}}{l_{1in} l_{1out}}\right) \tag{4.15}$$

For $w_{cin} = w_{cout} = w_c$ and $l_{1in} = l_{1out} = l_{1e}$ the specific permeance (see eqn (4.15)) of two end connections has the same form as that for a cylindrical stator, i.e.

$$\lambda_{1e} = 0.34 q_1 \left(1 - \frac{2}{\pi}\frac{w_c}{l_{1e}}\right) \tag{4.16}$$

Putting $w_c/l_{1e} = 0.64$, eqn (4.16) also gives satisfactory results for single-layer windings, i.e. the specific permeance of two end connections becomes

$$\lambda_{1e} \approx 0.2 q_1 \tag{4.17}$$

For double-layer, high-voltage windings the specific permeances of the inner and outer end connections are:

- inner end connection

$$\lambda_{1ein} \approx 0.21 q_1 (1 - \frac{2}{\pi}\frac{w_{cin}}{l_{1in}}) k_{w1}^2 \tag{4.18}$$

- outer end connection

$$\lambda_{1eout} \approx 0.21q_1(1 - \frac{2}{\pi}\frac{w_{cout}}{l_{1out}})k_{w1}^2 \tag{4.19}$$

where the stator winding factor k_{w1} for the fundamental space harmonic $\nu = 1$ is according to eqn (2.10). The specific permeance of two end connections

$$\lambda_{1e} = \lambda_{1ein} + \lambda_{1eout} = 0.42q_1\left(1 - \frac{1}{\pi}\frac{w_{cin}l_{1out} + w_{cout}l_{1in}}{l_{1in}l_{1out}}\right)k_{w1}^2 \tag{4.20}$$

In general, the following approximate relation may be used for most of the windings:

$$\lambda_{1e} \approx 0.3q_1 \tag{4.21}$$

The specific permeance of the differential leakage flux is

$$\lambda_{1d} = \frac{m_1 q_1 \tau k_{w1}^2}{\pi^2 g' k_{sat}}\tau_{d1} \tag{4.22}$$

where g' is the equivalent air gap according to eqns (2.110) or (2.111) and k_{sat} is the saturation factor of the magnetic circuit. The differential leakage factor τ_{d1} is

$$\tau_{d1} = \frac{1}{k_{w1}^2}\sum_{\nu>1}(\frac{k_{w1\nu}}{\nu})^2 \tag{4.23}$$

The curves of the differential leakage factor τ_{d1} are given in publications dealing with design of induction motors, e.g. [106, 121, 159].

In practical calculations it is convenient to use the following formula

$$\tau_{d1} = \frac{\pi^2(10q_1^2 + 2)}{27}\left[\sin\left(\frac{30^{\circ}}{q_1}\right)\right]^2 - 1 \tag{4.24}$$

The specific permeance between the heads of teeth:

$$\lambda_{1t} \approx \frac{5g/b_{14}}{5 + 4g/b_{14}} \tag{4.25}$$

should be added to the differential specific permeance λ_{1d} of slotted stator windings.

4.6 Performance Characteristics

Specifications of *Mavilor* AFPM motors (Figs. 4.1 and 4.2) are shown in Table 4.1. Torque-speed characteristics of *Mavilor* AFPM motors are plotted in Fig. 4.7. AFPM servo motors MA-3 to MA-30 are designed with Class F insulation and IP-65 protection. Table 4.2 shows specifications of brakes for MA-3 to MA-30 servo motors.

Table 4.3 shows specifications of *GenSmart*^{TM} AFPM synchronous generators with amorphous alloy stator cores manufactured by *LE Incorporated*, Indianopolis, IN, U.S.A. *GenSmart*^{TM} AFPM synchronous generators can be used both as stationary and mobile generators.

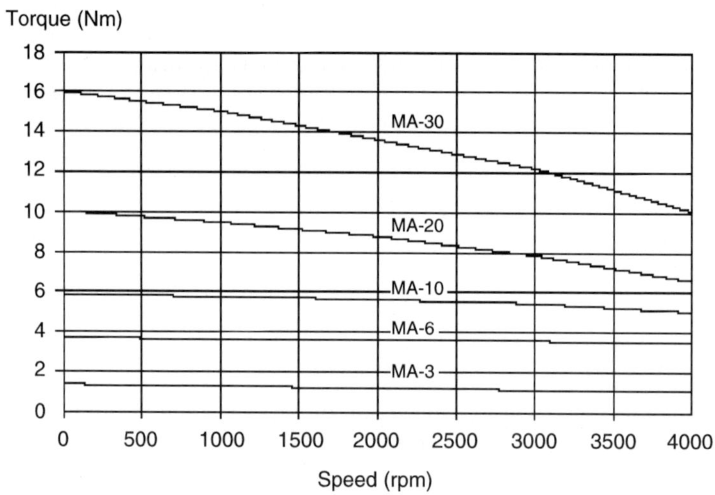

Fig. 4.7. Torque-speed curves of AFPM brushless servo motors manufactured by *Mavilor*, Barcelona, Spain.

4.7 Performance Calculation

4.7.1 Sine-Wave AFPM Machine

The electromagnetic power of a sine-wave (synchronous) AFPM motor is expressed by eqn (2.91) which can also be brought to the following form:

$$P_{elm} = m_1[I_a E_f \cos\Psi + \frac{1}{2}(X_{sd} - X_{sq})I_a^2 \sin 2\Psi] \tag{4.26}$$

Hence, the developed torque is

Table 4.1. AFPM brushless servomotors manufactured by *Mavilor*, Barcelona, Spain

	MA-3	MA-6	MA-10	MA-20	MA-30
Maximum speed, rpm	9000	6000	6000	6000	6000
Stall torque, Nm	1.3	3.6	5.8	10.0	16.0
Stall current, A	2.2	4.2	6.8	10.3	16.5
Peak torque, Nm	5.2	28.5	40.7	69.8	96.0
Torque density, Nm/kg	0.7	0.8	1.1	1.2	1.6
EMF constant, Vs/rad	0.3	0.5	0.5	0.6	0.6
Torque constant, Nm/A	0.6	0.9	0.9	1.0	1.0
Cogging torque, Nm	< 0.2	< 0.1	< 0.1	< 0.2	< 0.3
Winding resistance, Ω	10.2	5.3	2.2	1.4	0.5
Winding inductance, mH	25.0	11.6	0.4	7.0	4.0
Rotor moment of inertia, kgm^2 $\times 10^{-3}$	0.04	0.30	2.10	0.80	1.60
Mechanical time constant, ms	2.0	3.8	3.6	2.1	1.6
Electrical time constant, ms	2.5	2.2	7.5	5.0	7.5
Thermal time constant, s	1500	1500	1800	1500	1500
Thermal resistance, °C/W	1.1	0.6	0.5	0.4	0.4
Mass, kg				34	36
Radial load, N	218	225	225	390	390

Table 4.2. Specifications of built-in brakes for AFPM brushless servomotors manufactured by *Mavilor*, Barcelona, Spain

	MA-3	MA6	MA-10	MA-20	Ma-30
Holding torque, Nm	1	4	4	8	8
d.c. voltage, V	4.8	24	24	24	24
Input power, W	5	22	22	22	22
Moment of inertia, kgm^2 $\times 10^{-3}$	0.08	0.3	0.3	0.3	0.3
Disengagement response time, ms	7	30	30	30	30
Engagement response time, ms	5	7	7	7	7
Mass, kg	0.3	0.8	0.8	0.8	0.8

$$T_d = \frac{P_{elm}}{2\pi n_s}$$

$$= \frac{m_1 p}{2}\left[\sqrt{2}N_1 k_{w1}\Phi_f I_a \cos\Psi + \frac{1}{2}(L_{sd} - L_{sq})I_a^2 \sin 2\Psi\right] \quad (4.27)$$

For buried and interior PM rotors $X_{sd} > X_{sq}$ and $L_{sd} > L_{sq}$, which implies that an additional reluctance power and torque component is developed by the machine. For surface PM rotors the reluctance dependent components of power and torque are almost not existent as $X_{sd} \approx X_{sq}$ and $L_{sd} \approx L_{sq}$. In eqn (4.26) the EMF E_f can be calculated by means of eqn (2.36). The current I_a and angle Ψ in eqns (4.26) and (4.27) are either specified, or determined according to Section 2.7 using voltage phasor $\mathbf{V_1}$ projections on the d and q

Table 4.3. *GenSmart*^{TM} AFPM synchronous generators manufactured by LE, Inc., Indianapolis, IN, U.S.A.

	28-G22	90-G32	120-G49
Rated output power, kW	28	90	120
Rated speed, rpm	3600		2500
Maximum speed, rpm	4200		3000
Voltage, V		480 or 208	
Stator insulation		Class H	
Efficiency, %	92.2	94.9	95.3
Frame external diameter, mm	216	315	485
Frame external length, mm	150	186	200
Stator core material		Amorphous alloy	
Cooling		liquid or air cooled	
Ambient temperature, °C		-50 to 60	
Maximum allowable temperature, °C		125 (rise from 40°C)	
Enclosure		Totally enclosed	
Mounting		Flange or foot	
Overtemperature control		Embedded thermistor	
Mass, kg	21	44	115
Power density, kW/kg	1.333	2.045	1.043

axes. The reactances and inductances are calculated as described in Sections 2.9 and 4.5.2. Hence, with all the parameters known, the performance of the AFPM brushless motor in terms of the electromagnetic power and developed torque can be calculated. Note that, for the generating mode, the second term in eqns (4.26) and (4.27) (dependent on the difference between the *d*- and *q*-axis synchronous inductances) is written with the minus (−) sign.

The electric power available, after losses, of an AFPM machine with the stator ferromagnetic core is calculated as

- the input power for motoring mode

$$P_{in} = P_{elm} + \Delta P_{1w} + \Delta P_{1Fe} + \Delta P_{2Fe} + \Delta P_{PM} + \Delta P_e \tag{4.28}$$

- the output power for generating mode

$$P_{out} = P_{elm} - \Delta P_{1w} - \Delta P_{1Fe} - \Delta P_{2Fe} - \Delta P_{PM} - \Delta P_e \tag{4.29}$$

where P_{elm} is according to eqn (4.26), ΔP_{1w} are the stator winding losses according to eqn (2.49), ΔP_{1Fe} are the stator core losses according to eqn (2.54), ΔP_{2Fe} are the rotor core losses according to eqn (2.66), ΔP_{PM} are the losses in the PMs according to eqn (2.61) and ΔP_e are the eddy current losses in the stator conductors according to eqns (2.68) or (2.69) (only for slotless stators). The shaft power P_{sh} (input power P_{in} for generators, output power P_{out} for motors) is calculated by adding to the electromagnetic power

P_{elm} (generators) or subtracting from the electromagnetic power (motors) the rotational losses ΔP_{rot} according to eqn (2.70).

The shaft torque is

- for a motor (developed by the motor itself)

$$T_{sh} = \frac{P_{out}}{2\pi n} \tag{4.30}$$

- for a generator (developed by the prime mover)

$$T_{sh} = \frac{P_{in}}{2\pi n} \tag{4.31}$$

and the efficiency is

$$\eta = \frac{P_{out}}{P_{in}} \tag{4.32}$$

4.7.2 Synchronous Generator

The phasor diagram of a salient pole synchronous generator with RL load is shown in Fig. 4.8. The input voltage projections on the d and q axes according to Fig. 4.8 are

$$V_1 \sin\delta = I_{aq}X_{sq} - I_{ad}R_1$$

$$V_1 \cos\delta = E_f - I_{ad}X_{sd} - I_{aq}R_1 \tag{4.33}$$

and

$$V_1 \sin\delta = I_{ad}R_L - I_{aq}X_L$$

$$V_1 \cos\delta = I_{aq}R_L + I_{ad}X_L \tag{4.34}$$

where $\mathbf{Z}_L = R_L + jX_L$ is the load impedance per phase across the output terminals. The d- and q-axis currents are

- from eqns (4.33)

$$I_{ad} = \frac{E_f X_{sq} - V_1(X_{sq}\cos\delta + R_1\sin\delta)}{X_{sd}X_{sq} + R_1^2} \tag{4.35}$$

$$I_{aq} = \frac{V_1(X_{sd}\sin\delta - R_1\cos\delta) + E_f R_1}{X_{sd}X_{sq} + R_1^2} \tag{4.36}$$

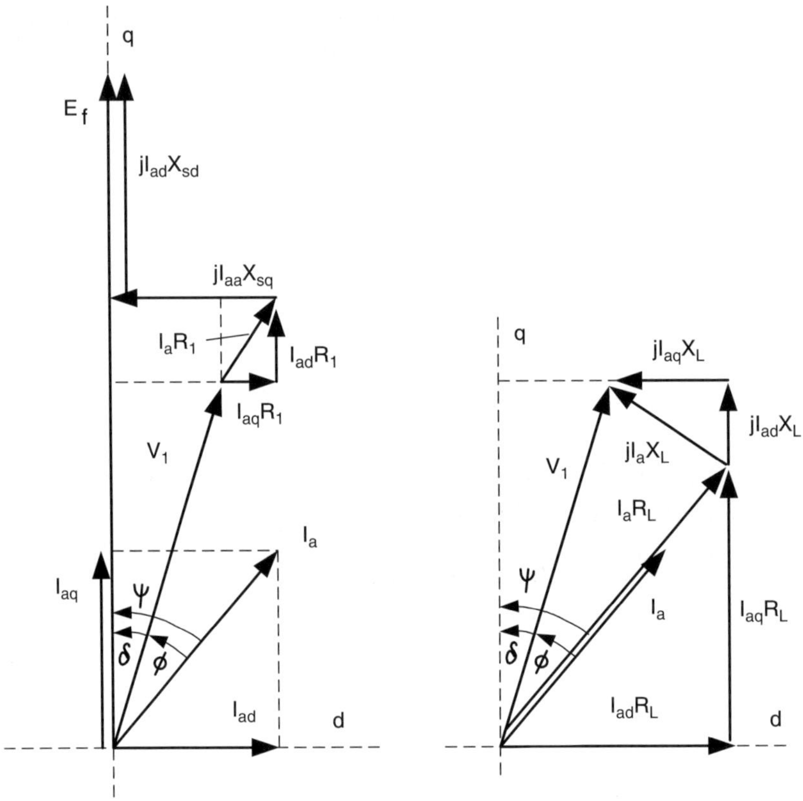

Fig. 4.8. Phasor diagram of an overexcited salient pole synchronous generator for RL load.

- from eqns (4.34)

$$I_{ad} = \frac{V_1(R_L \sin\delta + X_L \cos\delta)}{R_L^2 + X_L^2} \tag{4.37}$$

$$I_{aq} = \frac{V_1(R_L \cos\delta - X_L \sin\delta)}{R_L^2 + X_L^2} \tag{4.38}$$

The load angle δ between the voltage V_1 and EMF E_f can be determined, e.g. from the first eqn (4.34), i.e.

$$\delta = \arcsin\left(\frac{I_{ad}R_L - I_{aq}X_L}{V_1}\right) \tag{4.39}$$

Combining eqns (4.33) and (4.34), the d- and q-axis currents are independent of the load angle δ, i.e.

$$I_{ad} = \frac{E_f(X_{sq} + X_L)}{(X_{sd} + X_L)(X_{sq} + X_L) + (R_1 + R_L)^2} \tag{4.40}$$

$$I_{aq} = \frac{E_f(R_1 + R_L)}{(X_{sd} + X_L)(X_{sq} + X_L) + (R_1 + R_L)^2} \tag{4.41}$$

The angle Ψ between the current I_a and q-axis and the angle ϕ between the current I_a and voltage V_1 are respectively,

$$\Psi = \arccos\left(\frac{I_{aq}}{I_a}\right) = \arccos\left(\frac{I_{aq}}{\sqrt{I_{ad}^2 + I_{aq}^2}}\right) \tag{4.42}$$

$$\phi = \arccos\left(\frac{I_a R_L}{V_1}\right) = \arccos\left(\frac{R_L}{Z_L}\right) \tag{4.43}$$

where $I_a = \sqrt{I_{ad}^2 + I_{aq}^2}$ (see also eqn (2.89)). The output electrical power on the basis of phasor diagram (Fig. 4.8) and eqn (4.33) is

$$P_{out} = m_1 V_1 I_a \cos\phi = m_1 V_1 (I_{aq}\cos\delta + I_{ad}\sin\delta)$$

$$= m_1[E_f I_{aq} - I_{ad} I_{aq}(X_{sd} - X_{sq}) - I_a^2 R_1] \tag{4.44}$$

Including only the stator winding losses, the internal electromagnetic power of the generator is

$$P_{elm} = P_{out} + \Delta P_{1w} = m_1[E_f I_{aq} - I_{ad} I_{aq}(X_{sd} - X_{sq})] \tag{4.45}$$

4.7.3 Square-Wave AFPM Machine

The square (trapezoidal) wave AFPM brushless machine has already been introduced in Section 2.10.2. Square-wave machines are characterized by their trapezoidal or quasi-square back EMF waveforms with the conduction angle from 100 to 150 electrical degrees, depending on the construction. The trapezoidal back EMF waveforms can be obtained by proper design of the magnetic and electric circuits. Both slotted and slotless AFPM machines can be designed so as to obtain sinusoidal or trapezoidal back EMF waveforms, except the toroidal AFPM machine of Figs 2.4 and 2.5. The toroidal machine is almost always designed to have a trapezoidal back EMF waveform. The amplitude of the back EMF can be calculated by means of eqn (2.125) for a three phase Y-connected machine with two phase windings active.

The square-wave AFPM machine is always fed by means of a solid state inverter. The inverter supplies d.c. PWM voltages to the machine phase windings during the conduction intervals. The current square waveforms (see Fig. 1.3a) are described by the flat-topped value of the current $I_a^{(sq)}$. The developed torque T_d of the square-wave AFPM brushless machine can be calculated by means of eqn (2.127). Hence, the electromagnetic power P_{elm} of the machine can be found. Using the same loss and power balance equations as for the sine-wave machine, the input and output power can be calculated with the aid of eqn (4.28) and the output power $P_{out} = P_{elm} - \Delta P_{rot}$. Note that ΔP_{1w} must be calculated by using the true rms value of the current given by eqn (2.135).

4.8 Finite Element Calculations

Instead of using analytical or semi-empirical formulae, some of the equivalent circuit parameters and losses of the AFPM machine can be calculated accurately by means of the FEM analysis. The 2D and 3D FEM electromagnetic software packages are nowadays available at affordable prices and the field solution time has become attractively short. For example, most of the 2D FEM magnetostatic problems can be solved in seconds on today's personal computers.

The synchronous inductances L_{sd} and L_{sq} and the EMF E_f of an AFPM brushless machine are the critical parameters used in the performance calculations.

The first step is to specify on a 2D plane the outline, dimensions and materials of the AFPM machine. Due to the symmetry of an AFPM machine, it is usually sufficient to model only one pole pitch of the machine by applying negative periodic boundary conditions [261]. However, for AFPM machines using non-overlap windings, a greater portion of the machine (or sometime the entire machine) needs to be modelled [146]. The cross section is taken at the average radius of the active part of the stator winding, i.e. at $0.25(D_{out}+D_{in})$. Depending on the machine topology and material properties, either Neuman or Dirichlet conditions may be imposed on the remaining boundaries. The length of the winding, which is in the radial direction, is taken as the active length of the stator winding, i.e. $L_i = l_M = 0.5(D_{out} - D_{in})$ — eqn (3.34). After assigning different material properties to various regions of the model, the defined structure can then be meshed.

The second step is to specify the phase currents, i_A, i_B and i_C, in the machine according to the position of the rotor. To calculate the parameter L_{sd} or L_{sq} the phase currents must be assigned in such a way that the stator MMF can be aligned with the $d-$axis or $q-$axis respectively. In both cases the PMs have to be kept inactive. In the case of calculation of the EMF E_f equal to the no load voltage in generating mode, the phase currents are set to zero and only PMs are active.

The third step is to run the FEM solver and obtain for each of the three cases described above a solution for the air gap magnetic flux density in the axial direction of the machine. The fundamental component of the air gap magnetic flux density is determined by Fourier expansion of the axial component of the air gap flux density. Hence, the peak values of the fundamental air gap flux densities can be determined for each of the cases above: (1) B_{mad1} with only the d-axis MMF and thus only the d-axis current I_{ad}, (2) B_{maq1} with only the q-axis MMF and the $q-$axis current I_{aq}, and (3) B_{mg} with only the PMs active.

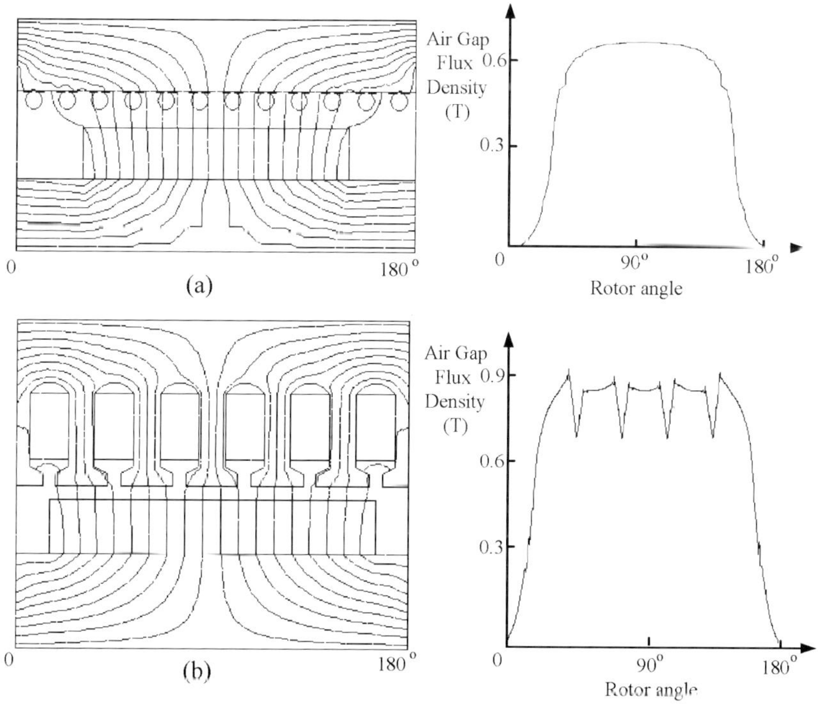

Fig. 4.9. Magnetic flux plot and air gap flux density with only PM excitation of (a) slotless and (b) slotted disc type machine with surface PMs.

Step four is to calculate by means of analytical equations the inductances and induced EMF. With B_{mad1} and B_{maq1} known from the FEM analysis, the flux linkages Ψ_d and Ψ_q can be calculated according to eqns (2.104) and (2.105). The inductances L_{sd} and L_{sq} can be calculated as described in Section 5.5.2. With B_{mg} known from the FEM analysis, the back EMF E_f can be found by first calculating the magnetic flux per pole Φ_f according to eqn (2.29) and then E_f according to eqn (2.36).

Fig. 4.9 shows the FEM flux plots and air gap flux densities due to PM excitation alone for two single-sided AFPM machine configurations with slotted and slotless stators. Fig. 4.10 shows the FEM flux plots and air gap flux densities for a single-sided interior AFPM machine with only the d- and q-axis stator winding excitation.

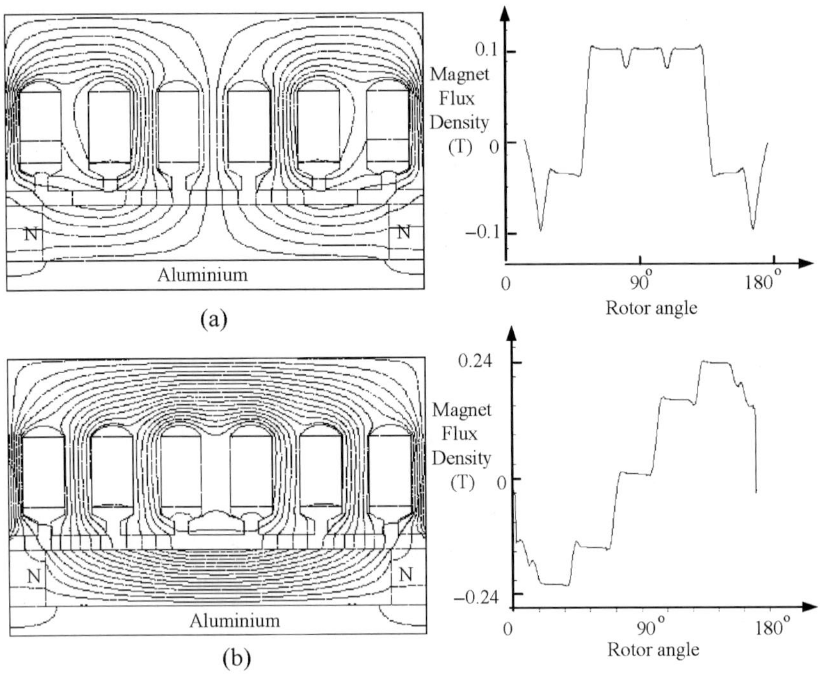

Fig. 4.10. Magnetic flux plot and air gap flux density of interior AFPM machine with (a) only d-axis and (b) only q-axis stator winding excitation.

It should be noted that using the cross section at the average radius of an AFPM machine to represent the entire machine in the FEM modelling has inherent limitations. As explained in [214], for AFPM machines with profiled PM shapes and varied tooth width along the radius the 3D effects become more prominent and thus cannot be disregarded in FEM modelling. A quasi-3D FEM modelling scheme has been proposed in [214], in which the AFPM machine is virtually divided into a number of sub-machines. The overall performance of the machine is obtained by combining the performance of each sub-machine. A similar approach that uses a multilayer and multislice modelling technique for coreless stator AFPM machines has been reported in [262, 263].

Numerical Example 4.1

A three-phase, 2.2-kW, 50-Hz, 380-V (line-to-line), Y-connected, 750-rpm, $\eta = 78\%$, $\cos\phi = 0.83$, double-sided disc PM synchronous motor has the following dimensions of the magnetic circuit: PM outer diameter $D_{out} = 0.28$ m, PM inner diameter $D_{in} = 0.16$ m, thickness of the rotor (PMs) $2h_M = 8$ mm, single-sided mechanical clearance $g = 1.5$ mm. A surface configuration of uniformly distributed PMs has been used. The rotor does not have any soft ferromagnetic material. The rotor outer and inner diameters correspond to the outer and inner outline of PMs and the stator stack. The dimensions of rectangular semi-closed slots (Fig. 4.6b) are: $h_{11} = 11$ mm, $h_{12} = 0.5$ mm, $h_{13} = 1$ mm, $h_{14} = 1$ mm, $b_{11} = 13$ mm and $b_{14} = 3$ mm. The number of stator slots (one unit) is $s_1 = 24$, the number of armature turns of a single stator per phase is $N_1 = 456$, the diameter of the stator copper bare conductor is 0.511 mm (AWG 24), the number of stator parallel wires is $a_w = 2$ and the air gap magnetic flux density is $B_{mg} = 0.65$ T. The rotational losses are $\Delta P_{rot} = 80$ W and the core and PM losses are $\Delta P_{1Fe} + \Delta P_{PM} = 0.05P_{out}$. There are two winding layers in each stator slot. The twin-stator Y-connected windings are fed in parallel.

Find the motor performance at the load angle $\delta = 11^{\circ}$. Compare the calculations obtained from the circuital approach with the FEM results.

Solution

The phase voltage is $V_1 = 380/\sqrt{3} = 220$ V. The number of pole pairs is $p = f/n = 50 \times 60/750 = 4$ and $2p = 8$. The minimum slot pitch is

$$t_{1min} = \frac{\pi D_{in}}{s_1} = \frac{\pi \times 0.16}{24} = 0.0209 \text{ m} \approx 21 \text{ mm}$$

The width of the slot is $b_{12} = 13$ mm, that means the narrowest tooth width $c_{1min} = t_{1min} - b_{12} = 21 - 13 = 8$ mm. The magnetic flux density in the narrowest part of the stator tooth

$$B_{1tmax} \approx \frac{B_{mg} t_{1min}}{c_{1min}} = \frac{0.65 \times 21}{8} = 1.7 \text{ T}$$

which is rather low.

The average diameter and average pole pitch according to eqn (1.14) are

$$D = 0.5(0.28 + 0.16) = 0.22 \text{ m}$$

$$\tau = \frac{\pi \times 0.22}{8} = 0.0864 \text{ m}$$

Because the number of slots per pole per phase (2.2)

$$q_1 = \frac{s_1}{2pm_1} = \frac{24}{8 \times 3} = 1$$

the winding factor $k_{w1} = k_{d1}k_{p1} = 1 \times 1 = 1$ as expressed by eqns (2.8), (2.9) and (2.10).

The magnetic flux according to eqn (2.26) and EMF induced by the rotor excitation system according to eqns (2.36) are respectively

$$\Phi_f = \frac{2}{\pi} \times 0.65 \frac{\pi}{8}[(0.5 \times 0.28)^2 - (0.5 \times 0.16)^2] = 0.002145 \text{ Wb}$$

$$E_f = \pi\sqrt{2}\, 50 \times 456 \times 1 \times 0.002145 = 217.3 \text{ V}$$

It has been assumed that $B_{mg1} \approx B_{mg}$.

Now it is necessary to check the electric loading, current density and fill (space) factor of the stator slot. For two parallel wires $a_w = 2$, the number of conductors per coil of a double-layer phase winding according to eqn (2.3) is

$$N_c = \frac{a_w N_1}{(s_1/m_1)} = \frac{2 \times 456}{(24/3)} = 114$$

Thus the number of conductors in a single slot N_{sl} is equal to the (number of layers) $\times$ (number of conductors per coil N_c) $= 2 \times 114 = 228$ or according to eqn (2.5)

$$N_{sl} = \frac{1 \times 2 \times 456}{4 \times 1} = 228$$

The rated input current in a single stator is

$$I_a = \frac{P_{out}}{2m_1 V_1 \eta \cos\phi} = \frac{2200}{2 \times 3 \times 220 \times 0.78 \times 0.83} = 2.57 \text{ A}$$

The stator line current density (peak value) on the basis of eqn (2.93) is

$$A_m = \frac{4\sqrt{2}\, m_1 I_a N_1}{\pi(D_{out} + D_{in})} = \frac{3\sqrt{2} \times 456 \times 2.57}{0.0864 \times 4} = 14\,452.2 \text{ A/m}$$

which is rather a low value even for small PM a.c. motors. The cross section of the stator (armature) conductor

$$s_a = \frac{\pi d_a^2}{4} = \frac{\pi 0.511^2}{4} = 0.207 \text{ mm}^2$$

gives the following current density under rated conditions

$$j_a = \frac{2.57}{2 \times 0.197} = 6.29 \text{ A/mm}^2$$

This is an acceptable value of the current density for disc-type a.c. machines rated from 1 to 10 kW.

For the class F enamel insulation of the armature conductors, the diameter of the wire with insulation is 0.548 mm. Hence, the total cross sectional area of all conductors in the stator slot is

$$228 \times \frac{\pi \times 0.548^2}{4} \approx 54 \text{ mm}^2$$

The cross section area of a single slot is approximately $h_{11}b_{12} = 11 \times 13 = 143$ mm^2. The slot fill factor $54/143 = 0.38$ shows that the stator can be easily wound, since the average slot fill factor for low voltage machines with round stator conductors is about 0.4.

The average length of the stator end connection for a disc rotor a.c. machine is $l_{1e} \approx 0.154$ m. The average length of the stator turn according to eqn (2.39) is

$$l_{1av} = 2(L_i + l_{1e}) = 2(0.06 + 0.154) = 0.428 \text{ m}$$

where $L_i = 0.5(D_{out} - D_{in}) = 0.5(0.28 - 0.16) = 0.06$ m. The stator winding resistance per phase at a temperature of 75°C (hot motor) according to eqn (2.40) is

$$R_1 = \frac{N_1 l_{1av}}{a_w \sigma s_a} = \frac{456 \times 0.428}{47 \times 10^6 \times 2 \times 0.1965} = 10.57\ \Omega$$

Carter's coefficient is calculated on the basis of eqns (1.2) and (1.3), i.e.

$$k_C = \left(\frac{28.8}{28.8 - 0.012 \times 1.5}\right)^2 = 1.001$$

$$\gamma = \frac{4}{\pi}\left[\frac{3}{2 \times 11}\arctan\left(\frac{3}{2 \times 11}\right) - \ln\sqrt{1 + \left(\frac{3}{2 \times 11}\right)^2}\right] = 0.012$$

where $t_1 = \pi D/s_1 = \pi \times 0.22/24 = 0.0288$ m $= 28.8$ mm. The nonferromagnetic air gap in calculation of Carter's coefficient is $g' \approx 2g + 2h_M = 2 \times 1.5 + 8 = 11$ mm. Since there are two slotted surfaces of twin stator cores, Carter's coefficient has been squared.

The stator (one unit) leakage reactance has been calculated according to eqn (4.8) in which $l_{1in}\lambda_{1ein}/L_i + l_{1out}\lambda_{1eout}/L_i \approx l_{1e}\lambda_{1e}/L_i$, i.e.

$$X_1 = 4 \times 0.4\pi \times 10^{-6}\pi \times 50 \times \frac{456^2 \times 0.06}{4 \times 1} \times (0.779 + \frac{0.154}{0.06} \times 0.218$$

$$+0.2297 + 0.9322) = 6.158\ \Omega$$

where

- the coefficient of slot leakage permeance – eqn (4.9) is

$$\lambda_{1s} = \frac{11}{3 \times 13} + \frac{0.5}{13} + \frac{2 \times 1}{13 + 3} + \frac{1}{3} = 0.779$$

- the coefficient of end connection leakage permeance – eqns (4.16) in which the average coil span $w_c = \tau$ is

$$\lambda_{1e} \approx 0.34 \times 1 \left(1 - \frac{2}{\pi}\frac{0.0864}{0.154}\right) = 0.218$$

- the coefficient of differential leakage permeance – eqns (4.22) and (4.24) is

$$\lambda_{1d} = \frac{3 \times 1 \times 0.0864 \times 1^2}{\pi^2 \times 0.011 \times 1.001} 0.0966 = 0.2297$$

$$\tau_{d1} = \frac{\pi^2(10 \times 1^2 + 2)}{27}\left(\sin\frac{30^{\circ}}{1}\right)^2 - 1 = 0.0966$$

- the coefficient of tooth-top leakage permeance – eqn (4.25) is

$$\lambda_{1t} = \frac{5 \times 11/3}{5 + 4 \times 11/3} = 0.9322$$

According to eqns (2.121) and (2.122) in which $k_{fd} = k_{fq} = 1$, the armature reaction reactances for surface type PM rotors and unsaturated machines are

$$X_{ad} = X_{aq} = 2 \times 3 \times 0.4 \times \pi \times 10^{-6} \times 50 \times \left(\frac{456 \times 1}{4}\right)^2$$

$$\times \frac{(0.5 \times 0.28)^2 - (0.5 \times 0.16)^2}{1.001 \times 0.011} = 5.856\ \Omega$$

where the air gap, in the denominator, for the armature flux should be equal to $g' \approx 2\times1.5+8 = 11$ mm ($\mu_{rrec} \approx 1$). The synchronous reactances according to eqns (2.79) and (2.80) are

$$X_{sd} = X_{sq} = 6.158 + 5.856 = 12.01\ \Omega$$

The armature currents are calculated on the basis of eqns (2.87), (2.88) and (2.89). For $\delta = 11^{\circ}$ ($\cos\delta = 0.982$, $\sin\delta = 0.191$) the current components are: $I_{ad} = -1.807$ A, $I_{aq} = 1.96$ A and $I_a = 2.666$ A.

The input power absorbed by one stator is expressed by eqn (2.90). The input power according to eqn (2.90) absorbed by two stators in parallel is twice as much, i.e.

$$P_{in} = 2 \times 3 \times 220 \times (1.88 \times 0.982 - (-1.82) \times 0.191) = 2986.8\ \text{W}$$

The input apparent power absorbed by two stators is

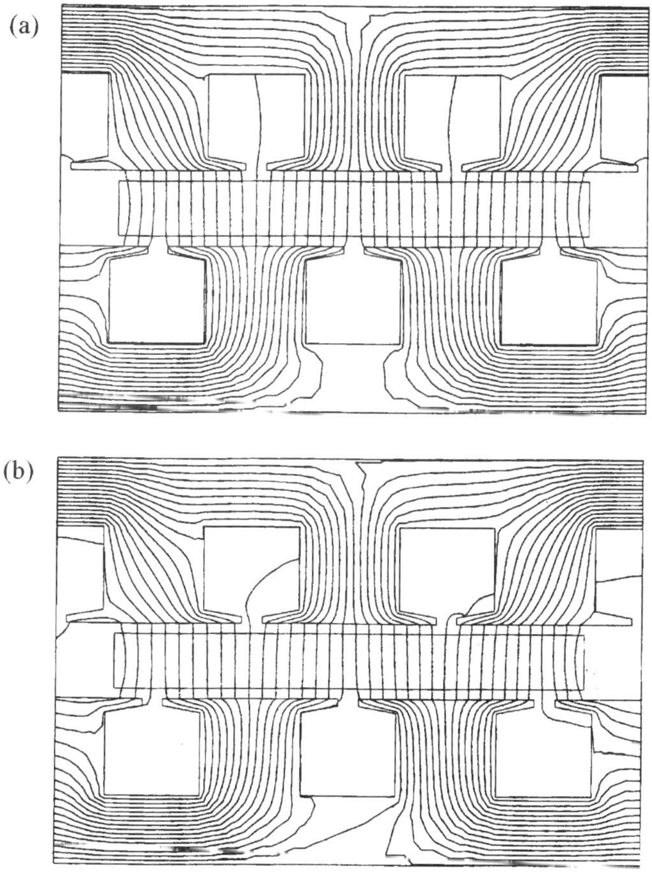

Fig. 4.11. Magnetic flux plots in the double sided AFPM motor: (a) zero armature current, (b) rated armature current. *Numerical example 4.1.*

$$S_{in} = 2 \times 3 \times 220 \times 2.62 = 3509.7 \text{ VA}$$

The power factor is

$$\cos\phi = \frac{2986.8}{3509.7} = 0.851$$

The losses in two stator windings according to eqn (2.49), in which $k_{1R} \approx 1$, are calculated as

$$\Delta P_{1w} = 2 \times 3 \times 2.62^2 \times 10.57 = 431.4 \text{ W}$$

The output power assuming that $\Delta P_{1Fe} + \Delta P_{PM} = 0.05 P_{out}$ is

$$P_{out} = \frac{1}{1.05}(P_{in} - \Delta P_{1w} - \Delta P_{rot}) = \frac{1}{1.05}(2986.8 - 431.4 - 80.0) = 2357.5 \text{ W}$$

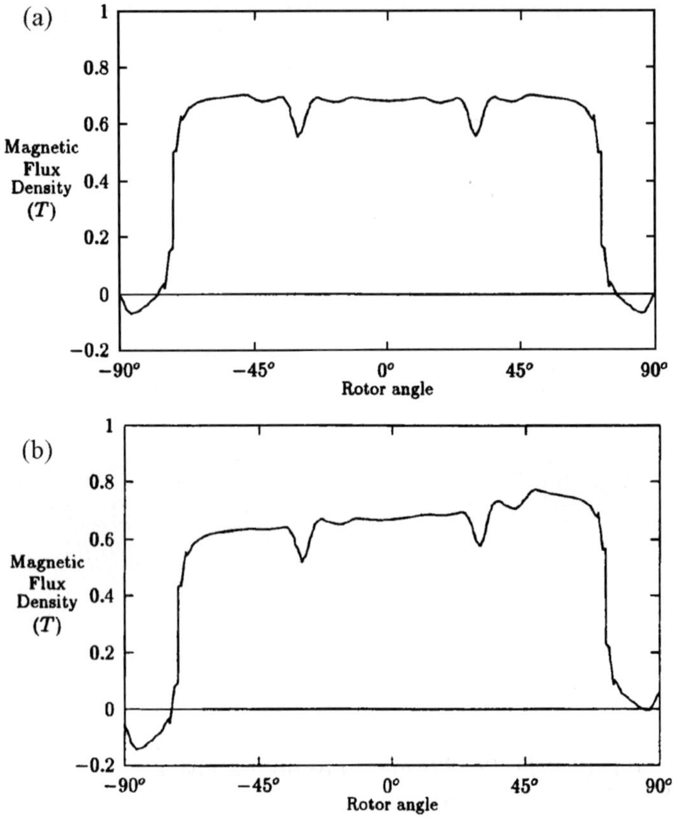

Fig. 4.12. Magnetic flux density distribution along the pole pitch: (a) zero armature current, (b) rated armature current. *Numerical example 4.1.*

Thus

$$\Delta P_{1Fe} + \Delta P_{PM} = 0.05 \times 2357.5 = 117.9 \text{ W}$$

The motor efficiency is

$$\eta = \frac{2357.5}{2986.8} = 0.789 \qquad \text{or} \qquad \eta = 78.9\%$$

The shaft torque is

$$T_{sh} = \frac{2357.5}{2\pi(750/60)} = 30.02 \text{ Nm}$$

The electromagnetic power is

$$P_{elm} = P_{in} - \Delta P_{1w} - \Delta P_{1Fe} - \Delta P_{PM} = 2986.8 - 431.4 - 117.9 = 2437.5 \text{ W}$$

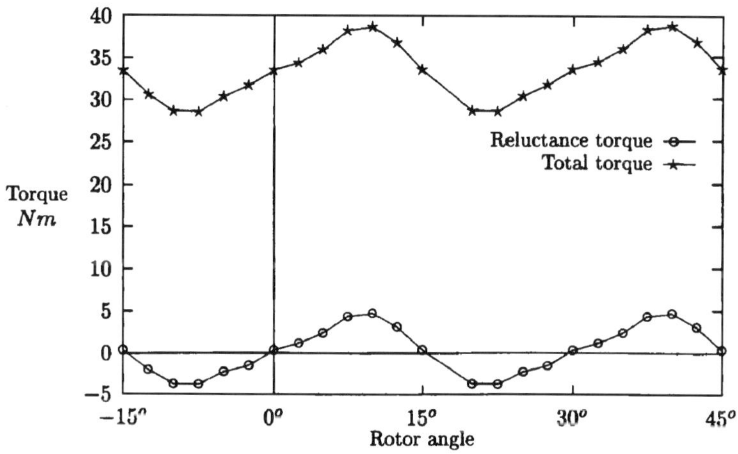

Fig. 4.13. Torque as a function of the disc rotor position. *Numerical example 4.1.*

The electromagnetic torque developed by the motor is

$$T_d = \frac{2437.5}{2\pi \times 750/60} = 31.04 \text{ Nm}$$

The results of the FEM analysis are shown in Figs 4.11, 4.12 and 4.13. The FEM gives higher values of the average developed torque than the analytical approach. Eqns (2.121), (2.122) and (4.8) do not give accurate values of X_{sd}, X_{sq} and X_1. The electromagnetic torque plotted in Fig. 4.13 against the rotor position has a significant cogging effect component with the period equal to double slot pitch, i.e. 30°.

Numerical Example 4.2

A single-sided AFPM machine has the same stator and dimensions of PMs as those in Numerical example 4.1. The rotor SmCo PMs with $B_r = 1.05$ T and $H_c = 720\ 000$ A/m are fixed to a solid steel disc. The air gap (mechanical clearance) is $g = 1.5$ mm, the height of PM is $h_M = 4$ mm, the coefficient of leakage flux of PM is $\sigma_{lM} = 1.1$, the *pole shoe-to-pole pitch* ratio is $\alpha_i = 0.78$ and the approximation coefficient of the shape of the magnetic flux density under PMs in eqn (1.13) is $\alpha = 0$.

Calculate the magnetic flux density distribution in the air gap at $D = 0.5(D_{in} + D_{out})$ and attractive force between the stator and rotor at zero stator current (no armature reaction).

Solution

1. Air gap permeance

According to eqns (1.14) and (2.92) the average diameter $D = 0.22$ m and average pole pitch $\tau = 0.086$ m. The number of pole pairs $p = (50/750)/60 = 4$. The average slot pitch $t_1 = 2p\tau/s_1 = 2 \times 4 \times 0.086/24 = 0.0288$ m. The relative recoil permeability expressed by eqn (3.4) is

$$\mu_{rrec} = \frac{1}{\mu_0}\frac{B_r}{H_c} = \frac{1}{\mu_0} \times \frac{1.05}{720\ 000} = 1.161$$

The equivalent air gap calculated on the basis of eqn (2.110) is practically independent of Carter's coefficient k_C (large equivalent air gap) and equal to $g' = 4.95$ mm. Carter's coefficient according to eqns (1.2) and (1.3) is very small, i.e., $k_C = 1.01$. The *slot opening–to–the air gap* ratio and *slot opening–to–the slot pitch* according to eqn (2.60) are $\kappa = 0.606$ and $\Gamma = 0.104$.

The relative permeance of smooth air gap is expressed by the same equation as eqn (2.108), i.e., $\lambda_0 = \mu_0/g' = \mu_0/0.00495 = 0.00025$ H/m^2. Amplitudes of slot harmonics a_ν are given by eqns (4.5) and (4.6). To obtain the relative slot permeance (4.4) it is enough to sum 25 slot harmonics, i.e.,

$$\lambda_{sl}(x) = -\sum_{\nu=1}^{25} a_\nu cos(\nu\frac{2\pi}{t_1}x)$$

The relative permeance of the air gap is

$$\lambda_g(x) = \lambda_0 + \lambda_{sl}(x)$$

For example, for $x = t_1 = 0.0288$ m the relative permeance of the air gap is $\lambda_{(x=t_1)} = 0.00022$ H/m^2.

2. Magnetic flux density in the air gap

The normal component of the magnetic flux density excited by PMs in smooth (without slots) air gap can be calculated on the basis of eqn (3.13). Including the PM leakage flux coefficient $\sigma_{lM} = 1.1$ and Carter's coefficient $k_C = 1.01$

$$B_g = \frac{1}{\sigma_{lM}}\frac{B_r}{1 + \mu_{rrec}k_C g/h_M} = \frac{1}{1.1}\frac{1.05}{(1 + 1.161 \times 1.01 \times 0.0015/0.004)} = 0.663\ \text{T}$$

The coefficient b_ν of the shape of the magnetic flux density distribution is expressed by eqn (1.13). For example, for $\nu = 1$ $b_1 = 1.228$, for $\nu = 3$ $b_3 = -0.296$ and for $\nu = 5$ $b_5 = 0.064$, where the pole face average width $b_p = \alpha_i\tau = 0.78 \times 0.086 = 0.0674$ m, $c_p = 0$ and $b_t = 0.0095$.

The distribution of the normal component of the magnetic flux density $B_{zPM}(x)$ excited in the smooth air gap by the rotor PMs is expressed by eqn (4.2). It is enough to include maximum $\nu = 1, 3.5, ...65$ slot harmonics.

The distribution of the magnetic flux density due to slot openings $B_{sl}(x)$ is expressed by eqn (4.3). The resultant magnetic flux density distribution in the air gap including stator slotting $B_g(x)$ is given by eqn (4.1). The flat-top value of the resultant magnetic flux density $B_g(x = 0.5\tau) = 0.671$ T. The difference between $B_g = 0.663$ T according to eqn (3.13) is negligible. Fig. 4.14 shows the magnetic flux density waveforms $B_g(x)$ and $B_{sl}(x)$ and harmonic flux densities $B_{g1}(x)$, $B_{g3}(x)$ and $B_{g3}(x)$.

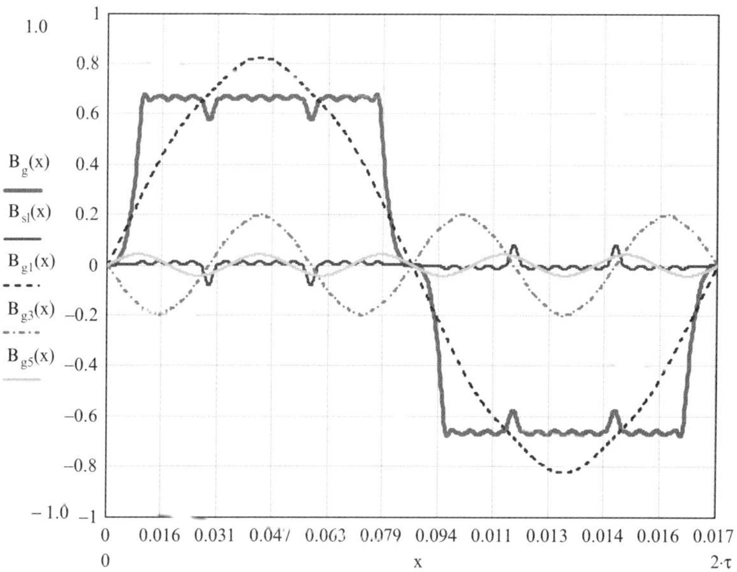

Fig. 4.14. Resultant air gap magnetic flux density waveform, slot component, fundamental, third and fifth harmonics (two pole pitches) *Numerical example 4.2.*

3. Normal attractive force between the stator and rotor

The active surface area of all PMs calculated on the basis of eqn (2.63) is $S_{PM} = 0.0323$ m^2. The magnetic pressure on the stator and rotor active surfaces is given by eqn (4.7). Thus, the normal attractive force

$$F_z(x) = p_z(x)S_{PM} = 0.0323p_z(x)$$

For the center axis of PM (d-axis), the normal attractive force is $F_z(x = 0.5\tau) = 5797$ N. The distribution of the normal attractive force $F_z(x)$ at the distance $0.5D$ from the center around the machine circumference $2p\tau$ is plotted in Fig. 4.14.

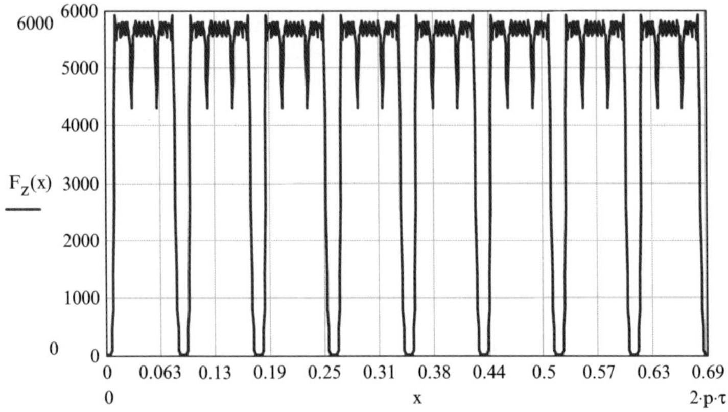

Fig. 4.15. Distribution of normal attractive force between stator and rotor. *Numerical Example 4.2.*

Numerical Example 4.3

A three-phase, Y-connected, 50 Hz, 5.5-kW AFPM synchronous generator with slotted stator has the following parameters of the steady-state equivalent circuit: $R_1 = 0.1\ \Omega$, $X_{sd} = 2.3\ \Omega$ and $X_{sq} = 2.2\ \Omega$. The generator is loaded with a three-phase Y-connected resistance of $R_L = 2.2\ \Omega$ and inductance of $L_L = 0.0007$ H connected in series. The EMF per phase at no load induced by the rotor PM excitation system is $E_f = 100$ V and rotational losses are $\Delta P_{rot} = 90$ W.

Find the stator currents, electromagnetic power, output power and efficiency. Stator core losses and losses in PMs have been ignored.

Solution

The load reactance and module of load impedance are

$$X_L = 2\pi f L_L = 2\pi \times 50 \times 0.0007 = 0.22\ \Omega$$

$$Z_L = \sqrt{2.2^2 + 0.22^2} = 2.211\ \Omega$$

The stator currents according to eqns (4.40), (4.41) and (2.89) are

$$I_{ad} = \frac{100 \times (2.2 + 0.22)}{(2.3 + 0.22) \times (2.2 + 0.22) + (0.1 + 2.2)^2} = 21.25\ \text{A}$$

$$I_{aq} = \frac{100 \times (0.1 + 2.2)}{(2.3 + 0.22) \times (2.2 + 0.22) + (0.1 + 2.2)^2} = 20.20\ \text{A}$$

$$I_a = \sqrt{21.25^2 + 20.20^2} = 29.32 \text{ A}$$

The terminal voltage (across the load impedance) is

$$V_1 = I_a Z_L = 29.32 \times 2.211 = 64.82 \text{ V}$$

$$V_{1L} = \sqrt{3}\, V_1 = \sqrt{3} \times \; 64.82 = 112.3 \text{ V}$$

The power factor, load angle δ and angle Ψ according to eqns (4.43), (4.39) and (4.42) respectively are

$$\cos\phi = \frac{29.32 \times 2.2}{64.82} = 0.995 \qquad\qquad \phi = 5.71^{\circ}$$

$$\Psi = \arccos\left(\frac{20.20}{29.32}\right) = 46.45^{\circ}$$

$$\delta = \arcsin\left(\frac{21.25 \times 2.2 - 20.20 \times 0.22}{64.82}\right) = 40.74^{\circ}$$

$$\delta = \Psi - \phi = 46.45^{\circ} - 5.71^{\circ} = 40.74^{\circ}$$

The electromagnetic power according to eqn (4.45) is

$$P_{elm} = 3 \times [100 \times 20.20 - 21.25 \times 20.20 \times (2.3 - 2.2)] = 5930.3 \text{ W}$$

The stator winding losses according to eqn (2.49) is

$$\Delta P_{1w} = 3 \times 29.32^2 \times 0.1 = 257.8 \text{ W}$$

The output power is

$$P_{out} = m_1 V_1 I_a \cos\phi = 3 \times 64.82 \times 29.32 \times 0.995 = 5672.5 \text{ W}$$

$$P_{out} = P_{elm} - \Delta P_{1w} = 5930.3 - 257.8 = 5672.4$$

The input power is

$$P_{in} = P_{elm} + \Delta P_{rot} = 5930.3 + 90 = 6020.3 \text{ W}$$

The efficiency is

$$\eta = \frac{5672.5}{6020.3} = 0.957$$

5

AFPM Machines Without Stator Cores

5.1 Advantages and Disadvantages

Depending on the application and operating environment, stators of AFPM machines may have ferromagnetic cores or be completely coreless. A coreless stator AFPM machine has an internal stator and twin external PM rotor (Fig. 1.4d). PMs can be glued to the rotor backing steel discs or nonmagnetic supporting structures. In the second case PMs are arranged in Halbach array (Fig. 3.15) and the machine is completely coreless. The electromagnetic torque developed by a coreless AFPM brushless machine is produced by the open space current-carrying conductor–PM interaction (Lorentz force theorem).

Coreless configurations eliminate any ferromagnetic material, i.e. steel laminations or SMC powders from the stator (armature), thus eliminating the associated eddy current and hysteresis core losses. Because of the absence of core losses, a coreless stator AFPM machine can operate at higher efficiency than conventional machines. On the other hand, owing to the increased nonmagnetic air gap, such a machine uses more PM material than an equivalent machine with a ferromagnetic stator core.

Typical coil shapes used in the winding of a coreless stator are shown in Figs 3.16 and 3.17.

In this chapter AFPM brushless machines with coreless stator and conventional PM excitation, i.e. PM fixed to backing steel discs, will be discussed.

5.2 Commercial Coreless Stator AFPM Machines

Bodine Electric Company, Chicago, IL, U.S.A. manufactures 178-mm (7-inch) and 356-mm (14-inch) diameter *e-TORQ*™ AFPM brushless motors with coreless stator windings and twin external PM rotors with steel backing discs (Fig. 5.1a). The coreless stator design eliminates the so called *cogging (detent) torque*, improves low speed control, yields linear torque-current characteristics due to the absence of magnetic saturation and provides peak torque up to ten times the rated torque. Motors can run smoothly at extremely low speeds, even when powered by a standard solid state converter. In addition, the high peak torque capability can allow, in certain applications, the elimination of costly gearboxes and reduce the risk of lubricant leaks.

Fig. 5.1. AFPM brushless *e-TORQ*™ motor with coreless stator windings: (a) general view; (b) motor integrated with wheel of a solar powered car. Photo courtesy of *Bodine Electric Company*, Chicago, IL, U.S.A.

The 356-mm diameter *e-TORQ*™ motors have been used successfully by students of North Dakota State University for direct propulsion of a *solar powered car* participating in 2003 American Solar Challenge (Fig. 5.1b). A well designed solar-powered vehicle needs a very efficient and very light electric motor to convert the maximum amount of solar energy into mechanical energy at minimum rolling resistance. Coreless AFPM brushless motors satisfy these requirements.

Small ironless motors may have *printed circuit stator windings* or *film coil windings*. The film coil stator winding has many coil layers while the printed circuit winding has one or two coil layers. Fig. 5.2 shows an ironless brushless motor with film coil stator winding manufactured by *EmBest*, Soeul, South Korea. This motor has single-sided PM excitation system at one side of the stator and backing steel disc at the other side of the stator. Small film coil motors can be used in computer peripherals, computer hard disc drives (HDDs) [139, 140], mobile phones, pagers, flight recorders, card readers, copiers, printers, plotters, micrometers, labeling machines, video recorders and medical equipment.

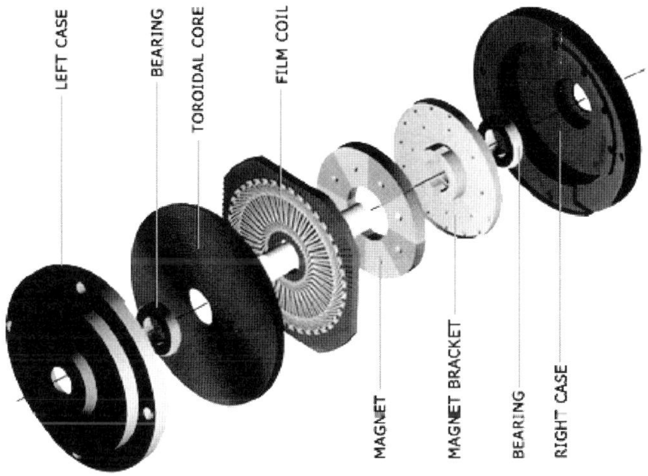

Fig. 5.2. Exploded view of the AFPM brushless motor with film coil coreless stator winding and single-sided rotor PM excitation system. Courtesy of *Embest*, Soeul, South Korea.

5.3 Coreless Stator AFPM Microgenerators

Fig. 5.3 shows an AFPM microgenerator developed at Georgia Institute of Technology, Atlanta, GA, U.S.A. [12, 13]. The stator uses interleaved, electroplated copper windings that are dielectrically isolated from a 1-mm thick NiFeMo substrate by a 5 μm polyimide layer. The rotor consists of a 2 to 12 pole, 500 μm thick, annular PM (OD = 10 mm, ID = 5 mm) and a 500 μm thick Hiperco 50 (FeCoV alloy) ring as a return path for the magnetic flux. SmCo PMs and Hiperco alloy have been used because microgenerators will be operating in high temperature environment, i.e., powered by gas-fueled turbine engines. Specifications are given in Table 5.1 [12].

A 16 W of mechanical–to–electrical power conversion and delivery of 8 W of d.c. power to a resistive load at a rotational speed of 305 krpm has been demonstrated [12]. The power density of the AFPM microgenerator was 59 W/cm^3.

5.4 Performance Calculation

5.4.1 Steady-State Performance

To calculate the steady-state performance of a coreless stator AFPM brushless machine it is essential to consider the equivalent circuits.

The *steady-state per phase equivalent circuit* of a coreless stator AFPM brushless machine may be represented by the circuit shown in Fig. 5.5, where

Fig. 5.3. Experimental tests on AFPM microgenerators by spinning a small magnet with the aid of air-powered drill above a mesh of coils fabricated on a chip. Photo courtesy of Georgia Institute of Technology, Atlanta, GA, U.S.A.

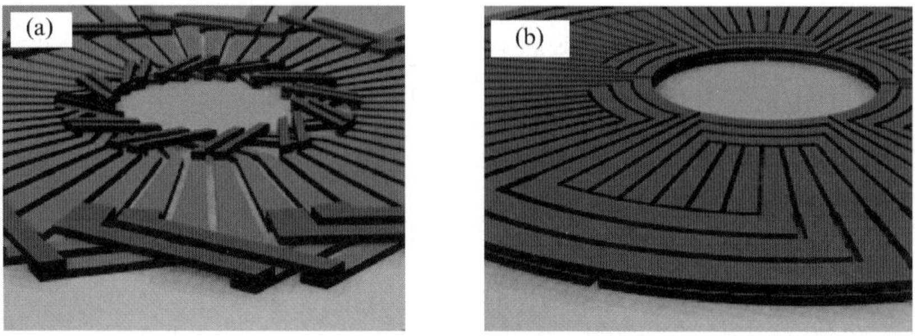

Fig. 5.4. Stator winding patterns for 8-pole microgenerator: (a) 2 turns per pole; (b) 3-turns pole. Photo courtesy of Georgia Institute of Technology, Atlanta, GA, U.S.A.

R_1 is the stator resistance, X_1 is the stator leakage reactance, E_f is the EMF induced in the stator winding by the rotor PM excitation system, E_i is the *rms* value of the internal phase voltage, V_1 is the terminal voltage and I_a is the *rms* stator current. The shunt resistance R_e is the stator eddy current loss resistance, which is defined in the same way as the core loss resistance [126] for slotted PM brushless motors.

The stator currents I_{ad}, I_{aq} and I_a for a given load angle δ (corresponding to the slip s in induction machines) can be calculated on the basis of eqns (2.87), (2.88) and (2.89) respectively. If $I_{ad} = 0$, all the stator current $I_a = I_{aq}$

Table 5.1. Specifications of AFPM microgenerator fabricated at the School of Electrical and Computer Engineering, Georgia Institute of Technology, Atlanta, GA, U.S.A.

Number of poles	2 to 12
Number of turns per pole	1 to 6
Outer end turn extension	0 to 2.5 mm
Inner end turn extension	0 to 2.5 mm
Magnet outer radius, mm	5.0
Magnet inner radius, mm	2.5
Stator radial conductor outer radius, mm	4.75
Stator radial conductor inner radius, mm	2.75
Hiperco ring thickness, μm	500
Magnet thickness, μm	500
Radial conductor thickness, μm	200
End turn thickness, μm	80
Substrate thickness, μm	1000
Air gap, μm	100
Power electronics equivalent resistance, mΩ	100

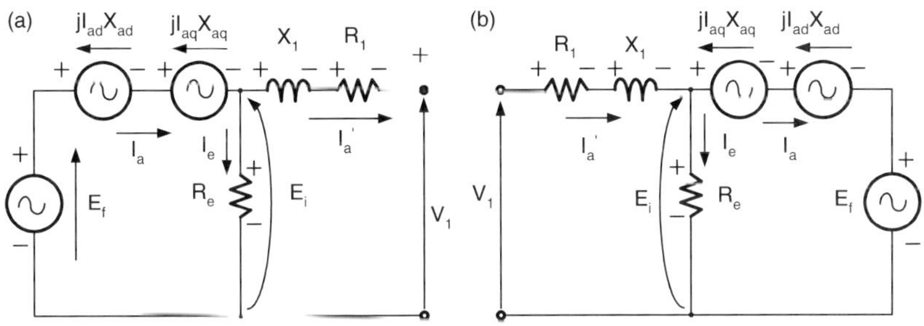

Fig. 5.5. Per phase equivalent circuit of an AFPM machine with a coreless stator: (a) generator arrow system; (b) consumer (motor) arrow system. Eddy current losses are accounted for by the shunt resistance R_e.

produces the electromagnetic torque and the load angle $\delta = \phi$. The angle ϕ between the stator current I_a and terminal voltage V_1 is determined by the power factor $\cos\phi$.

If we ignore the losses in PMs and losses in rotor backing steel discs, the input power can then be calculated as follows

- for the motoring mode (electrical power)

$$P_{in} = P_{elm} + \Delta P_{1w} + \Delta P_e \tag{5.1}$$

- for the generating mode (mechanical shaft power)

$$P_{in} = P_{elm} + \Delta P_{rot} \tag{5.2}$$

where P_{elm} is the electromagnetic power according to eqn (2.91) in which $\Delta P_{1Fe} = 0$, ΔP_{1w} is the stator winding loss according to eqn (2.49), ΔP_e are the eddy current losses in the stator conductors according to eqn (2.68) or eqn (2.69) and ΔP_{rot} are the rotational losses according to eqn (2.70).

Similarly, the output power is:

- for the motoring mode (shaft power)

$$P_{out} = P_{elm} - \Delta P_{rot} \tag{5.3}$$

- for the generating mode (electrical power)

$$P_{out} = P_{elm} - \Delta P_{1w} - \Delta P_e \tag{5.4}$$

The shaft torque is given by eqn (4.30) for motoring mode or eqn (4.31) for generating mode, and the efficiency is expressed by eqn (4.32).

5.4.2 Dynamic Performance

For a salient pole synchronous machine without any rotor winding the voltage equations for the stator circuit are

$$v_{1d} = R_1 i_{ad} + \frac{d\psi_d}{dt} - \omega\psi_q \tag{5.5}$$

$$v_{1q} = R_1 i_{aq} + \frac{d\psi_q}{dt} + \omega\psi_d \tag{5.6}$$

in which the linkage fluxes are defined as

$$\psi_d = (L_{ad} + L_1) i_{ad} + \psi_f = L_{sd} i_{ad} + \psi_f \tag{5.7}$$

$$\psi_q = (L_{aq} + L_1) i_{aq} = L_{sq} i_{aq} \tag{5.8}$$

In the above equations (5.5) to (5.8) v_{1d} and v_{1q} are *d*- and *q*-axis components of the terminal voltage, ψ_f is the maximum flux linkage per phase produced by the excitation system, R_1 is the armature winding resistance, L_{ad}, L_{aq} are the *d*- and *q*-axis components of the armature self-inductance, $\omega = 2\pi f$ is the angular frequency of the armature current, i_{ad}, i_{aq} are the *d*- and *q*-axis components of the armature current. The resultant armature inductances $L_{sd} = L_{ad} + L_1$ and $L_{sq} = L_{aq} + L_1$ are referred to as synchronous inductances. In a three-phase machine $L_{ad} = (3/2)L'_{ad}$ and $L_{aq} = (3/2)L'_{aq}$ where L'_{ad} and L'_{aq} are self-inductances of a single phase machine. The excitation linkage flux $\psi_f = L_{fd} I_f$ where L_{fd} is the maximum value of the mutual inductance between the armature and field winding. In the case of a PM excitation, the fictitious field current is $I_f = H_c h_M$.

Putting eqns (5.7) and (5.8) into eqns (5.5) and (5.6), the stator voltage equations in the *d* and *q*-axis can be written as

$$v_{1d} = R_1 i_{ad} + \frac{di_{ad}}{dt} L_{sd} - \omega L_{sq} i_{aq} \quad (5.9)$$

$$v_{1q} = R_1 i_{aq} + \frac{di_{aq}}{dt} L_{sq} + \omega L_{sd} i_{ad} + \omega \psi_f \quad (5.10)$$

For the steady state operation $(d/dt)L_{sd}i_{ad} = (d/dt)L_{sq}i_{aq} = 0$, $\mathbf{I}_a = I_{ad} + jI_{aq}$, $\mathbf{V}_1 = V_{1d} + jV_{1q}$, $i_{ad} = \sqrt{2}I_{ad}$, $i_{aq} = \sqrt{2}I_{aq}$, $v_{1d} = \sqrt{2}V_{1d}$, $v_{1q} = \sqrt{2}V_{1q}$, $E_f = \omega L_{fd} I_f/\sqrt{2} = \omega\psi_f/\sqrt{2}$ [94]. The quantities ωL_{sd} and ωL_{sq} are known as the $d-$ and q-axis synchronous reactances respectively.

The instantaneous power to the motor input terminals is [94, 106]

$$p_{in} = \frac{m_1}{2}(v_{1d} i_{ad} + v_{1q} i_{aq}) \quad (5.11)$$

The input power of a motor (5.11) is equivalent to the output power of a generator. The electromagnetic power of a three-phase machine is [94, 106]

$$p_{elm} = \frac{3}{2}\omega[\psi_f + (L_{sd} - L_{sq})i_{ad}]i_{aq} \quad (5.12)$$

The electromagnetic torque of a three phase motor with p pole pairs is

$$T_d = p\frac{p_{elm}}{\omega} = \frac{3}{2}p[\psi_f + (L_{sd} - L_{sq})i_{ad}]i_{aq} \quad (5.13)$$

The relationships between i_{ad}, i_{aq} and phase currents i_{aA}, i_{aB} and i_{aC} are

$$i_{ad} = \frac{2}{3}\left[i_{aA}\cos\omega t + i_{aB}\cos\left(\omega t - \frac{2\pi}{3}\right) + i_{aC}\cos\left(\omega t + \frac{2\pi}{3}\right)\right] \quad (5.14)$$

$$i_{aq} = -\frac{2}{3}\left[i_{aA}\sin\omega t + i_{aB}\sin\left(\omega t - \frac{2\pi}{3}\right) + i_{aC}\sin\left(\omega t + \frac{2\pi}{3}\right)\right] \quad (5.15)$$

The reverse relations, obtained by simultaneous solution of eqns (5.14) and (5.15) in conjunction with $i_{aA} + i_{aB} + i_{aC} = 0$, are

$$i_{aA} = i_{ad}\cos\omega t - i_{aq}\sin\omega t$$

$$i_{aB} = i_{ad}\cos\left(\omega t - \frac{2\pi}{3}\right) - i_{aq}\sin\left(\omega t - \frac{2\pi}{3}\right) \quad (5.16)$$

$$i_{aC} = i_{ad}\cos\left(\omega t + \frac{2\pi}{3}\right) - i_{aq}\sin\left(\omega t + \frac{2\pi}{3}\right)$$

5.5 Calculation of Coreless Winding Inductances

5.5.1 Classical Approach

The synchronous inductance L_s consists of the armature reaction (mutual) inductance L_a and the leakage inductance L_1. For a machine with magnetic asymmetry, i.e. with a difference in reluctances in the d and q axes, the synchronous inductances in the $d-$ and q-axis, L_{sd} and L_{sq}, are written as sums of the armature reaction inductances (mutual inductances), L_{ad} and L_{aq}, and leakage inductance L_1, i.e.

$$L_{sd} = L_{ad} + L_1 \qquad L_{sq} = L_{aq} + L_1 \tag{5.17}$$

The armature reaction inductances are expressed by eqns (2.116), (2.117) and (2.118), in which air gaps in the d- and q-axis are given by eqns (2.112) and (2.113). The armature reaction reactances are given by eqns (2.121) and (2.122). Table 2.1 compares armature reaction equations for cylindrical and disc-type machines.

The leakage inductance is expressed analytically as a sum of three components, i.e.

$$L_1 = L_{1s} + L_{1e} + L_{1d} = 2\mu_0 \frac{N_1^2 L_i}{pq_1}(\lambda_{1s} + \frac{l_{1e}}{L_i}\lambda_{1e} + \lambda_{1d}) \tag{5.18}$$

where $L_i = 0.5(D_{out} - D_{in})$ is the active length of a coil equal to the radial length l_M of the PM, q_1 is the number of coil sides per pole per phase (equivalent to the number of slots) according to eqn (2.2), $l_{1e} = 0.5(l_{1in} + l_{1out})$ is the average length of the single-sided end connection, L_{1s} and λ_{ls} are the inductance and specific permeance for the leakage flux about radial portions of conductors (corresponding to slot leakage in classical machines) respectively, L_{1e} and λ_{1e} are the inductance and specific permeance for the leakage flux about the end connections respectively, and L_{1d} and λ_{1d} are the inductance and specific permeance for differential leakage flux (due to higher space harmonics) respectively.

It is difficult to derive an accurate analytical expression for λ_{1s} for a coreless electrical machine. The specific permeances λ_{1s} and λ_{1e} can roughly be estimated from the following semi-analytical equation:

$$\lambda_{1s} \approx \lambda_{1e} \approx 0.3q_1 \tag{5.19}$$

The specific permeance for the differential leakage flux can be found in a similar way as for an induction machine [121], using eqn (4.22) in which $k_{sat} \approx 1$ and $g' \approx 2[(g+0.5t_w)+h_M]$. The thickness of the stator winding is t_w and the distance from the stator disc surface to the PM active surface is g (mechanical clearance). It is not difficult to show that $L_{1d} = L_a\tau_{d1}$ [121].

5.5.2 FEM Approach

Unlike conventional slotted AFPM machines, there is no clear definition for main and leakage inductances in a coreless or slotless machine as discussed in [14, 85, 133]. Using the 2D FEM analysis both mutual and leakage fluxes can be taken into account. The only remaining part is the end winding leakage flux.

With the 2D finite element solution the magnetic vector potential $\mathbf{A}$ has only a z component, i.e. $\mathbf{A} = A(x, y) \cdot \hat{a}_z$ where $\hat{a}_z$ is the unit vector in z direction (axial direction). The total stator flux of a phase winding ψ_{ABC} that excludes the end-winding flux leakage can be readily calculated by using Stokes theorem, i.e.

$$\psi_{ABC} = \int_s \mathbf{B} \cdot \mathbf{dS} = \int_s \nabla \times \mathbf{A} \cdot \mathbf{dS} = \oint_c \mathbf{A} \cdot \mathbf{dl} \tag{5.20}$$

As an approximation the flux linkage of a phase coil can be calculated by working out the difference between maximum magnetic vector potential values of each coil side. In the case that the coil is not very thin, magnetic vector potential varies in the coil cross-section area. Therefore, the average magnetic vector potential values should be used. For first-order triangular elements, the flux linkage of a coil with N_1 turns, area S and length l is given by [144]

$$\psi = N_1 \sum_{j=1}^{n} \frac{\Delta_j}{S} \left[\frac{\zeta}{3} \sum_{i=1}^{3} A_{ij} \right] l \tag{5.21}$$

where A_{ij} is the nodal value of the magnetic vector potential of the triangular element j, $\zeta = +1$ or $\zeta = -1$ indicates the direction of integration either into the plane or out of the plane, Δ_j is the area of the triangular element j, and n is the total number of elements of the in-going and out-going areas of the coil. It follows that for an AFPM machine with only one pole modelled, the total flux linkage of a phase winding is

$$\psi_{ABC} = \frac{2\,p\,N_1\,l}{a_p S} \sum_{j=1}^{u} \left[\frac{\Delta_j \zeta}{3} \sum_{i=1}^{3} A_{ij} \right] \tag{5.22}$$

where u is the total number of elements of the meshed coil areas of the phase in the pole region and a_p is the number of parallel circuits (parallel current paths) of the stator winding.

From a machine design perspective, it is important to find the fundamental components of the total flux linkages. For a coreless stator AFPM machine with usually unsaturated rotor yoke, the flux linkage harmonics due to iron stator slots and magnetic saturation are absent. Owing to the large air gap, the stator winding MMF space harmonics are negligible in most cases. The most important harmonics needed to account for are those due to the flat-shaped PMs.

Given these considerations, the flux linkage wave of an AFPM machine is nearly sinusoidal, though for a concentrated parameter (non-distributed) winding, an appreciable 3rd and less significant 5th and 7th harmonics are still present in the total flux linkage waveform. If the 5th, 7th and higher harmonics are ignored, the fundamental total phase flux linkages can be calculated by using the technique given in [144], i.e.

$$[\psi_{ABC1}] \approx [\psi_{ABC}] - [\psi_{ABC3}] \tag{5.23}$$

where the co-phasal 3rd harmonic flux linkage, including the higher order triple harmonics, can be obtained from:

$$\psi_{A3} = \psi_{A3} = \psi_{C3} \approx \frac{1}{3}(\psi_A + \psi_B + \psi_C) \tag{5.24}$$

With the fundamental total phase flux linkages and rotor position known, the d and q-axis flux linkages are calculated using Park's transformation as follows [94]:

$$[\psi_{dq0}] = [K_p]\,[\psi_{ABC1}] \tag{5.25}$$

where [94]

$$K_p = \frac{2}{3}\begin{bmatrix} \cos\theta & \cos(\theta - \frac{2\pi}{3}) & \cos(\theta + \frac{2\pi}{3}) \\ -\sin\theta & -\sin(\theta - \frac{2\pi}{3}) & -\sin(\theta + \frac{2\pi}{3}) \\ \frac{1}{2} & \frac{1}{2} & \frac{1}{2} \end{bmatrix} \tag{5.26}$$

With a constant rotor angular velocity ω the angle $\theta = \omega t$. In the 2D FEM model the d-axis and q-axis synchronous inductances do not include the component due to end connection leakage flux [106], i.e.

$$L'_{sd} = L_{ad} + L'_1 = L_{ad} + L_{1s} + L_{1d} \tag{5.27}$$

$$L'_{sq} = L_{aq} + L'_1 = L_{aq} + L_{1s} + L_{1d} \tag{5.28}$$

The inductances L'_{sd} and L'_{sq} result from eqns (5.7) and (5.8), i.e.

$$L'_{sd} = \frac{\psi_d - \psi_f}{i_{ad}} \quad \text{and} \quad L'_{sq} = \frac{\psi_q}{i_{aq}} \tag{5.29}$$

The end winding inductance L_{1e} can be calculated from numerical evaluation of the energy stored in the end connections [184] or simply by using an approximate equation resulting from eqn (5.18), i.e.

$$L_{1e} = 2\mu_0 \frac{N_1^2 l_{1e}}{p q_1} \lambda_{1e} \tag{5.30}$$

Finally,

$$L_{sd} = L'_{sd} + L_{1e} \quad \text{and} \quad L_{sd} = L'_{sd} + L_{1e} \tag{5.31}$$

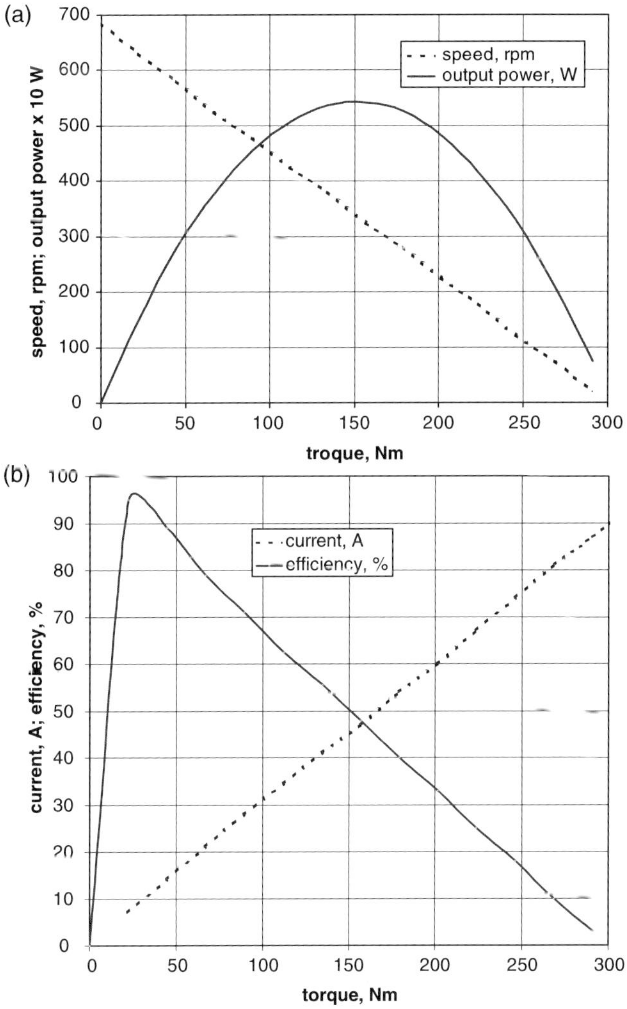

Fig. 5.6. Steady state characteristics of of 356-mm, 1-kW, 170-V *e-TORQ*™ AFPM motor; (a) output power P_{out} and speed n versus shaft torque T_{sh}; (b) phase current and efficiency versus shaft torque T_{sh}. Data captured at 22°C for totally enclosed non-ventilated motor. Courtesy of *Bodine Electric Company*, Chicago, IL, U.S.A.

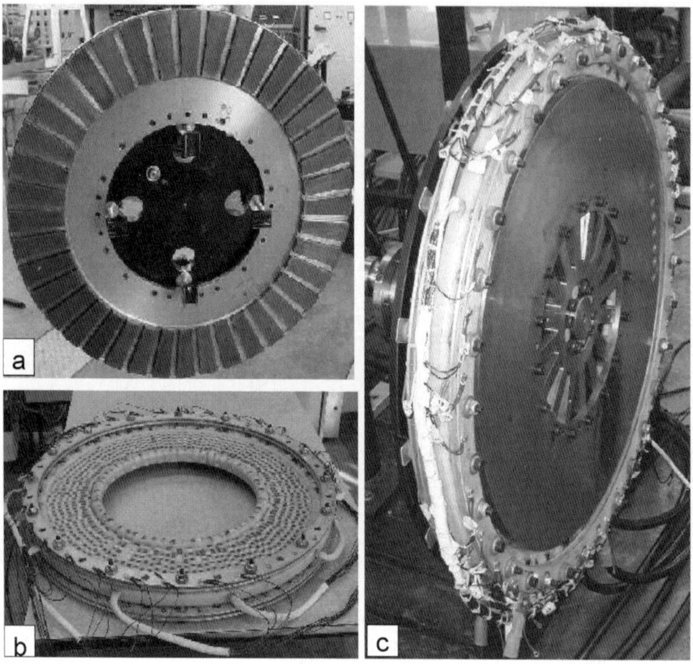

Fig. 5.7. Air-cooled single-stage synchronous AFPM machine: (a) rotor disc with surface mounted PM segments, (b) coreless stator with busbars, and (c) the assembled machine. Courtesy of the *University of Stellenbosch*, South Africa.

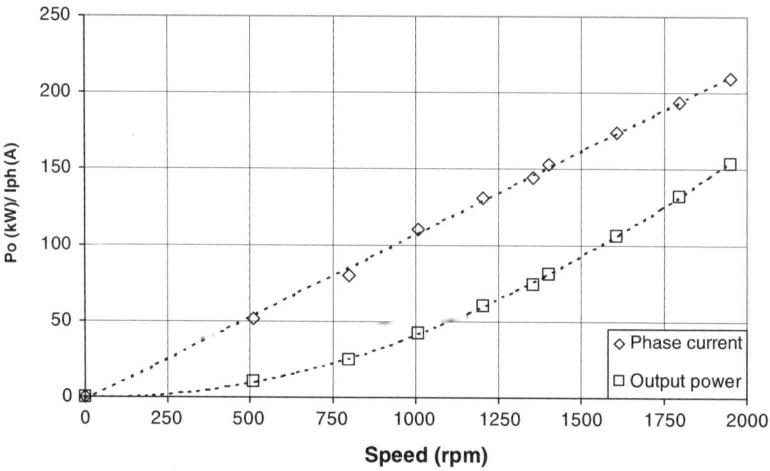

Fig. 5.8. Output power and phase current versus speed for generating mode of a 160-kW AFPM brushless generator. Specifications are given in Table 5.3.

Table 5.2. Specifications of *e-TORQ*™ AFPM motors with coreless stators manufactured by *Bodine Electric Company*, Chicago, IL, U.S.A.

Parameter	178-mm AFPM motor Low EMF constant	178-mm AFPM motor High EMF constant	356-mm AFPM motor Low EMF constant	356-mm AFPM motor High EMF constant
Output power, kW	0.7	0.57	1.0	0.26
Number of poles	8	8	16	16
d.c. bus voltage, V	170	300	170	300
Speed, rpm	3000	1500	300	70
Peak speed, rpm	3500	2200	700	400
Torque, Nm	2.26	2.83	31.1	33.9
Peak torque, Nm	22.6	13.67	152.1	84.4
Current, A	5	2	13	5
Peak current, A	50	10.5	64	12.5
Efficiency, %	81	75	84	77
Torque constant, Nm/A	0.4859	1.299	2.38	6.814
EMF constant, V/krpm	50	137	249	713
Winding resistance, Ω	2.2	14.3	1.33	43
Winding inductance, mH	1.4	10.5	3.6	29.4
Viscous friction, Nm/rad/s	9.9×10^{-5}	0.00019	0.00669	0.012
Static friction, Nm	0.00728	0.0378	0.02357	0.1442
Electrical time constant, ms	0.6364	0.734	2.71	0.684
Mechanical time constant, ms	4.538	4.026	4.5	17.78
Moment of inertia, kgm^2	0.00525	0.00525	0.21	0.21
Mass of active parts, kg	6.17	6.17	30.87	30.87
Power density, W/kg	113.5	92.4	32.39	8.42

5.6 Performance Characteristics

Specifications of *e-TORQ*™ AFPM brushless motors manufactured by *Bodine Electric Company*, Chicago, IL, U.S.A. are given in Table 5.2. The EMF constant, torque constant, winding resistance and winding inductance are line-to-line quantities. Motors are designed for a maximum winding continuous temperature of 130°C. Steady-state performance characteristics of a 356-mm, 1-kW, 170-V e-TORQ™ motor are shown in Fig. 5.6.

Fig. 5.7 shows an air-cooled 160-kW AFPM brushless generator with a coreless stator built at the University of Stellenbosch, South Africa [260]. The stator winding consists of sixty single-layer trapezoidal-shape coils, which have the advantage of being easy to fabricate and have relatively short overhangs (Fig. 3.17). The winding coils are held together to form a disc-type stator by using composite material of epoxy resin and hardener. Sintered NdFeB PMs with $B_r \approx 1.16$ T and maximum allowable working temperature around 130°C have been used. The detailed design data are given in Table 5.3.

The output power and phase current at different speeds are shown in Fig. 5.8. Owing to very low stator winding inductance per phase, the output voltage varies almost linearly with the load current.

Table 5.3. Specifications of a three-phase, 160-kW, 1950-rpm AFPM brushless synchronous generator with ironless stator core built at the *University of Stellenbosch*, South Africa

Design data	
Output power P_{out}, kW	160
Speed n, rpm	1950
Number of phases m_1	3 (Wye)
Rated line voltage, V	435
Rated phase current, A	215
Frequency f, Hz	100
Number of stator modules	1
Number of pole pairs p	20
Number of stator coils (3 phases)	60
Number of turns per phase	51
Wire diameter, mm	0.42
Number of parallel wires a_w	12
Axial height of PM per pole h_M, mm	10.7
Axial thickness of the winding t_w, mm	15.7
Air gap (one side) g, mm	2.75
Air gap magnetic flux density B_{mg} under load, T	0.58
Current density, A/mm^2	7.1
PM outer diameter D_{out}, mm	720
$k_d = D_{in}/D_{out}$ ratio	0.69
Class of insulation	F
Winding temperature rise, °C	56
Cooling system	Self air-cooled
Type of winding	Single layer trapezoidal

It has been found from both experimental tests and calculations that for a typical sine-wave AFPM machine with a coreless stator, the ratio of the d- and q-axis inductances of the phase winding is near unity, i.e. $L_{sd}/L_{sq} \approx 1$. Thus, the analysis of the AFPM brushless machine with a coreless stator may be done in a similar way as that of the three-phase cylindrical machine with surfaced PMs [128, 157].

5.7 Performance of Coreless Non-Overlap Winding AFPM Machines

The use of non-overlap coils in AFPM electrical machines has certain advantages, amongst other things,

- easier coil manufacture and stator assembly;
- reduced stator winding cost due to less number of coils and simple winding structure;
- smaller overall diameter of the machine due to shorter end-turn length;

- lower stator winding losses due to shorter average length of turn;
- higher slot fill factor in the case of slotted stator with ferromagnetic core.

A drawback in general of non-overlap coil machines is the lower output torque due to a lower winding factor expressed by eqns (2.14) and (2.15) [146]. Recent studies, however, shows that non-overlap coil PM machines with high pole numbers can have high winding factors and good output torque [67, 166, 176, 177, 229]. These studies focused on radial flux PM machines with coreless stators. The characterization of an AFPM brushless generator with a coreless non-overlap winding with concentrated parameters is described in [255]. Research work on AFPM machines with coreless non-overlap concentrated parameter windings has been reported in [27, 146, 213]. Coreless AFPM machines with non-overlap windings do not have the problem of cogging torque and of putting the coils into slots. Hence, there is more freedom in the layout and assembly of the winding. Furthermore, non-overlap coil PM machines with ferromagnetic stator core are known for their additional losses in the current conducting PMs and rotor ferromagnetic yoke due to higher harmonic flux pulsations. In coreless non-overlap winding AFPM machines this is almost completely absent due to the low armature reaction effect. There are, thus, no disadvantages in using non-overlap coils with concentrated parameters in coreless stator AFPM machines, except for the lower torque performance in the case of some winding and PM configurations.

In Fig. 5.9 the layouts of overlap and non overlap windings are shown. Congruent with these layouts, three air-cored AFPM stators (1 kW) with normal overlap, non-overlap and phase-group windings are shown in Fig. 5.10(a)-(c). These stators have been designed and built for the PM rotor discs of the AFPM machine under test shown in Fig. 5.10(d). For the tests an a.c. electromechanical drive has been used as prime mover and a torque transducer to measure the shaft torque as shown in Fig. 5.10(d).

For the numeric modeling of the AFPM machine the 2D FEM analysis has been used. The 2D FEM models of the normal three-phase overlap winding and two types of non-overlap coil AFPM machines are shown in Fig. 5.11. Owing to the axial symmetry, it is only necessary to model half of the machine in the FEM analysis, i.e., one rotor disc and half a stator. For the normal overlap winding AFPM machine, it is possible to model just one pole-pitch of the machine by applying negative periodical condition on the left and right boundaries, as shown in Fig. 5.11(a). However, for the non-overlap winding AFPM machine, it is necessary to model two pole-pairs of the machine by using positive periodical conditions, as shown in Fig. 5.11(b). For the phase-group non-overlap winding AFPM machine, it is inevitable to model the complete machine as there is no symmetry (Fig. 5.11(c)).

The design data of the AFPM machines with different windings is given in Table 5.4. The per unit calculated and measured results of the different winding AFPM machines are given in Table 5.5. The analytically calculated, FEM-calculated and measured results of the overlap winding are throughout

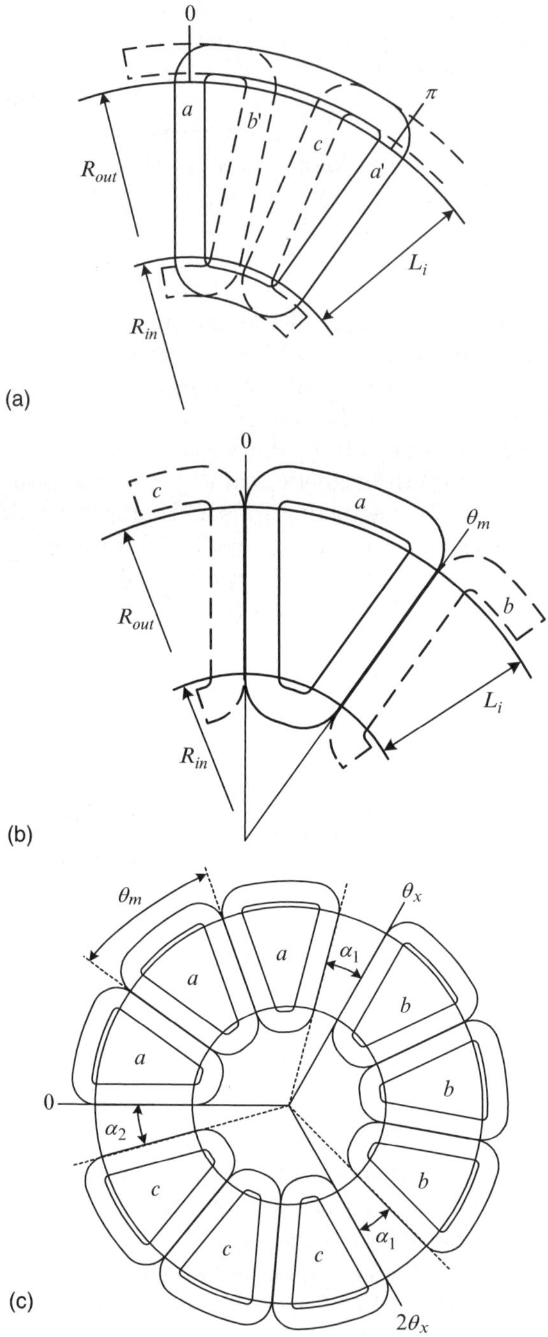

Fig. 5.9. Layout of different coreless windings with (a) normal overlap winding, (b) non-overlap winding, and (c) phase-group non-overlap winding.

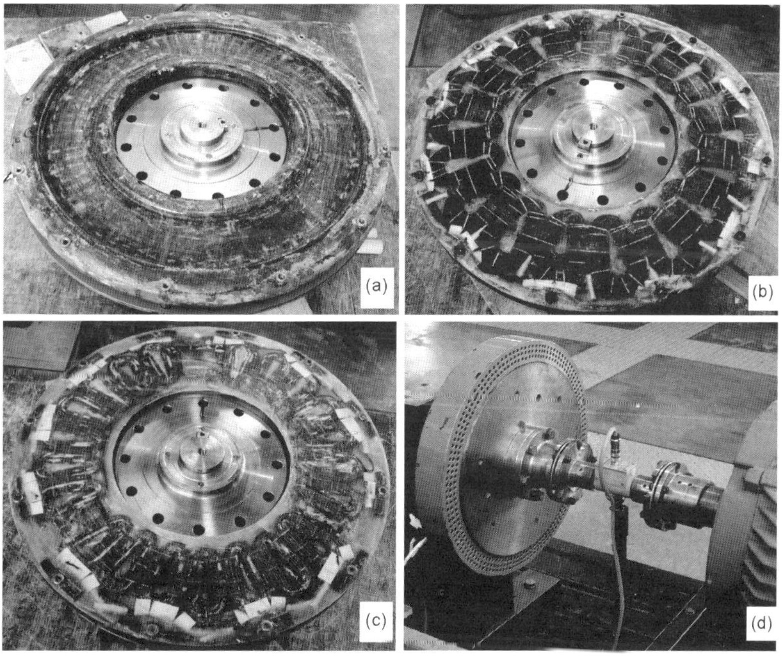

Fig. 5.10. Coreless AFPM machine stators with (a) normal overlap winding, (b) non-overlap winding, (c) phase-group non overlap winding, and (d) testing set-up of AFPM machine as a generator [146].

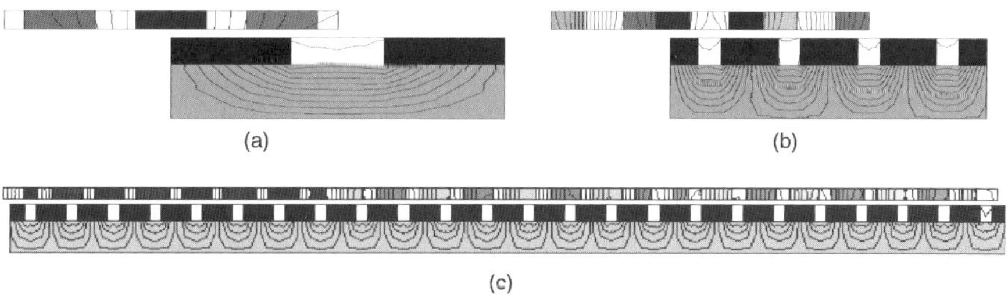

Fig. 5.11. 2D FEM models of coreless AFPM machine with (a) normal overlap, (b) non-overlap winding, and (c) phase-group non-overlap winding [146].

taken as the basis in the per unit calculation. The calculation and measurements were performed for the same copper loss of 120 W for each type of stator winding. It is clear from the results of Table 5.5 that the same torque per copper loss is obtained with the non-overlap winding as with the normal overlap windings. However, non-overlap winding needs significantly less copper.

Table 5.4. Design data of AFPM machines with different winding topologies

Design data	overlap	non-overlap	phase-group
Output power P_{out}, kW	1	1	1
Speed n, rpm	200	200	200
Number of phases m_1	3 (Wye)	3 (Wye)	3 (Wye)
Rated line voltage, V	19.88	19.56	18.43
Rated phase current, A	13.79	14.23	13.06
Frequency f, Hz	40	40	40
Number of stator modules	1	1	1
Number of pole pairs p	12	12	12
Number of stator coils (3 phases)	36	18	21
Number of turns per coil	11	25	23
Wire diameter, mm	0.8	0.85	0.7
Number of parallel wires a_w	7	6	6
Axial height of PM h_M, mm	6	6	6
Axial thickness of winding t_w, mm	8.4	8.4	8.4
Air gap (one side) g, mm	1.8	1.8	1.8
Air gap flux density B_{mg}, T	0.527	0.527	0.527
PM outer diameter D_{out}, mm	400	400	400
$k_d = D_{in}/D_{out}$ ratio	0.7	0.7	0.7
Class of insulation	F	F	F
Winding temperature rise, °C	9.9	12.1	8.3
Cooling system	Air-cooled	Air-cooled	Air-cooled

Table 5.5. Comparison of calculated and measured per unit results for AFPM brushless machines with $D_{in} = 280$ mm $D_{out} = 400$ mm and $p = 12$ [146]

			Torque			Mass of copper	
Type	z	Q_c	analytical	FEM	measured	analytical	measured
Overlap	-	36	1	1	1	1	1
Non-overlap	1	18	0.988	1.013	1.014	0.891	0.854
Phase-group	7	21	0.885	0.902	0.899	0.683	0.624

Fig. 5.12 shows the measured induced open circuit voltage of the normal overlap and the non-overlap windings. It is clear that an improved sinusoidal voltage waveform is obtained with the non-overlap winding.

5.8 Eddy Current Losses in the Stator Windings

5.8.1 Eddy Current Loss Resistance

For an AFPM machine with a coreless stator, associated stator iron losses are absent. The rotor discs rotate at the same speed as the main magnetic field, thus the core losses in the rotor discs due to the fundamental harmonic of the stator field also do not exist. However, the eddy current losses in the

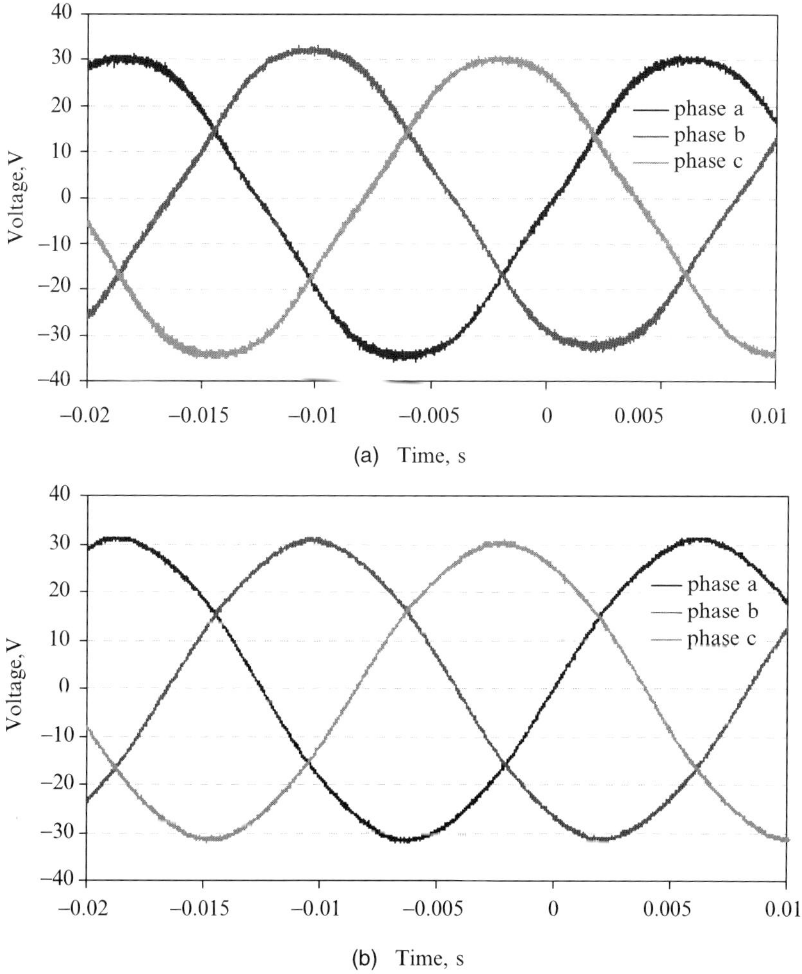

Fig. 5.12. Measured open circuit voltage waveform of AFPM machine with (a) overlap, and (b) non-overlap winding [146].

stator winding are significant due to the fact that the machine is designed with $2p \geq 6$ to operate at relatively high frequencies $f = pn$.

Eddy current losses ΔP_e in stator conductors are calculated according to eqn (2.68) or eqn (2.69). A more detailed method of the eddy current losses computation is discussed in [263]. Eddy current losses in the stator conductors can be accounted for in the same way as core losses in the stator stack [126]. The eddy-loss current I_e and its d- and q-axis components I_{ed} and I_{eq} shown in Fig. 5.5 is in phase with the internal phase voltage E_i across the shunt resistance R_e. The internal phase voltage E_i is generated by the resultant air gap flux and usually referred to as the air gap voltage [94]. Consequently, the shunt resistance in the equivalent circuit representing the eddy current loss is expressed as

$$R_e = m_1 I_e^2 R_e = \frac{m_1 E_i^2}{\Delta P_e} \tag{5.32}$$

where

$$I_e = \sqrt{I_{ed}^2 + I_{eq}^2} \tag{5.33}$$

On the basis of equivalent circuits (Fig. 5.5), the following equations can be written in the phasor form:

- for generators

$$\mathbf{E}_f = \mathbf{E}_i + j\mathbf{I}_{ad} X_{ad} + j\mathbf{I}_{aq} X_{aq} \tag{5.34}$$

$$\mathbf{I}'_a = \mathbf{I}_a + \mathbf{I}_e \tag{5.35}$$

- for motors

$$\mathbf{E}_f = \mathbf{E}_i - j\mathbf{I}_{ad} X_{ad} - j\mathbf{I}_{aq} X_{aq} \tag{5.36}$$

$$\mathbf{I}'_a = \mathbf{I}_a - \mathbf{I}_e \tag{5.37}$$

where $\mathbf{I}'_a$ is the stator current with eddy currents being accounted for and $\mathbf{I}_a$ is the stator current with eddy currents being ignored.

5.8.2 Reduction of Eddy Current Losses

In an AFPM machine with a coreless stator the winding is directly exposed to the air gap magnetic field (see Fig. 5.13). Motion of PMs over the coreless winding produces an alternating field through each conductor inducing eddy currents. The loss due to eddy currents in the conductors depends on both the geometry of the wire cross section and the amplitude and waveform of the flux density. In order to minimize the eddy current loss in the conductors, the stator winding should be designed in one of the following ways using:

- parallel wires with smaller cross sections instead of a one thick conductor;
- stranded conductors (Litz wires);
- coils made of copper or aluminum ribbon (foil winding).

In an AFPM machine with an ironless winding arrangement (as shown in Fig. 5.13a), in addition to its normal component, the air gap magnetic field has a tangential component, which can lead to serious additional eddy current loss (Fig. 5.13b). The existence of a tangential field component in the air gap discourages the use of ribbon conductors as a low cost arrangement. Litz wires

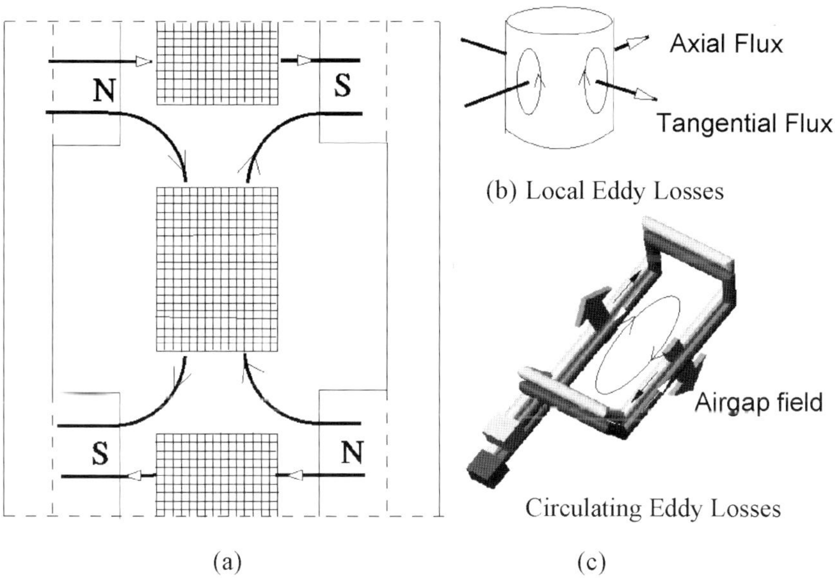

Fig. 5.13. Eddy currents in the coreless stator winding of an AFPM machine: (a) magnetic field distribution in a coreless stator; (b) eddy currents in a conductor; (c) circulating eddy currents among parallel connected conductors.

allow significant reduction of eddy current loss, but they are more expensive and have fairly poor filling factors.

As a cost effective solution, a bunch of parallel thin wires can be used. However, this may create a new problem, i.e. unless a complete balance of induced EMFs among the individual conducting paths is achieved, a circulating current between any of these parallel paths [161, 231, 263] may occur as shown in Fig. 5.13c, causing circulating eddy current losses.

When operating at relatively high frequency magnetic fields, these eddy current effects may cause a significant increase of winding losses, which are intensified if there are circulating currents among the parallel circuits. These losses will deteriorate the performance of the AFPM brushless machine. Predicting the winding eddy current losses with a good accuracy is therefore very important at the early stage of design of such machines.

Eddy current losses may be resistance limited when the flux produced by the eddy currents has a negligible influence on the total field [231]. In this case the conductor dimensions (diameter of thickness) are small when compared to the equivalent depth of penetration $\Delta = 1/k_\nu$ of the electromagnetic field – eqn (1.17).

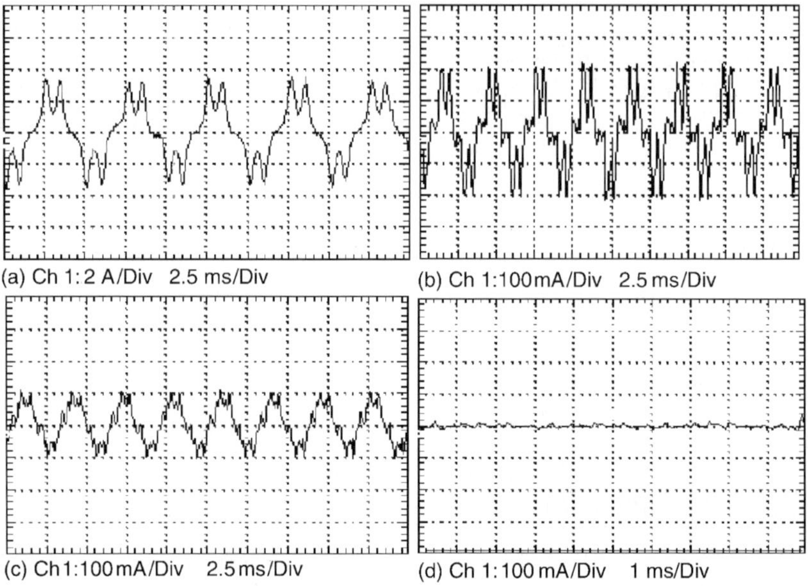

Fig. 5.14. Measured circulating current of (a) non-twisted coil, (b) slightly twisted coil, (c) moderately twisted coil, (d) heavily twisted coil.

5.8.3 Reduction of Circulating Current Losses

To minimise the circulating current in a coil made of parallel conductors, the normal practice is to twist or transpose the wires in such a fashion that each parallel conductor occupies all possible layer positions for the same length of the coil. The effect is to equalize the induced EMFs in all parallel conductors, and to allow them to be paralleled at the ends without producing eddy circulating currents between the parallel conductors.

Figure 5.14 illustrates the effectiveness of suppressing circulating eddy current by wire twisting. Four coils are made with the only difference that the parallel wires of each coil are: (a) non-twisted, (b) slightly twisted (10 to 15 turns per metre), (c) moderately twisted (25 to 30 turns per metre), and (d) heavily twisted (45 to 50 turns per metre) respectively. All the coils have been used to form a portion of an experimental stator, which is placed in the middle of the two opposing PM rotor discs. The machine operates at a constant speed (400 rpm in this case). The circulating currents between two parallel conductors have been measured and logged on a storage oscilloscope. It can be seen that the induced circulating current is greatly reduced even with a slightly twisted coil and can generally be ignored in a heavily twisted coil. These twisted wires can easily be manufactured in a cost effective way.

The fill (space) factors for the non-twisted and the heavily twisted coils are estimated as 0.545 and 0.5 respectively, which is slightly less than that of the

Litz wires (typically from 0.55 to 0.6). However, the saving in manufacturing costs is more important.

It should be noted that due to the low impedance of a coreless stator winding, circulating current could also exist among coil groups connected in parallel (parallel current paths) if a perfect symmetry of coils cannot be guaranteed.

5.8.4 Measurement of Eddy Current Losses

The resistance limited eddy loss in the stator of an AFPM machine may be experimentally determined by measuring the difference in input shaft powers of the AFPM machine at the same speed, first with the stator in, and then by replacing it with a dummy stator (no conductors). The dummy stator has the same dimensions and surface finish as the real stator and is meant to keep the windage losses the same.

A schematic representation of the experimental test setup for measuring eddy current losses in stator conductors is shown in Fig. 5.15. The shaft of the tested prototype and the shaft of the prime mover (driving machine) are coupled together via a torque meter. The stator is positioned in the middle of the two rotor discs with the outer end ring (see Fig. 5.15a) mounted on the outside supporting frame. Temperature sensors are also attached to the conductors in order to take into account temperature factor in the measurement.

Initially, the prototype (with the coreless stator placed in it) was driven by a variable speed motor for a number of different speeds. The corresponding torque measurements were taken. Replacing the real stator with a dummy one, the tests were repeated for the same speeds. Both the torque and the temperature values were recorded. The difference of the torques multiplied by the speed gives the eddy current loss. The eddy losses due to eddy-circulating current in the windings can also be determined by measuring the difference of input shaft powers, first with all the parallel circuits connected and then with disconnected parallel circuits.

Fig. 5.16 shows the measured and calculated resistance limited eddy losses of the prototype machine. It can be seen that the calculated eddy losses obtained by using the standard analytical formula – eqns (2.68) and (2.69) with only the B_{mz} component taken into account yield underestimated values (43% less). The discrepancy between the measured and calculated results becomes large at high speeds. Better accuracy may be achieved by including both the normal B_{mz} and tangential B_{mx} components of the magnetic flux density and implementing eqn (2.68) or eqn (2.69) into the 2D or 3D FEM [262].

5.9 Armature Reaction

The 3-D FEM analysis has been applied to the modelling of a coreless stator AFPM machine in [10]. Both no-load and load operations have been modelled. It has been shown that good accuracy can be expected even when the

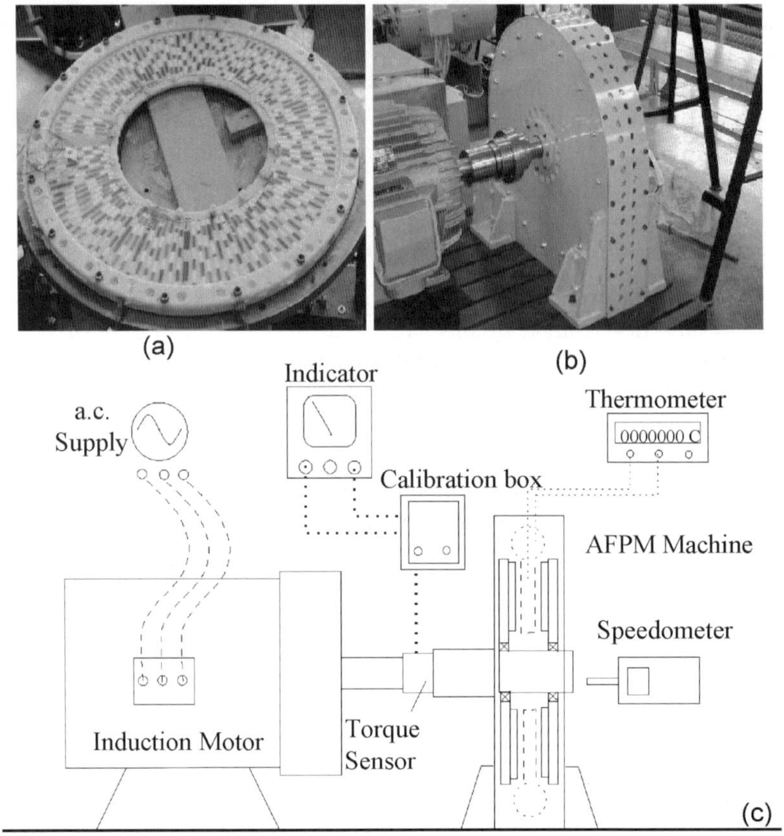

Fig. 5.15. Laboratory set for measuring eddy current losses in the stator winding: (a) specially designed stator; (b) experimental machine; (c) schematic of experimental set-up.

first order linear FEM solution is applied to the AFPM machine if the magnetic circuit is not saturated (large air gap). It has also been found that the armature reaction for an AFPM machine with a coreless stator is generally negligible.

Using a 40-pole, 766 Hz coreless stator AFPM machine as an example, the simulated effects of armature reaction to the air gap flux distribution has been demonstrated in Figs 5.17 and 5.18. It can be seen that the plateau of the axial field plot is somewhat tilted due to the interaction between the flux of the PMs and the armature flux generated by the rated current (Fig. 5.17). Similarly, the modified tangential field plot is compared with the original one in Fig. 5.18.

As the armature reaction flux is small in coreless AFPM machines with large air gaps, the air gap magnetic flux density maintains its maximum value very

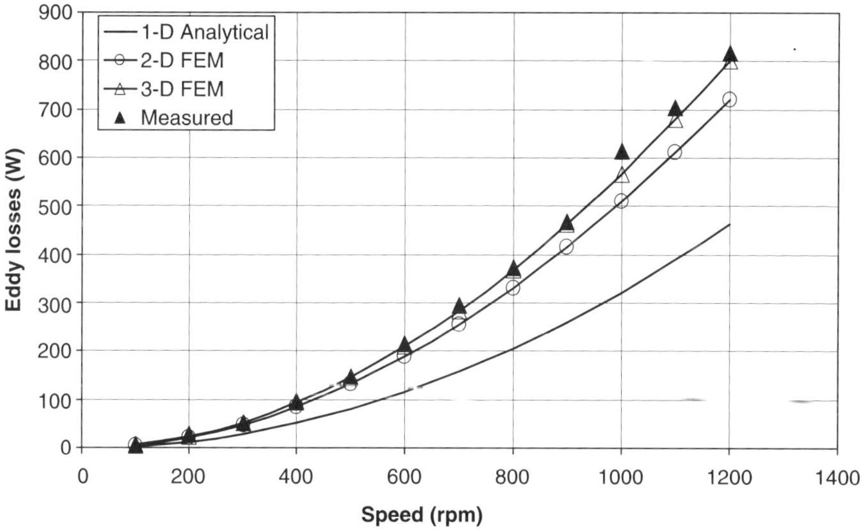

Fig. 5.16. Comparison of calculated eddy loss with measurements.

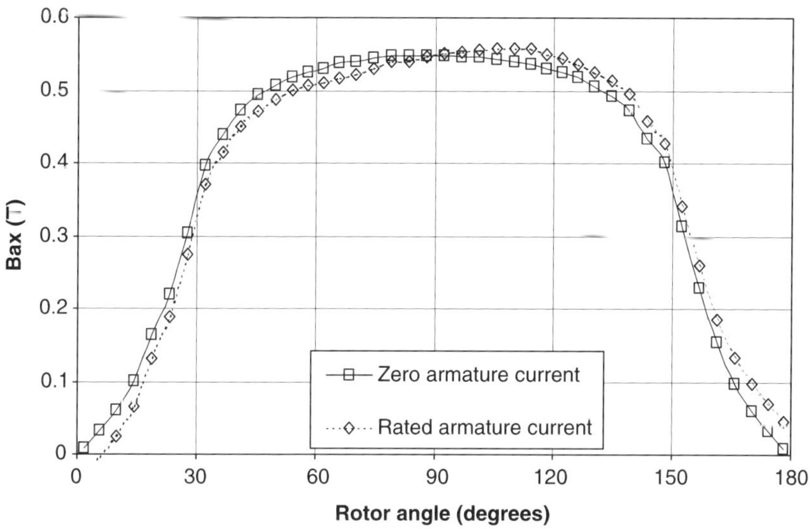

Fig. 5.17. Air gap axial field component at no-load and full load conditions.

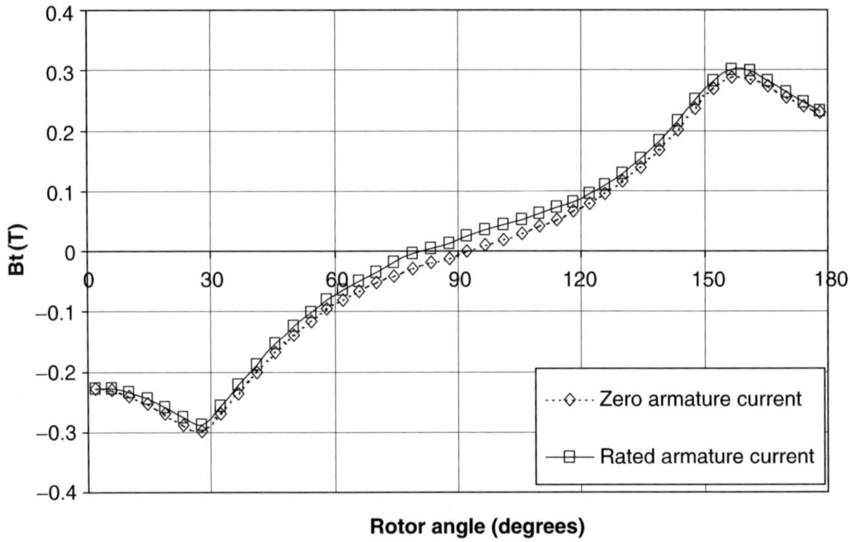

Fig. 5.18. Air gap tangential field component at no-load and full load conditions.

close to that at no load. The influence of the armature reaction on the eddy current losses is usually insignificant though it can be easily accounted for when the FEM modelling is used. The harmonic contents of the air gap flux density obtained with and without armature reaction are compared in Table 5.6. It is evident that, in this case, the change of the field harmonic composition due to the armature reaction is negligible.

Table 5.6. Harmonics content of the axial air-gap flux density with and without armature reaction

Harmonics	Rated armature current	Zero armature current
1st	0.60346784	0.60246033
3nd	0.03745651	0.03745759
5rd	0.04099061	0.04070525
7th	0.03502689	0.03514609
9th	0.01231397	0.01230032
11th	0.00304573	0.00262309
13th	0.00559543	0.00579739
15th	0.00314908	0.00312179
17th	0.00083159	0.00086812
19th	0.00307828	0.00272939

5.10 Mechanical Design Features

In the mechanical design of an AFPM brushless machine, obtaining a uniform air gap between the rotor disc and the stator is important. Therefore, the methods of fixing the rotor discs onto the shaft and the stator onto the enclosure (frame) are very important. Improper methods of fixing, or misalignment in the assembling of the stator and rotor will cause a nonuniform air gap, vibration, noise, torque pulsation and deterioration of electrical performance.

With regard to cooling in an air-cooled AFPM machine, the entry losses of the air flow can be quite high if the machine-air-inlet is poorly designed. It is important to reduce these losses without weakening the mechanical structure for better cooling.

To summarise, attention should be paid to the following aspects of the mechanical design:

- **Shaft**. The load torque, the first critical speed and the shaft dynamics should be taken into account in the shaft design.
- **Rotor**. (i) The deflection of the rotor disc due to the strong magnetic attraction force, (ii) the means of mounting and securing the magnets on the rotor discs to counteract the strong centrifugal force especially for high speed applications, and (iii) the balancing of the rotor discs.
- **Stator**. (i) The strength and rigidity of the resin reinforced stator and frame, and (ii) the positioning and spacing of the coils to ensure perfect symmetry.
- **Cooling**. For air cooled AFPM machines, the air inlet and air flow paths through the machine should be carefully designed in order to ensure a better mass flow rate and therefore better cooling.
- **Assembly**. An effective tool to facilitate the assembling and dismantling of the machine for maintenance.

5.10.1 Mechanical Strength Analysis

The deflection of the rotor discs due to the strong magnetic pull may have the following undesirable effects on the operation and condition of AFPM machines:

- closing the running clearance between the rotor disc and the stator;
- loose or broken PMs;
- reducing air-flow discharging area thus deteriorating the cooling capacity;
- nonuniform air gap causing a drift in electrical performance from the optimum.

For a double-sided AFPM machine with an internal coreless stator, the rotor discs account for roughly 50% of the total active mass of an AFPM machine. Hence, the optimal design of the rotor discs is of great importance to realise a design of high power–to–mass ratio. All these aspects require the mechanical stress analysis of the rotor disc.

Attraction Force between Rotor Discs

The attraction force between two parallel rotor discs can be calculated by using the *virtual work* method, i.e.

$$F_z = -\frac{dW}{dz} \approx -\frac{\Delta W}{\Delta z} = -\frac{W_2 - W_1}{z_2 - z_1} \tag{5.38}$$

where W is the total magnetic energy stored in the machine and Δx is the small variation of the air gap length. The accurate prediction of the attraction force is a prerequisite for the mechanical stress analysis. Hence, the respective magnetic stored energies W_1 and W_2 for air gap lengths z_1 and z_2 are usually calculated by using the FEM.

Analytically, the normal attractive force between two parallel discs with PMs can be expressed as

$$F_z \approx \frac{1}{2}\frac{B_{mg}^2}{\mu_0} S_{PM} \tag{5.39}$$

where the active area of PMs is according to eqn (2.63).

Optimum Design of Rotor Discs

The structure of the rotor steel discs may be optimised with the aid of an FEM structural program. It is important that the deflection of the rotor steel discs, due to the axial magnetic pull force, is not too dangerous for small running clearance between the coreless stator and PMs. Two important constraints should be considered, i.e. (i) the maximum allowable deflection and (ii) the maximum mechanical strength of the material used to fabricate the disc.

When choosing allowable deflection, one needs to make sure that the PMs do not experience any excessive forces due to bending of the disc that can potentially peel them off from the backing steel disc. Owing to the cyclical symmetry of the disc structures, it is sufficient to model only a section of the disc with symmetry boundary conditions applied. In the FEM analysis the axial magnetic force can be applied in the form of a constant pressure load over the total area that the PMs occupy. Usually, the axial-symmetric elements are preferred for modelling a relatively thick disc. However, it has been shown in [182] that the FEM modelling of rotor discs using both axial-symmetric-elements and shell-elements gives very close results.

Fig. 5.19 shows the finite element model of the analysed rotor disc using 4-node shell-elements, with symmetrical boundary conditions applied. The axial magnetic attraction force has been calculated as 14.7 kN and applied in the form of a constant 69.8 kPa pressure load. Specification of the investigated AFPM machine are given in Table 5.3. The stiffness provided by the magnets has not been included, to keep the design on the conservative side. As the rotor discs are mounted on the centre support hub, additional boundary constraints have been defined so that there is no axial displacement in the vicinity of the mounting bolts and contact area.

To find the suitable thickness of the rotor disc, which satisfies the critical strength requirements of the rotor disc with a low steel content, the linear FEM static analysis was performed for different thicknesses of the rotor disc.

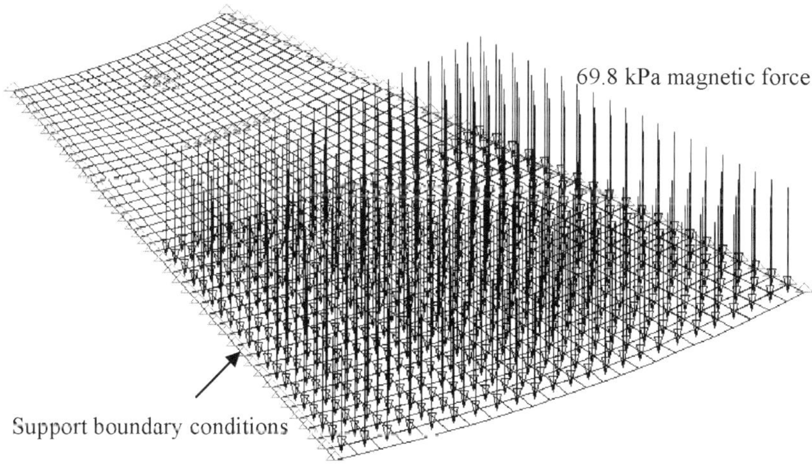

Fig. 5.19. FEM model for analysing the mechanical stress of the rotor disc.

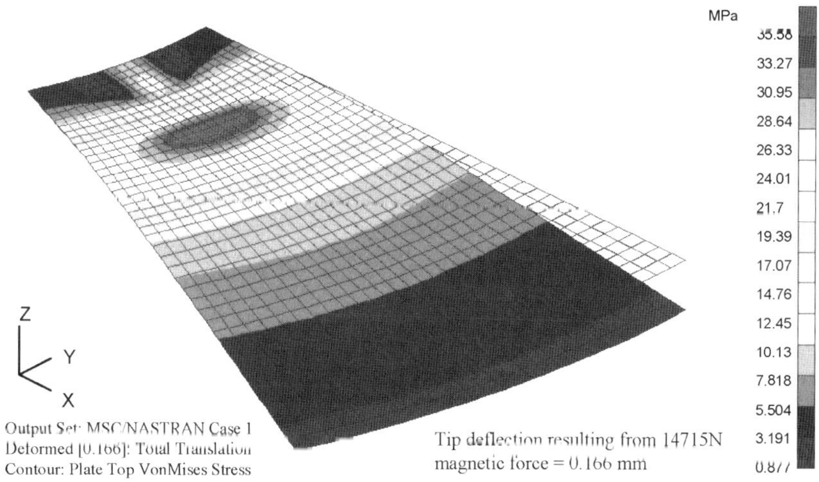

Fig. 5.20. Deflection (blown-up) and von Mises stress distribution of the rotor disc.

Based on the analysis, the rotor disc thickness was chosen as 17 mm with a maximum deflection of 0.145 mm. Fig. 5.20 shows the deflection (blown-up) and the von Mises stress distribution of the laboratory prototype of the 17-mm thick disc. The maximum stress of 35.6 MPa is much lower than the typical yield strength of mild steel, that is in the region of 300 MPa. It has been shown in literature [182] that the bending of the rotor disc decreases towards its outer periphery. The rotor disc may be machined in such a way that the disc becomes thinner towards the outer periphery. As shown in Table 5.7, the tapered disc uses approximately 10% less steel than the straight disc. The maximum deflection increases by only 0.021 mm with the tapered disc, which is negligible. This can effectively reduce the active mass of the machine without compromising the mechanical strength, but does not bring down the cost of the steel.

If manufacturing costs are taken into account for small production volumes, it is better to use a steel disc with uniform thickness. Obviously, a uniform disc means a heavier rotor disc. It can, however, be argued that the extra machining work needed for producing a tapered rotor disc may be too costly to justify the profit due to improvement in the dynamic performance. As long as the added mass can be tolerated, this option is viable for small production volumes and for laboratory prototypes.

Table 5.7. Comparison of different designs of rotor disc

Parameter	Straight disc	Tapered disc
Mass, kg	39.184	31.296
Maximum deflection, mm	0.145	0.166
Maximum von Mises stress, MPa	35.6	33.4

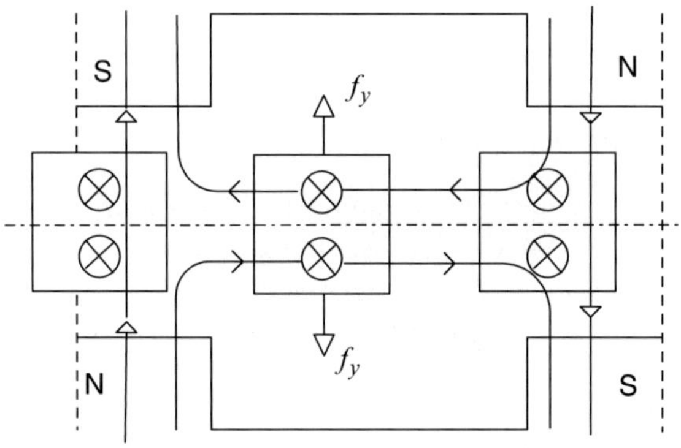

Fig. 5.21. Schematic diagram showing the axial force exerted on stator.

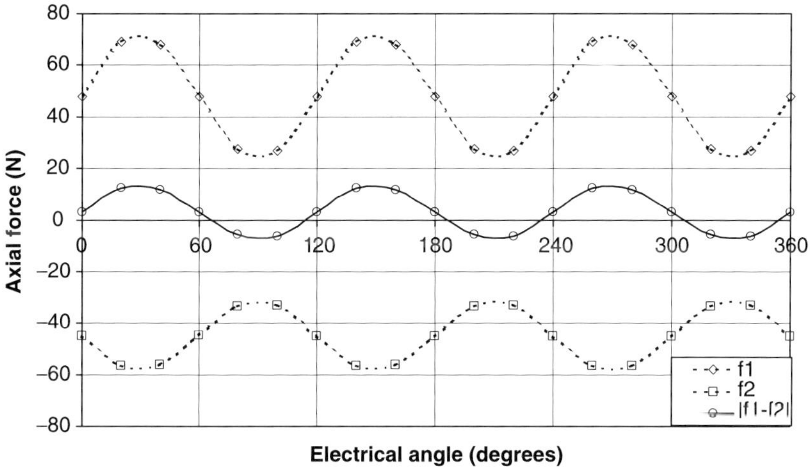

Fig. 5.22. Unbalanced axial force exerted on the stator.

5.10.2 Imbalanced Axial Force on the Stator

As a result of the interaction of alternating currents in conductors and the tangential component of the magnetic field, there is an axial magnetic force f_z on each half of the coil as shown in Fig. 5.21. When the stator is located in the middle of the air gap, the forces on each side of the stator should cancel each other.

Assuming that the coreless stator of the AFPM machine is slightly off centre, the axial forces on each side of the stator (f_1 and f_2) as shown in Fig. 5.22 will not be the same, resulting in an unbalanced force, $\Delta f = |f_1 - f_2|$, being exerted on the stator. This unbalanced force may cause extra vibration and thus have an adverse effect on the mechanical strength of the epoxy reinforced stator.

5.11 Thermal Problems

Owing to the excessive heat generated in the stator winding and the resultant thermal expansion, an epoxy encapsulated stator is subject to certain deformation. When the deformation is significant, it may cause physical contact between the stator and the magnets resulting in serious damage to the stator winding and the magnets.

Fig. 5.23 shows the thermal expansion of the stator at different temperatures. The rated current (16 A) has been conducted through a coil in the epoxy encapsulated stator. A thermal coupler and a roundout meter have been used to measure the coil temperature and surface deformation respectively. It

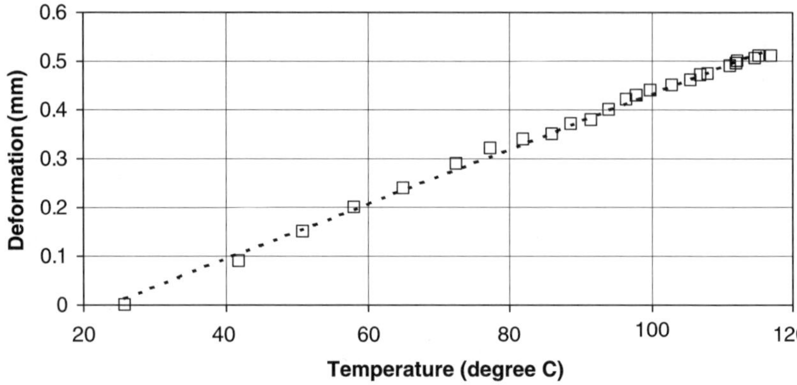

Fig. 5.23. Thermal expansion of the stator for different temperatures.

has been found that the relationship between the stator deformation and its temperature is almost linear. On average, the deflection of the stator surface is about 0.0056 mm/°C, resulting in a 0.5 mm deflection at the temperature of 117°C. The remaining air contents in the epoxy can also contribute to the heat transfer deterioration and temperature increase. To solve this problem, a more dedicated production process is suggested. The running clearance between the stator and the rotor should be kept reasonably large.

Numerical Example 5.1

A three-phase, Y-connected, 3000 rpm PM disc motor has a coreless stator and twin external rotor with surface PMs and backing steel discs. Sintered NdFeB PMs with $B_r = 1.2$ T and $\mu_{rrec} = 1.045$ have been used. The nonmagnetic distance between opposite PMs is $t = 11$ mm, the winding thickness is $t_w = 8$ mm and the height of PMs (in axial direction) is $h_M = 6$ mm. The outer diameter of PMs equal to the outer diameter of the stator conductors is $D_{out} = 0.22$ m and the parameter $k_d = 1/\sqrt{3}$. The number of poles is $2p = 6$, the number of single layer coil sides (equivalent to the number of slots) is $s_1 = 54$, the number of turns per phase is $N_1 = 234$, the number of parallel conductors $a_w = 2$, the diameter of wire $d_w = 0.912$ mm (AWG 19) and the coil pitch is $w_c = 7$ coil sides.

Find the motor steady state performance, i.e. output power, torque, efficiency and power factor assuming that the total armature current $I_a = 8.2$ A is torque producing ($I_{ad} = 0$). The saturation factor of the magnetic circuit is $k_{sat} = 1.02$, the motor is fed with sinusoidal voltage, the mass of the twin rotor $m_r = 3.4$ kg, the mass of shaft $m_{sh} = 0.64$ kg, the radius of shaft $R_{sh} = 15$ mm, the conductivity of copper conductors $\sigma = 47 \times 10^6$ S/m at 75°C, the specific density of copper conductors $\rho = 8800$ kg/m^3, the density of PMs $\rho_{PM} = 7700$ kg/m^3, the air density $\rho = 1.2$ kg/m^3, the dynamic viscosity of air $\mu = 1.8 \times 10^{-5}$ Pa s, the coefficient of bearing friction $k_{fb} = 1.5$ and the coefficient of distortion of the magnetic flux density $\eta_d = 1.15$. Losses in PMs, losses in steel rotor discs and the tangential component of the magnetic flux density in the air gap are negligible.

Solution

The number of coils per phase for a single layer winding is $n_c = s_1/(2m_1) = 54/(2 \times 3) = 9$. The number of turns per coil is $N_{ct} = N_1/n_c = 234/9 = 26$. The number of coil sides per pole per phase (equivalent to the number of slots per pole per phase) $q_1 = s_1/(2pm_1) = 54/(6 \times 3) = 3$.

The air gap (mechanical clearance) is $g = 0.5(t - t_w) = 0.5(11 - 8) = 1.5$ mm and the pole pitch measured in coil sides is $\tau_c = s_1/(2p) = 54/6 = 9$.

The input frequency at 3000 rpm is

$$f = n_s p = \frac{3000}{60} \times 3 = 150 \text{ Hz}$$

The magnetic voltage drop equation per pole pair is

$$4\frac{B_r}{\mu_0 \mu_{rrec}} h_M = 4\frac{B_{mg}}{\mu_0}\left[\frac{h_M}{\mu_{rrec}} + \left(g + \frac{1}{2}t_w\right)k_{sat}\right]$$

Hence

$$B_{mg} = \frac{B_r}{1 + [\mu_{rrec}(g + 0.5t_w)/h_M]k_{sat}}$$

$$= \frac{1.2}{1 + [1.045(1.5 + 0.5 \times 8)/6] \times 1.02} = 0.607 \text{ T}$$

The magnetic flux according to eqn (2.28) is

$$\Phi_f = \frac{1}{8}\frac{2}{\pi}\frac{\pi}{3} 0.607 \times 0.22^2 \left[1 - \left(\frac{1}{\sqrt{3}}\right)^2\right] = 0.001632 \text{ Wb}$$

The winding factor according to eqns (2.8), (2.9) and (2.10) is

$$k_{d1} = \frac{\sin \pi/(2 \times 3)}{3 \sin \pi/(2 \times 3 \times 3)} = 0.9598; \qquad \beta = \frac{w_c}{\tau_c} = \frac{7}{9}$$

$$k_{p1} = \sin\left(\frac{7}{9}\frac{\pi}{2}\right) = 0.9397 \qquad k_{w1} = 0.9598 \times 0.9397 = 0.9019$$

The EMF constant according to eqn (2.37) and torque constant according to eqn (2.34) are respectively

$$k_E = \pi\sqrt{2} \times 3 \times 234 \times 0.9019 \times 0.001632 = 4.591 \text{ V/(rev/s)} = 0.0765 \text{ V/rpm}$$

$$k_T = k_E \frac{m_1}{2\pi} = 4.591 \frac{3}{2\pi} = 2.192 \text{ Nm/A}$$

The EMF at 3000 rpm is

$$E_f = k_E n = 0.0765 \times 3000 = 229.5 \text{ V}$$

The electromagnetic torque at $I_a = I_{aq} = 8.2$ A is

$$T_d = k_T I_a = 2.192 \times 8.2 = 17.97 \text{ Nm}$$

The electromagnetic power is

$$P_{elm} = 2\pi n T_d = 2\pi \frac{3000}{60} \times 17.97 = 5646.8 \text{ W}$$

The inner diameter $D_{in} = D_{out}/\sqrt{3} = 0.22/\sqrt{3} = 0.127$ m, the average diameter $D = 0.5(D_{out} + D_{in}) = 0.5(0.22 + 0.127) = 0.1735$ m, the average pole pitch $\tau = \pi 0.1735/6 = 0.091$ m, the length of conductor (equal to the radial length l_M of the PM) $L_i = l_M = 0.5(D_{out} - D_{in}) = 0.5(0.22 - 0.127) = 0.0465$ m, the length of shorter end connection without inner bends $l_{1emin} = (w_c/\tau_c)\pi D_{in}/(2p) = (7/9)\pi \times 0.127/6 = 0.052$ m and the length of the longer end connection without outer bends $l_{1emax} = 0.052 \times 0.22/0.127 = 0.0896$ m.

The average length of the stator turn with 15-mm bends (Fig. 3.16) is

$$l_{1av} \approx 2L_i + l_{1emin} + l_{1emax} + 4 \times 0.015$$

$$= 2 \times 0.0465 + 0.052 + 0.0896 + 0.06 = 0.2943 \text{ m}$$

The stator winding resistance at 75°C according to eqn (2.40) is

$$R_1 = \frac{234 \times 0.2943}{47 \times 10^6 \times 2 \times \pi \times (0.912 \times 10^{-3})^2/4} = 1.122 \ \Omega$$

The maximum width of the coil at the diameter D_{in} is $w_w = \pi D_{in}/s_1 = \pi 0.127/54 = 0.0074$ m $= 7.4$ mm. The thickness of the coil is $t_w = 8$ mm. The number of conductors per coil is $N_c = a_w \times N_{ct} = 2 \times 26 = 52$. The maximum value of the coil packing factor is at D_{in}, i.e.

$$k_{fmax} = \frac{d_w^2 \times N_c}{t_w w_w} = \frac{0.912^2 \times 52}{8 \times 7.4} = 0.732$$

The stator current density is

$$j_a = \frac{8.2}{2 \times \pi 0.912^2/4} = 6.28 \text{ A/mm}^2$$

The stator winding losses according to eqn (2.49) are

$$\Delta P_{1w} = 3 \times 8.2^2 \times 1.122 = 226.2 \text{ W}$$

The eddy current losses in stator round conductors according to eqn (2.68) are

$$\Delta P_e = \frac{\pi^2}{4}\frac{47 \times 10^6}{8800} 150^2 \times 0.000912^2 \times 0.75 \times 0.607^2 \times 1.15^2 = 90.2 \text{ W}$$

where the mass of stator conductors (radial parts) is

$$m_{con} = \rho_{cu} m_1 a_w N_1 \left(\frac{\pi d_w^2}{4}\right)(2L_i)$$

$$= 8800 \times 3 \times 2 \times 234 \left(\frac{\pi 0.000912^2}{4}\right)(2 \times 0.0465) = 0.75 \text{ kg}$$

The friction losses in bearings according to eqn (2.71) are

$$\Delta P_{fr} = 0.06 \times 1.5(3.4 + 0.64)\frac{3000}{60} = 18.2 \text{ W}$$

The windage losses according to eqn (2.74) are

$$\Delta P_{wind} = \frac{1}{2} 7.353 \times 10^{-3} \times 1.2 \left(2\pi \frac{3000}{60}\right)^3 [(0.5 \times 0.115)^5 - (0.5 \times 0.015)^5] \approx 2.8 \text{ W}$$

where

$$R'_{out} \approx 0.5 D_{out} + 0.005 = 0.5 \times 0.22 + 0.005 = 0.115 \text{ m}$$

$$Re = 1.2 \frac{2\pi(3000/60)}{1.8 \times 10^{-5}} \times 0.115^2 = 2.77 \times 10^5$$

$$c_f = \frac{3.87}{\sqrt{2.77 \times 10^5}} = 7.353 \times 10^{-3}$$

The rotational (mechanical) losses according to eqn (2.70) are

$$\Delta P_{rot} = 18.2 + 2.8 = 21 \text{ W}$$

The output power is

$$P_{out} = P_{elm} - \Delta P_{rot} = 5646.8 - 21 = 5625.8\text{W}$$

The shaft torque is

$$T_{sh} = \frac{5625.8}{2\pi \times 3000/60} = 17.91 \text{ Nm}$$

The input power is

$$P_{in} = P_{elm} + \Delta P_{1w} + \Delta P_e = 5646.8 + 226.2 + 90.2 = 5963.2 \text{ W}$$

The efficiency is

$$\eta = \frac{5625.8}{5963.2} = 0.943$$

The leakage reactance can be approximately calculated taking into account the radial part of conductors, end connection and differential leakage fluxes, i.e.

$$\lambda_{1e} \approx 0.3q_1 = 0.3 \times 3 = 0.9 \qquad \lambda_{1s} \approx \lambda_{1e} = 0.9$$

$$\lambda_{1d} = \frac{3 \times 3 \times 0.091 \times 0.9019^2}{\pi^2(2 \times 0.0015 + 0.008)1.02} \times 0.011 = 0.066$$

where according to eqn (4.23), the differential leakage factor $\tau_{d1} = 0.011$ for $5 \leq \nu \leq 997$.

The stator leakage reactance is

$$X_1 = 4\pi \times 0.4\pi \times 10^{-6} \times 150 \frac{234^2 \times 0.0465}{3 \times 3} \left(0.9 + \frac{0.071}{0.0465} \times 0.9 + 0.066\right)$$

$$= 1.564\ \Omega$$

where the average length of one end connection is $l_{1e} = 0.5(l_{1emin} + l_{1emax}) = 0.5(0.052 + 0.0896) = 0.071$ m.

The equivalent air gap in the d axis according to eqn (2.112) is

$$g' = 2\left[(1.5 + 0.5 \times 8)1.02 + \frac{6}{1.045}\right] = 22.7 \text{ mm}$$

The equivalent air gap in the q axis according to eqn (2.113) is

$$g'_q = 2[(1.5 + 0.5 \times 8) + 6] = 23 \text{ mm}$$

Armature reaction reactances according to eqns (2.121) and (2.122) are

$$X_{ad} = 2 \times 3 \times 0.4 \times \pi \times 10^{-6} \times 150 \left(\frac{234 \times 0.9019}{3}\right)^2 \frac{1}{0.0227}$$

$$\times[(0.5 \times 0.22)^2 - (0.5 \times 0.127)^2] = 1.989\ \Omega$$

$$X_{aq} = X_{ad}\frac{22.7}{23.0} = 1.963\ \Omega$$

where $k_{fd} = k_{fq} = 1$ for surface configuration of PMs [106].

Synchronous reactances according to eqns (2.79) and (2.80) are respectively

$$X_{sd} = 1.543 + 1.989 = 3.532\ \Omega$$

$$X_{sq} = 1.543 + 1.963 = 3.506\ \Omega$$

The input phase voltage is

$$V_1 = \sqrt{(E_f + I_a R_1)^2 + (I_a X_{sq})^2}$$

$$= \sqrt{(229.5 + 8.2 \times 1.122)^2 + (8.2 \times 3.506)^2} = 240.5\ \text{V}$$

The line–to–line voltage $V_{1L-L} = \sqrt{3} \times 240.5 = 416.5$ V.

The power factor is

$$\cos\phi = \frac{E_f + I_a R_1}{V_1} = \frac{229.5 + 8.2 \times 1.122}{240.5} = 0.993 \approx 1.0$$

Numerical Example 5.2

For the coreless stator AFPM brushless motor described in *Numerical example 5.1* find the rotor moment of inertia, mechanical and electromagnetic time constants and axial magnetic attractive force between the backing steel discs of the twin rotor.

Solution

The following input data and results of calculation of *Numerical example 5.1* are necessary:

- input phase voltage $V_1 = 240.5$ V
- input frequency $f = 150$ Hz
- air gap magnetic flux density $B_{mg} = 0.607$ T
- Stator winding resistance per phase $R_1 = 1.122\ \Omega$

- d-axis synchronous reactance $X_{sd} = 3.532\ \Omega$
- q-axis synchronous reactance $X_{sq} = 3.506\ \Omega$
- torque constant $k_T = 2.192$ Nm/A
- Electromagnetic torque $T_d = 19.97$ Nm
- speed $n = 3000$ rpm
- power factor $\cos\phi = 0.993$
- PM outer diameter $D_{out} = 0.22$ m
- PM inner diameter $D_{in} = 0.127$ m
- shaft diameter $D_{sh} = 2R_{sh} = 0.03$ m
- number of pole pairs $p = 3$
- pole width–to–pole pitch ratio $\alpha_i = 2/\pi$
- axial height of PM (one pole) $h_M = 6$ mm
- mass of twin rotor without shaft $m_r = 3.4$ kg
- mass of shaft $m_{sh} = 0.64$ kg
- specific mass density of PMs $\rho_{PM} = 7700$ kg/m^3
- specific mass density of mild steel $\rho_{Fe} = 7850$ kg/m^3

1. Rotor moment of inertia

The active surface area of PMs (one side) according to eqn (2.63) is

$$S_{PM} = \frac{2}{\pi}\frac{\pi}{4}(0.22^2 - 0.127^2) = 0.01614 \text{ m}^2$$

The mass of all PMs is

$$m_{PM} = 2 \times 7700 \times 0.01614 \times 0.006 = 1.49 \text{ kg}$$

The mass of backing steel discs is

$$m_{Fe} = m_r - m_{PM} = 3.4 - 1.49 = 1.91 \text{ kg}$$

The shaft moment of inertia is

$$J_{sh} = m_{sh}\frac{D_{sh}^2}{8} = 0.64\frac{0.03^2}{8} = 0.072 \times 10^{-3} \text{ kgm}^2$$

The moment of inertia of PMs is

$$J_{PM} = m_{PM}\frac{D_{out}^2 + D_{in}^2}{8} = 1.49\frac{0.22^2 + 0.127^2}{8} = 0.012 \text{ kgm}^2$$

The moment of inertia of backing steel discs is

$$J_{Fe} = m_{Fe}\frac{D_{out}^2 + D_{sh}^2}{8} = 1.91\frac{0.22^2 + 0.03^2}{8} = 0.0118 \text{ kgm}^2$$

The resultant moment of inertia of the rotor is

$$J_r = J_{sh} + J_{PM} + J_{Fe} = 0.072 \times 10^{-3} + 0.012 + 0.0118 = 0.02386 \text{ kgm}^2$$

2. Mechanical and electromagnetic time constants

Since the d-axis current $I_{ad} = 0$, the angle between the q axis and stator current $\psi = 0$. For the power factor $\cos\phi = 0.993$ the angle between stator current and terminal voltage $\phi = 6.78^o$. The load angle is

$$\delta = \phi - \psi = 6.78 - 0 = 6.78^o$$

At the first instant of starting the EMF $E_f = 0$. Assuming the load angle at starting is the same as for nominal operation, the stator current according to eqns (2.87), (2.88) and (2.89) is

$$I_{ashd} = \frac{240.5(3.506 \times 0.993 - 1.122 \times 0.118) - 0 \times 3.506}{3.532 \times 3.506 + 1.122^2} = 59.04 \text{ A}$$

$$I_{ashq} = \frac{240.5(1.122 \times 0.993 + 3.532 \times 0.118) - 0 \times 1.122}{3.532 \times 3.506 + 1.122^2} = 27 \text{ A}$$

$$I_{ash} = \sqrt{59.04^2 + 27^2} = 64.92 \text{ A}$$

where $\sin\phi = 0.118$ ($\cos\phi = 0.993$). The electromagnetic starting torque according to eqn (2.33) is

$$T_{dst} = k_T I_{ashq} = 2.192 \times 27.0 = 59.2 \text{ Nm}$$

The no load speed assuming linear torque-speed curve according to eqn (2.133) is

$$n_0 = \frac{3000}{1 - 19.97/59.2} = 4526.9 \text{ rpm} = 75.44 \text{ rev/s}$$

The mechanical time constant is

$$T_{mech} = J_r \frac{2\pi n_0}{T_{dst}} = 0.02386 \frac{2\pi 75.44}{59.2} = 0.191 \text{ s}$$

The d- and q-axis synchronous inductances are

$$L_{sd} = \frac{X_{sd}}{2\pi f} = 3.5322\pi 150 = 0.00375 \text{ H}$$

$$L_{sq} = \frac{X_{sq}}{2\pi f} - 3.5062\pi 150 = 0.00372 \text{ H}$$

The electromagnetic time constant of the stator winding is

$$T_{elm} = \frac{L_{sd}}{R_1} = \frac{0.00375}{1.122} = 0.00334 \text{ s}$$

The mechanical–to–electromagnetic time constant ratio is

$$\frac{T_{mech}}{T_{elm}} = \frac{0.191}{0.00334} = 57.2$$

3. Axial attractive force between backing steel discs of the twin rotor

Assuming that the two twin backing steel discs are perfectly parallel, the axial magnetic attractive force between them can be found on the basis of eqn (5.39)

$$F_z = \frac{0.606^2}{2\mu_0} 0.01614 = 2365.5 \text{ N}$$

The magnetic pressure is

$$p_z = \frac{F_z}{S_{PM}} = \frac{2365.5}{0.01614} = 146\,601.2 \text{ Pa} \approx 146.6 \text{ kPa}$$

The backing steel disc thickness is

$$d_{Fe} = \frac{m_{Fe}}{2\rho_{Fe}\pi(D_{out}^2 - D_{in}^2)/4}$$

$$= \frac{1.91}{2 \times 7850\pi(0.22^2 - 0.127^2)/4} = 0.0033 \text{ m} = 3.3 \text{ mm}$$

and outer diameter $D_{out} = 0.22$ m can provide adequate stiffness of the disc under the action of 146.6 kPa magnetic pressure in the axial direction.

6

AFPM Machines Without Stator and Rotor Cores

6.1 Advantages and Disadvantages

With the availability of high energy magnets the stators and rotors of AFPM brushless machines can be fabricated without ferromagnetic cores [107, 150, P83, 225]. A completely coreless design reduces the mass and increases the efficiency of the machine as compared with a conventional design. Besides this, a coreless AFPM brushless machine does not produce any normal attractive force between the stator and rotor. It also does not produce torque pulsations at zero current state.

There is a limit to the increase of the electromagnetic torque that can be achieved by enlarging the machine diameter. Factors limiting the single disc design are listed in Section 2.1.6. A reasonable solution for larger torques are double or triple-disc machines.

The disc-type PM brushless machines without stator and rotor cores were first manufactured commercially in the late 1990s for servo mechanisms and industrial electromechanical drives [150], solar powered electrical vehicles [225] as well as micromotors for computer peripherals and vibration motors for mobile phones [93].

6.2 Topology and Construction

The *AFPM brushless machine without any ferromagnetic core* is shown in Fig. 6.1. The machine consists of a twin rotor (3) with rare earth PMs (2) and nonmagnetic supporting structure. The steel-free stator (armature) winding (1) is located between two identical parts of the rotor. The stator polyphase winding fixed to the frame (6) is assembled as "flower petals" (Fig. 3.16) [P83]. Multi-turn coils are arranged in overlapping layers around the shaft-axis of the machine. The whole winding is then embedded in a high mechanical integrity plastic or resin. Since the topology shown in Fig 6.1 does not use any ferromagnetic core with slots, the machine is free of cogging (detent) torque

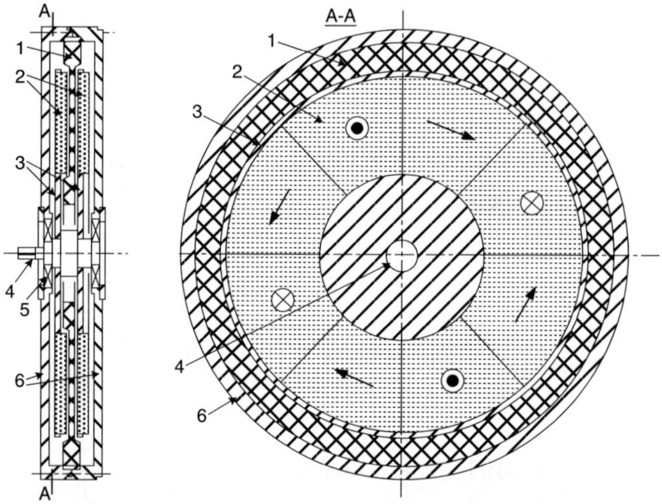

Fig. 6.1. Disc type coreless PM brushless machine: 1 — coreless stator (armature) winding, 2 — PMs, 3 — twin rotor, 4 — shaft, 5 — bearing, 6 — frame.

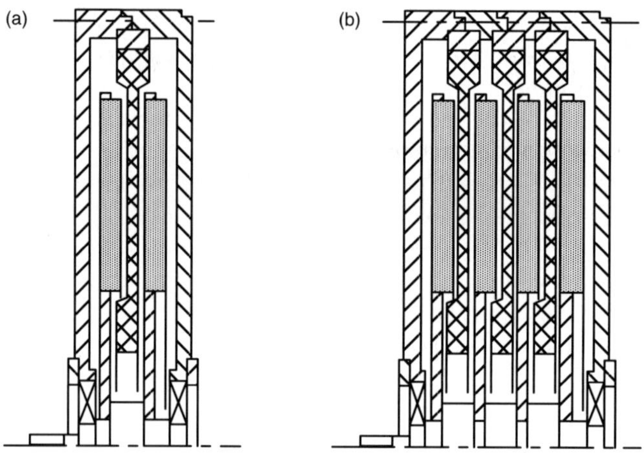

Fig. 6.2. Segmental construction of a disc type coreless PM brushless machine: (a) single module (segment), (b) three-module assembly.

and core losses. The only eddy current losses are losses in the stator winding conductors and metallic parts (if they exist) that reinforce the stator coreless winding.

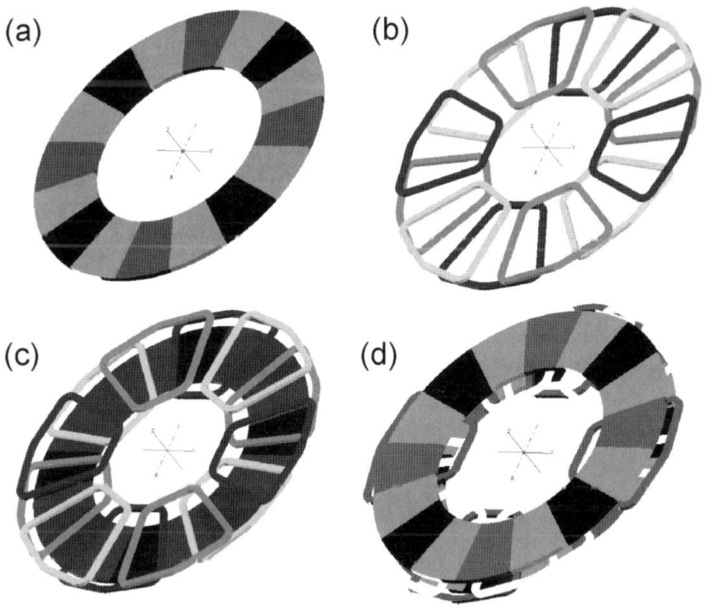

Fig. 6.3. Construction of a three-phase, 8-pole AFPM brushless machines with PMs arranged in Halbach array: (a) PM ring; (b) stator winding; (c) one half of the twin rotor; (d) stator winding and complete twin rotor.

The coreless machine designed as a segmental (modular) machine is shown in Fig. 6.2. The output shaft power can easily be adjusted to desired level by adding more modules.

To obtain a high power (or high torque) density motor, the magnetic flux density in the air gap should be as high as possible. This can be achieved by using PMs arranged in "Halbach array" (Figs 3.15, 6.3 and Fig 6.4). The magnetic flux density distribution excited by Halbach arrays is described by eqns (3.42) to (3.46). In practice, the angle between magnetization vectors of adjacent magnets is 90°, 60° or 45° (Fig. 6.5).

6.3 Air Gap Magnetic Flux Density

Fig. 6.6 shows the results of the 2D FEM modelling of the magnetic field in the air gap of a coreless AFPM brushless machine. NdFeB magnets with remanent flux density $B_r = 1.2$ T and coercivity $H_c = 950$ kA/m have been

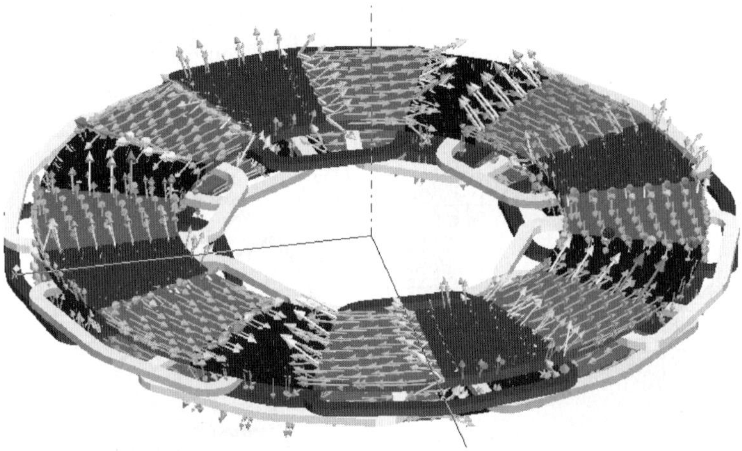

Fig. 6.4. 3D magnetic flux density distribution excited by an 8-pole twin rotor with PMs arranged in Halbach array.

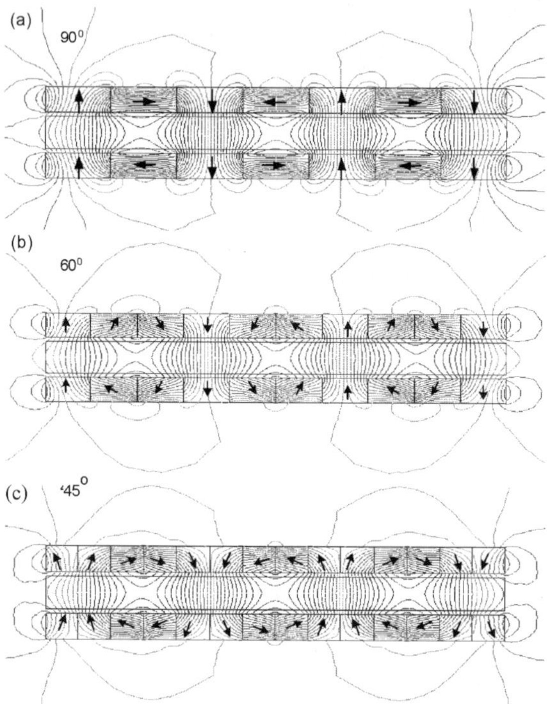

Fig. 6.5. Arrangement of the twin rotor PMs in 90°, 60°, and 45° Halbach array.

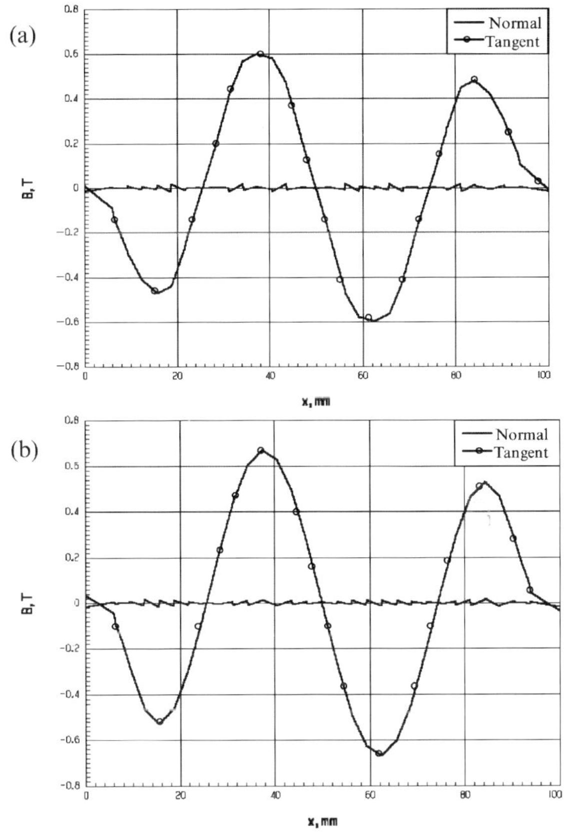

Fig. 6.6. Normal and tangential components of the magnetic flux density in the centre of the air gap of a double-sided coreless AFPM brushelss machine excited by Halbach arrays of PMs: (a) 90°; (b) 45°.

considered. The thickness of each PM has been assumed 6 mm, the coreless stator winding thickness is 10 mm and one-sided air gap thickness equals to 1 mm.

With the aid of Halbach array a high peak value (over 0.6 T) of the normal component of the magnetic flux density has been excited. This value is sufficient to obtain a high electromagnetic torque. The peak value of the flux density can be even higher for the optimized magnetic circuit of an AFPM machine. In practice, 60° and 45° Halbach arrays produce similar peak magnetic flux density (Fig. 3.22). The peak value of the normal component of the magnetic flux density is higher than that in a standard arrangement of surface PMs. The backing ferromagnetic discs added to the double-sided PM structure with a large nonmagnetic gap do not increase the flux density as much as Halbach array excitation does.

6.4 Electromagnetic Torque and EMF

The electromagnetic torque of a coreless AFPM brushless machines with Halbach array PM excitation can be calculated on the basis of eqns (2.32), (2.33), (2.34), (2.127) and (2.128). Similarly, the EMF can be calculated according to eqns (2.36), (2.37), (2.125) and (2.126).

6.5 Commercial Coreless AFPM Motors

Coreless AFPM machine architecture known as *segmented electro-magnetic array* (SEMA) has been proposed by *Lynx Motion Technology*, Greenville, IN, U.S.A. for electric motor, actuator and generator applications requiring high power density and high efficiency [151]. Applications that require an electromechanical drive with high power density, high efficiency and low torque ripple include precision motion control, naval propulsion systems and acoustically sensitive applications. SEMA technology may also find applications in distributed generation systems and energy storage systems, such as *flywheel motor–generator systems*. Motors with high torque and efficiency are demanded for gearless electromechanical drives. The use of direct drive motors without speed reducers eliminates gear noise, oil leaks, positioning errors from backlash and lack of torsional stiffness.

Fig. 6.7. Exploded view of the SEMA AFPM brushless machine. Courtesy of *Lynx Motion Technology*, Greenville, IN, U.S.A.

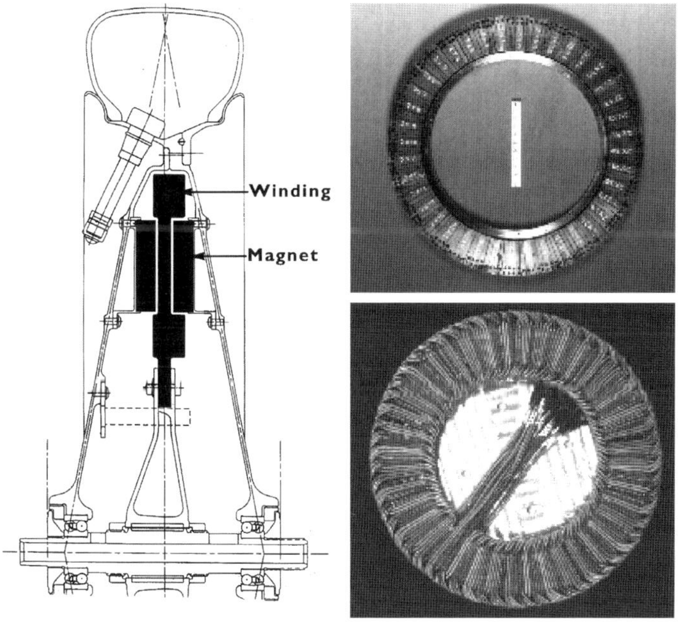

Fig. 6.8. In-wheel motor, PM ring and stator winding of a coreless AFPM brushless motor for solar-powered vehicles. Courtesy of *CSIRO*, Lindfield, NSW, Australia.

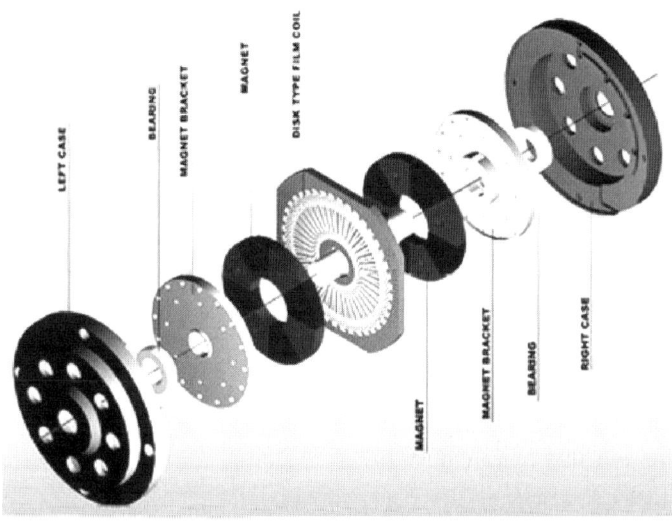

Fig. 6.9. Exploded view of the AFPM brushless motor with film coil coreless stator winding and twin PM rotor. Courtesy of *EmBest*, Soeul, South Korea.

The stator coreless and toothless design (Fig. 6.7) not only eliminates the cogging torque but also increases the area available for conductors. It also boosts peak torque capability and allows PMs to be more effectively utilized [151]. The stator coils are fully potted in high-strength, thermally conductive epoxy resin. Such construction gives the machine structural integrity and effectively damps the high frequency vibration when the motor is PWM inverted-fed. Machines with coreless stator winding perform excellently at constant speed, variable speed and at reversals.

CSIRO Telecommunications and Industrial Physics, Lindfield, NSW, Australia manufactures in-wheel coreless AFPM brushless motors for solar powered vehicles [225]. Solar-powered vehicles compete in the World Solar Challenge solar car race organized every two years in Australia between Darwin and Adelaide which has become a well-known international event.

CSIRO provides in-wheel AFPM brushless motors both with surface PMs glued to steel discs and with PMs arranged in Halbach array. The motor structure, PM disc and stator winding are shown in Fig. 6.8 [225].

Fig. 6.9 shows a coreless brushless motor with film coil stator winding and twin PM rotor manufactured by *EmBest*, Soeul, South Korea. The coreles stator has a foil winding at both sides. The 8-pole PM rotor is designed as a twin external rotor.

6.6 Case Study: Low-Speed AFPM Coreless Brushless Motor

A low speed, three-phase, Y-connected AFPM brushless motor rated at 10 kW, 750 rpm, 28.5 A has been designed and investigated. The rotor does not have any ferromagnetic core and consists of trapezoidal coils embedded in a high mechanical integrity resin. Average quality sintered NdFeB PMs with $B_r = 1.2$ T and $H_c = 950$ kA/m have been used. The double-disc construction is similar to that shown in Fig. 6.2b. The design data and calculated parameters are given in Table 6.1. All parameters except the air gap magnetic flux density B_{mg} and synchronous reactance $X_{sd} = X_{sq} = X_s$ have been calculated analytically. The magnetic flux density and synchronous reactance has been calculated using the 2D FEM. The resistance $R_1 = 0.175\ \Omega$ and synchronous reactance $X_s = 0.609\ \Omega$ per phase are for two stator modules connected electrically in series. The EMF constant $k_E = 5.013$ Vs and torque constant $k_T = 2.394$ Nm/A have been calculated for one stator segment.

6.6.1 Performance Characteristics

The current density 4.175 A/mm^2 in the stator winding at rated load is low and a totally enclosed motor can even operate without any forced air cooling. The recommended current density for enclosed a.c. motors with class F of insulation rated up to 10 kW is from 4.5 to 7 A/mm^2 [70].

Table 6.1. Design data and parameters of a three-phase, 10-kW, 750-rpm disc type AFPM brushless machine with ironless stator core

Design data	
Output power P_{out}, W	10,000
Speed n, rpm	750
Number of phases m_1	3 (Wye)
Input current I, A	28.4
Input frequency f, Hz	100
Number of stators (modules)	2
Number of pole pairs p	8
Number of coils (3 phases)	24
Number of turns per phase	100 (one module)
Coil pitch	1 slot
Wire diameter, mm	6×1.2
Axial thickness of PM h_M, mm	6
Axial thickness of the winding t_w, mm	10
Air gap (one side) g, mm	1
Air gap magnetic flux density B_{mg} under load, T	0.58
Current density, A/mm^2	4.175
$k_d = D_{in}/D_{out}$ ratio [5]	$1/\sqrt{3}$
Rotor outer diameter D_{out}, mm	360
Winding packing factor at $r = 0.5D_{in}$	0.65
Winding temperature, °C	75
Cooling system	natural
Class of insulation	F

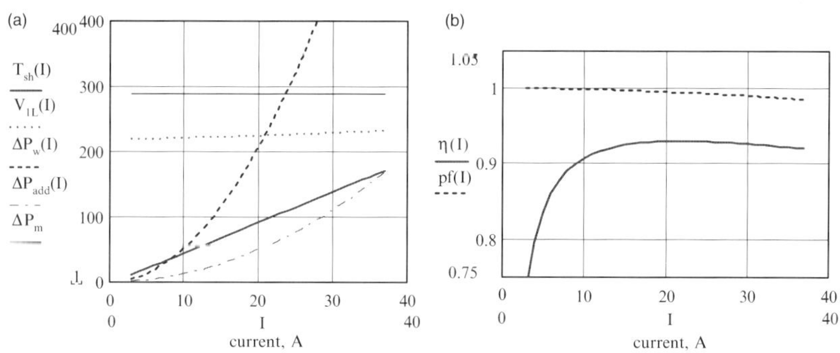

Fig. 6.10. Calculated performance characteristics of a 10 kW coreless AFPM brushless motor according to Table 6.1: (a) shaft torque T_{sh}, line-to-line voltage V_{1L}, winding losses ΔP_w, additional losses ΔP_{add}, and mechanical losses ΔP_m versus stator current at $n = 750$ rpm = const; (b) efficiency η and power factor pf versus stator current at $n = 750$ rpm = const.

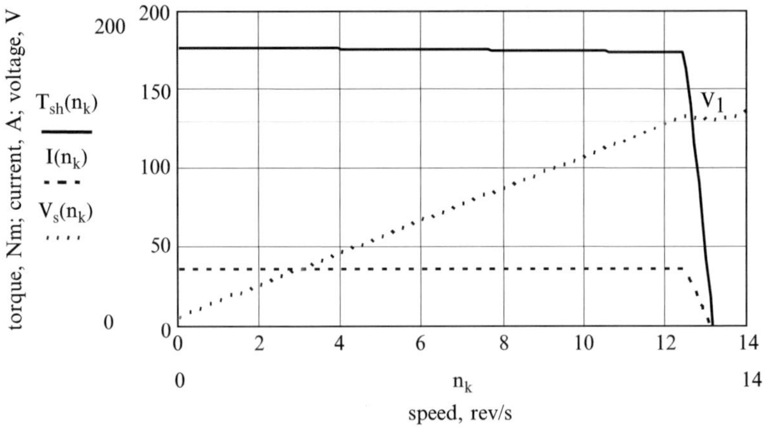

Fig. 6.11. Shaft torque, current and phase voltage versus speed of a 10 kW coreless AFPM brushless motor according to Table 6.1.

The calculated characteristics are shown in Figs. 6.10 and 6.11. It can be seen from Fig. 6.11 that the maximum–to–rated current ratio for torque–speed characteristic is 1.3. In Fig. 6.11 the voltage increases linearly up to approximately rated speed and then is kept constant.

6.6.2 Cost Analysis

The cost model has been presented in [106]. Average prices of materials for electrical machine construction are as follows (in US dollars: \$): silicon sheet steel c_{Fe} = 1.25 \$/kg, isolated copper conductor (magnet wire) c_{Cu} = 5.51 \$/kg, sintered NdFeB PMs c_{PM} = 15 \$/kg and steel bar for the shaft c_{steel} = 2.6 \$/kg. The active materials are conductors, sheet steel and PMs. The total volume of the steel bar to the volume of shaft is 1.94. The coefficient taking into account the cost of machining of the shaft is 2.15. The cost of labour involved in the assembly of active parts has not been estimated.

Table 6.2 lists the mass and cost of components of the 10 kW, 750 rpm coreless motor of disc construction. Table 6.3 lists the mass and cost of components of the equivalent cylindrical motor with laminated stator and rotor cores. Components independent of the machine shape are the encoder, terminal leads, terminal board, and name plate. The 10 kW coreless AFPM brushless motor costs \$613.05 (PMs cost \$185.66) while a classical PM brushless motor of cylindrical construction with the same ratings costs only \$510.01 (PMs cost \$23.55). The contribution of PMs to the total costs of a coreless motor is predominant.

6.6.3 Comparison With Cylindrical Motor With Laminated Stator and Rotor Cores

Table 6.4 compares the design data and performance of a 10 kW, 750 rpm coreless AFPM brushless motor and the equivalent 10 kW, 750 rpm PM brushless motor with laminated stator and rotor cylindrical cores. In both motors the number of phases is 3, number of poles is $2p = 16$, and input frequency is 100 Hz.

The analysis and the FEM simulation have shown the advantages of the coreless disc type PM brushless motors in terms of performance and mass of active materials. They have higher efficiency and power density when compared with conventional PM machines with laminated cores.

The efficiency of the 10-kW AFPM brushless motor is 1.2% higher (0.925 versus 0.914) and power density (output power to mass of active materials) is 82% higher (444.25 W/kg versus 243.55 W/kg) because the mass of active materials is 45% lower than that of an equivalent PM brushless motor with cylindrical ferromagnetic core. At zero current state the normal forces between the stator and rotor do not exist. However, the normal forces between the rotor PM discs are large.

Table 6.2. Mass and cost of components of a disc type coreless PM brushless motor rated at 10-kW, 750-rpm

Item	Mass, kg	Cost, $
Winding	10.13	55.84
Stator plastic material	1.01	7.09
PMs	12.38	185.66
Shaft	5.48	59.45
Frame, end bells, and bearings		165.00
Components independent of the machine shape		140.00
Total	22.51 (active parts)	613.05

Table 6.3. Mass and cost of components of a cylindrical PM brushless motor with laminated core rated at 10-kW, 750-rpm

Item	Mass, kg	Cost, $
Winding	8.51	46.9
Core	30.98	72.49
PMs	1.57	23.55
Shaft	5.72	62.07
Frame, end bells, and bearings		165.00
Components independent of the machine shape		140.00
Total	41.06 (active parts)	510.01

Table 6.4. Comparison of parameters and performance of 10-kW, 750-rpm PM brushless motors

Parameter	Coreless disc type	Cylindrical with laminated core
Input voltage (line-to-line), V	227.6	220.0
Shaft torque, Nm	131	133
Efficiency, %	92.5	91.4
Power factor	0.991	0.96
Air gap magnetic flux density, T	0.58	0.78
Stator winding current density, A/mm^2	4.175	4.247
Stator winding resistance per phase at 75°C, Ω	0.173	0.037
Synchronous reactance, Ω	0.609	0.758
Mass of PMs, kg	12.38	1.57
Mass of active materials, kg	22.51	41.06
Power density (only active materials), W/kg	444.25	243.55
Cost excluding labour, U.S.$	613.05	510.01

These new types of machines are cogging torque free. Owing to the lack of ferromagnetic cores the hysteresis and eddy current losses in those parts do not exist.

The cost of materials and components of the disc type 10 kW coreless motor is 20% higher than that of a cylindrical motor with laminated core. On the other hand, the cost of tooling will be much lower as a coreless motor does not require stamping and stacking tools.

The high cost of PMs limits commercial applications of coreless AFPM brushless motors to small size machines, special servo drives, airborne apparatus (lightweight construction) and electromechanical drives where the cogging torque, torque ripple and normal forces must be reduced to zero.

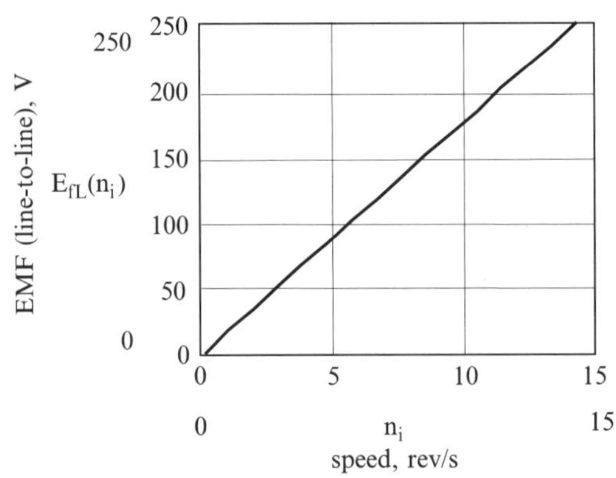

Fig. 6.12. Open circuit characteristic: line-to-line EMF versus speed.

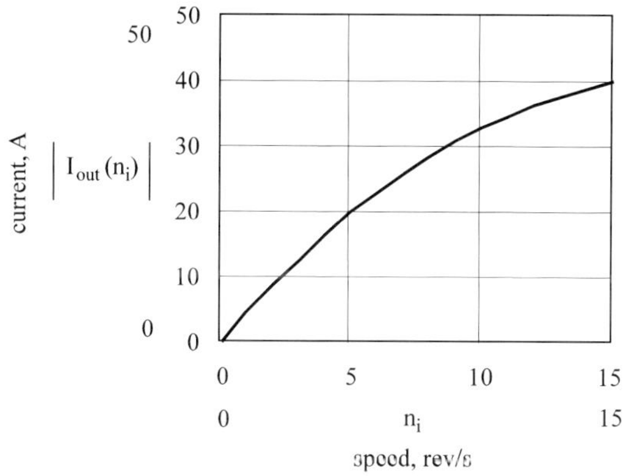

Fig. 6.13. Current versus speed for generating mode.

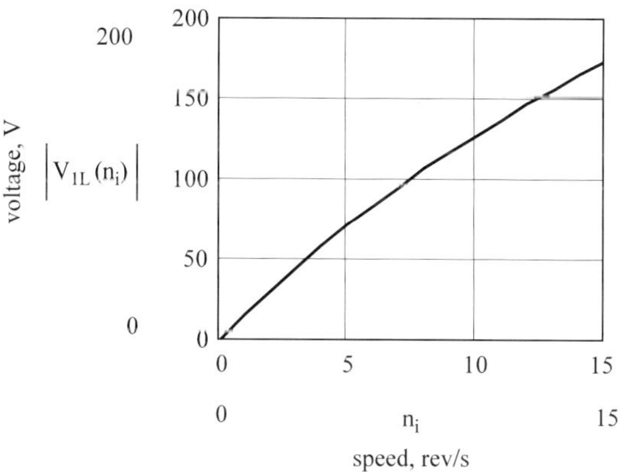

Fig. 6.14. Line voltage versus speed for generating mode.

6.7 Case Study: Low-Speed Coreless AFPM Brushless Generator

The AFPM brushless machine (Table 6.1) has been analysed as a stand alone a.c. generator. The open circuit characteristic, i.e. EMF (line-to-line) as a function of speed is shown in Fig. 6.12. This characteristic is linear.

The AFPM generator has been connected to the inductive load $\mathbf{Z}_L = R_L + jX_L = 1 + j0.628\ \Omega$. The current–speed characteristic for generating mode is shown in Fig. 6.13.

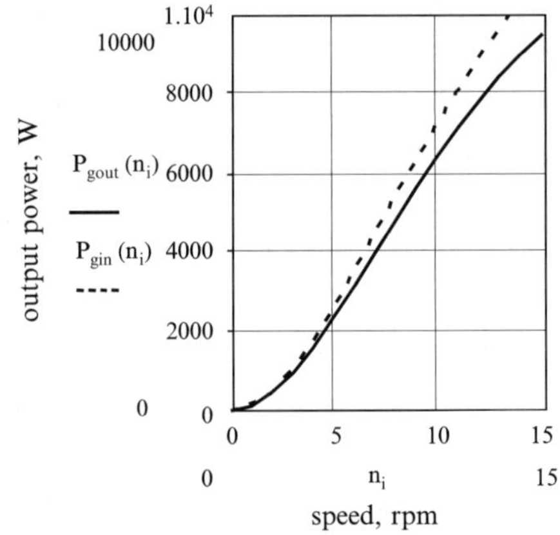

Fig. 6.15. Output power P_{gout} and input power P_{gin} for generating mode.

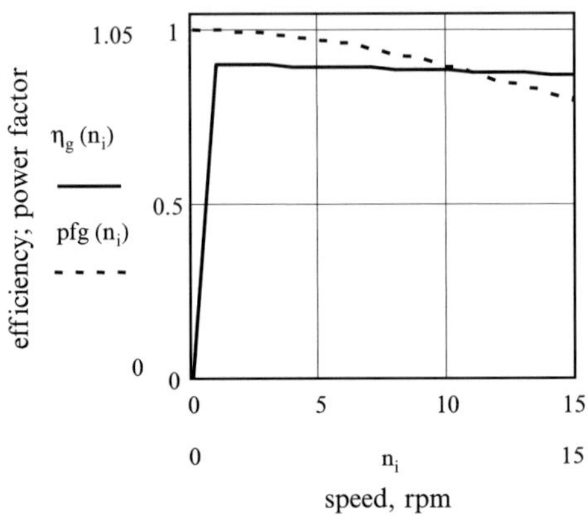

Fig. 6.16. Efficiency η_g and power factor pf_g for generating mode.

The line voltage versus speed is shown in Fig.6.14, output and input active power in Fig. 6.15, and efficiency and power factor in Fig. 6.16. The voltage is a nonlinear function of speed and depends on the value of the load impedance $\mathbf{Z}_L$ and power factor. For inductive load (RL) the power factor decreases as the speed increases.

6.8 Characteristics of Coreless AFPM Machines

Table 6.5 shows specifications of coreless AFPM brushless motors for solar powered vehicles with surface PMs glued to a steel yoke and PMs arranged in Halbach array (*CSIRO*, Australia) respectively. The steady-stare efficiency curves for surface PMs and Halbach array are plotted in Fig. 6.17. The maximum efficiency is obtained when the inverter applies a sinusoidal voltage synchronized to the rotor position. The inductance of the stator winding is low and most inverters require additional series inductance.

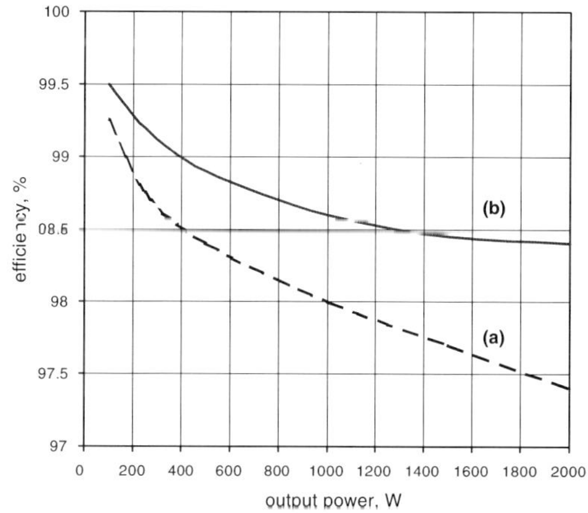

Fig. 6.17. Efficiency as a function of output power of a 3-phase, 1.8-kW coreless AFPM brushless motors for solar-powered vehicles: (a) rotor with surface PMs; (b) rotor with PMs arranged in Halbach array. Courtesy of *CSIRO*, Lindfield, NSW, Australia.

The E225 coreless AFPM motor (Table 6.6) has been developed by *Lynx Motion Technology*, Greenville, IN, U.S.A. for air-handling applications, in which system efficiency is a high priority. The family of *Lynx* 225-mm diameter motor designs offers a broad range of possible system configurations for speed and position servo applications requiring both high efficiency and high torque. Performance ratings of the E225 motor are based on air-cooled 105°C winding at 25°C ambient air temperature. The E813 coreless AFPM machines has been designed for power generation applications (distributed generation systems) where high efficiency of the system is important. Performance ratings of the E813 machine are based on 130°C winding at 40°C ambient air temperature. Both E225 and E813 machines have extremely linear torque constant k_T independent of speed. Other applications of *Lynx* coreless machines include servo applications, precision robotics, marine propulsion, generators and weapons turrets.

When coupled with a three-phase H-bridge inverter, *Lynx* motors offer a highly efficient electromechanical drive with the high peak torque capability required for the most demanding high-transient applications, e.g. fuel-cell air compressors.

Table 6.5. Comparison of three-phase 1.8-kW coreless AFPM brushless motors with surface and Halbach PM excitation manufactured by *CSIRO*, Australia

Parameter	Surface PMs and steel yoke	Halbach array
Output power, W	1800	1800
Efficiency, %	97.4	98.2
Number of poles	40	40
Nominal speed, rpm	1060	1060
Maximum speed, rpm	2865	2865
Nominal torque, Nm	16.2	16.2
Maximum continuous torque at 1060 rpm and maximum winding temperature, Nm	31	39
Absolute maximum torque, Nm	50.2	50.2
Torque constant per phase at sinusoidal excitation	0.39	0.56
Nominal phase EMF, V	43	62
EMF constant per phase at sinusoidal current, Vs/rad	0.39	0.56
Nominal phase current, A	13.9	9.6
Winding losses, W	43.9	27.6
Eddy current losses in the stator winding, W	2.6	2.7
Windage losses, W	2.1	2.1
Total losses, W	48.6	32.4
Winding temperature rise, °C	22	14
Overload winding temperature rise, °C	64	40
Maximum winding temperature, °C	110	110
Air gap at each side, mm	2	2
Resistance per phase, Ω	0.0757	0.0997
Mass of complete motor, kg	16.2	13.2
Mass of frameless motor, kg	10.7	7.7

Numerical Example 6.1

The coreless AFPM motor with the same dimensions as that described in *Numerical Example 5.1* has been redesigned to obtain the speed of 1000 rpm at $I_a = 8.2$ A ($I_{ad} = 0$). The number of pole pairs has been increased to $p = 12$, the number of coil sides to $s_1 = 72$ and PMs have been arranged in 90° Halbach array with $n_M = 4$ PM pieces per wavelength. The number of stator winding turns per phase is $N_1 = 240$, the number of parallel wires $a_w = 8$, the diameter of wire $d_w = 0.455$ mm (AWG 25) and the coil pitch measured in coils sides $w_c = 3$. Since both stator and rotor do not have any steel magnetic circuit, the magnetic saturation does not exist and $k_{sat} = 1$. All other parameters and assumptions are the same as those in *Numerical example 5.1*.

Table 6.6. Specifications of coreless AFPM motors manufactured by *Lynx Motion Technology*, Greenville, IN, U.S.A.

Parameter	E225 AFPM motor	E813 AFPM motor
Number of poles	12	28
d.c. bus voltage, V	155	850
Terminal voltage, V	117 rms phase voltage H-bridge	574 peak line-to-line
Speed, rpm	6000	2750
Torque, Nm	9.9	450
Peak torque, Nm	97.4	900
Current, A	18.7	119
Peak current, A	180	238
Output power, kW	6.2	130
Torque constant, Nm/A	0.53	3.78
EMF constant per phase, V/krpm	19	209
Electrical time constant, μs	0.348	1.66
Winding resistance at 100°C, Ω	0.204	0.0265
Winding inductance, μH	71	44
Rotor moment of inertia, kgm^2	0.0184	9.16
Mass of motor, kg	8.4	295
External diameter, m	0.225	0.813

Find the motor steady state performance, i.e. output power, torque and efficiency at rated current $I_a = I_{aq} = 8.2$ A.

Solution

The following calculated parameters are the same as those in *Numerical example 5.1*: $g = 1.5$ mm, $D_{in} = 0.127$ m, $D = 0.1735$ m, $L_i = l_M = 0.0465$ m, $R'_{out} = 0.115$ m (for windage losses), $R_{sh} = 0.015$ m, $m_r = 3.4$ kg, $m_{sh} = 0.64$ kg and $m_{con} = 0.75$ kg.

The number of coils per phase for a single layer winding is $n_c = s_1/(2m_1) = 72/(2 \times 3) = 12$. The number of turns per coil is $N_{ct} = N_1/n_c = 240/12 = 20$. The number of coil sides per pole per phase (equivalent to the number of slots per pole per phase) $q_1 = s_1/(2pm_1) = 72/(24 \times 3) = 1$. The pole pitch measured in coil sides is $\tau_c = s_1/(2p) = 72/24 = 3$ and the average pole pitch $\tau = \pi D/(2p) = \pi 0.1735/24 = 0.023$ m

The input frequency at 1000 rpm

$$f = n_s p = \frac{1000}{60} \times 12 = 200 \text{ Hz}$$

The wavelength l_a at the average diameter and constant β for Halbach array is

$$l_a = 2\tau = 2 \times 0.023 = 0.046 \text{ m} \qquad \beta = \frac{2\pi}{l_a} = \frac{2\pi}{0.046} = 138.3 \text{ 1/m}$$

The magnetic flux density in the air gap according to eqn (3.42) is

$$B_{mg} \approx B_{m0} = 1.2[1 - \exp(-138.3 \times 0.006)]\frac{\sin(\pi/4}{\pi/4} = 0.609 \text{ T}$$

The magnetic flux according to eqn (2.28) is

$$\Phi_f = \frac{1}{8}\frac{2}{\pi}\frac{\pi}{12}0.609 \times 0.22^2 \left[1 - \left(\frac{1}{\sqrt{3}}\right)^2\right] = 0.00041 \text{ Wb}$$

The winding factor according to eqns (2.8), (2.9) and (2.10) is

$$k_{d1} = \frac{\sin \pi/(2 \times 3)}{1 \sin \pi/(2 \times 3 \times 1)} = 1; \qquad \beta = \frac{w_c}{\tau_c} = \frac{7}{9}$$

$$k_{p1} = \sin\left(\frac{3}{3}\frac{\pi}{2}\right) = 1 \qquad k_{w1} = 1 \times 1 = 1$$

The EMF constant according to eqn (2.37) and torque constant according to eqn (2.34) is

$$k_E = \pi\sqrt{2} \times 12 \times 240 \times 1 \times 0.00041 = 5.24 \text{ V/(rev/s)} = 0.0873 \text{ V/rpm}$$

$$k_T = k_E\frac{m_1}{2\pi} = 5.24\frac{3}{2\pi} = 2.5 \text{ Nm/A}$$

The EMF at 1000 rpm is

$$E_f = k_E n = 0.0873 \times 1000 = 87.34 \text{ V}$$

The electromagnetic torque at $I_a = I_{aq} = 8.2$ A is

$$T_d = k_T I_a = 2.5 \times 8.2 = 20.5 \text{ Nm}$$

The electromagnetic power is

$$P_{elm} = 2\pi n_s T_d = 2\pi\frac{1000}{60} \times 20.5 = 2148.6 \text{ W}$$

The length of the stator shorter end connection without inner bends $l_{1emin} = (w_c/\tau_c)\pi D_{in}/(2p) = (3/3)\pi \times 0.127/24 = 0.017$ m and the length of the stator longer end connection without outer bends $l_{1emax} = 0.017 \times 0.22/0.127 = 0.0288$ m (Fig. 3.16).

The average length of the stator turn with 15-mm bends is

$$l_{1av} \approx 2L_i + l_{1emin} + l_{1emax} + 4 \times 0.015$$

$$= 2 \times 0.0465 + 0.017 + 0.0288 + 0.06 = 0.1984 \text{ m}$$

The stator winding resistance at 75°C according to eqn (2.40) is

$$R_1 = \frac{240 \times 0.1984}{47 \times 10^6 \times 8 \times \pi \times (0.455 \times 10^{-3})^2/4} = 0.7789\ \Omega$$

The maximum width of the coil at the diameter D_{in} is $w_w = \pi D_{in}/s_1 = \pi 0.127/72 = 0.0055$ m $= 5.5$ mm. The thickness of the coil is $t_w = 8$ mm. The number of conductors per coil is $N_c = a_w \times N_{ct} = 8 \times 20 = 160$. The maximum value of the coil packing factor is at D_{in}, i.e.

$$k_{fmax} = \frac{d_w^2 \times N_c}{t_w w_w} = \frac{0.455^2 \times 160}{8 \times 5.5} = 0.731$$

The stator current density is

$$j_a = \frac{8.2}{8 \times \pi 0.455^2/4} = 6.45 \text{ A/mm}^2$$

The stator winding losses according to eqn (2.49) are

$$\Delta P_w = 3 \times 8.2^2 \times 0.7789 = 157.1 \text{ W}$$

The eddy current losses in stator round conductors according to eqn (2.68) are

$$\Delta P_e = \frac{\pi^2}{4}\frac{47 \times 10^6}{8800} 200^2 \times 0.000455^2 \times 0.75 \times 0.609^2 \times 1.15^2 = 41.1 \text{ W}$$

The friction losses in bearings according to eqn (2.71) are

$$\Delta P_{fr} = 0.06 \times 1.5(3.4 + 0.64)\frac{1000}{60} = 6.1 \text{ W}$$

The windage losses according to eqn (2.74) are

$$\Delta P_{wind} = \frac{1}{2} 0.013 \times 1.2 \left(2\pi \frac{1000}{60}\right)^3 [(0.5 \times 0.115)^5 - (0.5 \times 0.015)^5]$$

$$\approx 0.18 \text{ W}$$

where

$$Re = 1.2\frac{2\pi(1000/60)}{1.8 \times 10^{-5}} \times 0.115^2 = 9.233 \times 10^4$$

$$c_f = \frac{3.87}{\sqrt{9.233 \times 10^4}} = 0.013$$

The rotational (mechanical) losses according to eqn (2.70) are

$$\Delta P_{rot} = 6.1 + 0.18 = 6.28 \text{ W}$$

The output power is

$$P_{out} = P_{elm} - \Delta P_{rot} = 2148.6 - 6.28 = 2142.3\text{W}$$

The shaft torque is

$$T_{sh} = \frac{2142.3}{2\pi \times 1000/60} = 20.46 \text{ Nm}$$

The input power is

$$P_{in} = P_{elm} + \Delta P_w + \Delta P_e = 2148.6 + 157.1 + 41.1 = 2346.7 \text{ W}$$

The efficiency is

$$\eta = \frac{2142.3}{2346.7} = 0.912$$

Numerical Example 6.2

A three-phase, 15-kW, 3600-rpm coreless AFPM motor with Halbach array PM twin rotor has the stator winding resistance per phase $R_1 = 0.24\ \Omega$, stator winding leakage reactance $X_1 = 0.92\ \Omega$ and armature reaction reactances $X_{ad} = X_{aq} = 0.88\ \Omega$. The input phase voltage is $V_1 = 220$ V, power factor $\cos\phi = 0.96$, angle between the stator current and q axis $\Psi = 5°$, EMF constant $k_E = 3.363$ V/(rev/s), eddy current losses in stator conductors $\Delta P_e = 184$ W and rotational losses $\Delta P_{rot} = 79$ W.

Find the internal voltage E_i excited by the resultant air gap flux and the stator d- and q-axis current adjusted for eddy current losses.

Solution

The phasor diagram showing the EMFs E_f and E_i is drawn in Fig. 6.18. The EMF E_f excited by the rotor magnetic flux

$$E_f = k_E n = 3.363 \times \frac{3600}{60} = 201.78 \text{ V}$$

The synchronous reactance is

$$X_{sd} = X_{sq} = X_{ad} + X_1 = X_{aq} + X_1 = 0.88 + 0.92 = 1.8\ \Omega$$

The angle between the stator current and input voltage $\phi = \arccos 0.96 = 16.26^{\circ}$. The load angle between the voltage and EMF E_f is $\delta = \phi - \Psi = 16.26 - 5 = 11.26^{\circ}$. The stator currents according to eqns (2.87), (2.88) and (2.89) are:

$$I_{ad} = \frac{220(1.8\cos 11.26^{\circ} - 0.24\sin 11.26^{\circ}) - 201.78 \times 1.8}{1.8 \times 1.8 + 0.24^2} = 4.5 \text{ A}$$

$$I_{aq} = \frac{220(0.24\cos 11.26^{\circ} + 1.8\sin 11.26^{\circ}) - 201.78 \times 0.24}{1.8 \times 1.8 + 0.24^2} = 24.5 \text{ A}$$

$$I_a = \sqrt{4.5^2 + 24.5^2} = 24.9 \text{ A}$$

The stator winding losses according to eqn (2.49) are

$$\Delta P_{1w} = 3 \times 24.9^2 \times 0.24 = 445.6 \text{ W}$$

The input power is

$$P_{in} = m_1 V_1 I_a \cos\phi = 3 \times 220 \times 24.9 \times 0.96 = 15\ 763.0 \text{ W}$$

The electromagnetic power according to eqn (5.1) is

$$P_{elm} = 15\ 763 - 445.6 - 184 = 15\ 317.3 \text{ W}$$

The output power according to eqn (5.3) is

$$P_{out} = 15\ 317.3 - 79 = 15\ 238.3 \text{ W}$$

The efficiency is

$$\eta = \frac{15\ 238.3}{15\ 763.0} = 0.967$$

The internal voltage E_i excited by the resultant air gap flux according to the phasor diagram (Fig. 6.18) is

$$E_i = \sqrt{(E_f + I_{ad}X_{ad})^2 + (I_{aq}X_{aq})^2}$$

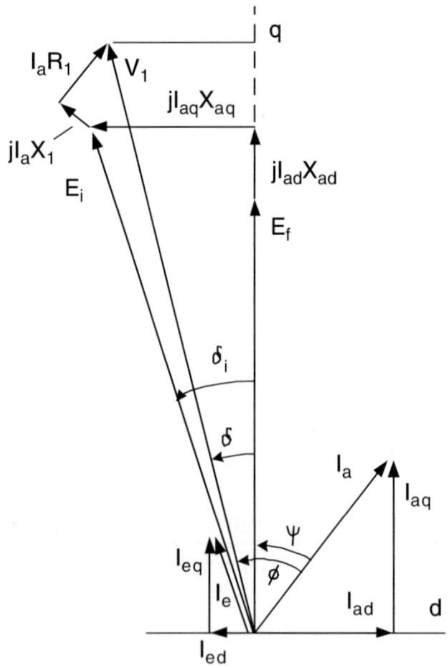

Fig. 6.18. Phasor diagram of a coreless AFPM synchronous motor (eddy current I_e is in phase with the internal voltage E_i). *Numerical example 6.2.*

$$= \sqrt{(201.78 + 4.5 \times 0.88)^2 + (24.5 \times 0.88)^2} = 206.87 \text{ V}$$

The angle δ_i between the EMF E_i and q-axis is

$$\delta_i = \arctan\left(\frac{I_{aq}X_{aq}}{E_f + I_{ad}X_{ad}}\right) = \arctan\left(\frac{24.5 \times 0.88}{201.78 + 4.5 \times 0.88}\right) = 5.97^{\circ}$$

The eddy current shunt resistance according to eqn (5.32) is

$$R_e = 3\frac{206.87^2}{184} = 697.75 \ \Omega$$

The current in the vertical branch of the equivalent circuit (Fig. 5.5) found on the basis of the phasor diagram (Fig. 6.18)

$$I_{ed} = -\frac{E_i}{R_e}\sin\delta_i = -\frac{206.87}{697.75}\sin 5.97^{\circ} = -0.031 \text{ A}$$

$$I_{eq} = \frac{E_i}{R_e}\cos\delta_i = \frac{206.87}{697.75}\cos 5.97^{\circ} = 0.295 \text{ A}$$

$$I_e = \sqrt{I_{ed}^2 + I_{eq}^2} = \sqrt{(-0.031)^2 + 0.295^2} = 0.296 \text{ A}$$

or

$$I_e = \frac{E_i}{R_e} = \frac{206.87}{697.75} \cos 5.97^{\text{o}} = 0.296 \text{ A}$$

The stator current adjusted for eddy current losses found on the basis of the phasor diagram (Fig. 6.18)

$$I'_{ad} = \frac{E_i \cos \delta_i - E_f}{X_{ad}} - \frac{E_i \sin \delta_i}{R_e}$$

$$= \frac{206.87 \cos 5.97^2 - 201.78}{0.88} - \frac{206.87 \sin 5.97^2}{697.75} \approx 4.5 \text{ A}$$

$$I'_{aq} = \frac{E_i \sin \delta_i}{X_{aq}} - \frac{E_i \cos \delta_i}{R_e} = \frac{206.87 \sin 5.97^2}{0.88} - \frac{206.87 \cos 5.97^2}{697.75} \approx 24.8 \text{ A}$$

$$I'_a = \sqrt{(I'_{ad})^2 + (I'_{aq})^2} = \sqrt{4.5^2 + 24.8^2} \approx 25.2 \text{ A}$$

7

Control

In the previous chapters it was shown that there is a direct relationship between the rotor speed and the frequency of the phase voltages and currents of the AFPM machine. It was also shown that the magnitude of the induced phase voltages is directly related to the frequency and hence the rotor-speed of the AFPM machine. Therefore, for variable speed operation of the AFPM machine both the frequency and magnitude of the supply voltage must be adjustable. This requires a solid-state converter to be used between the fixed a.c. or d.c. supply and the terminals of the AFPM machine as shown in Fig. 7.1. To have good position and/or speed control of the converter-fed AFPM machine drive, the torque and therefore the current of the machine must be controlled. For current control, information of the phase current as well as the position of the rotor is necessary. Thus, both the current and rotor position of the AFPM machine is sensed and fed back to the controller as shown in Fig. 7.1. The controller in its turn controls the supply voltage and frequency of the AFPM machine via the solid-state converter (inverter).

This chapter focuses on all the control aspects of the AFPM machine drive of Fig. 7.1. The control of the two types of AFPM machines considered in previous chapters, namely the trapezoidal and sinusoidal AFPM machines, is dealt with separately. Finally, position sensorless control is briefly discussed and some numerical examples are given.

7.1 Control of Trapezoidal AFPM Machine

Trapezoidal AFPM machines are surface-mounted PM machines and are characterised by their *trapezoidal or quasi-square EMF waveforms*. Example of these EMF waveforms are shown in Figs 7.2a and 7.2b, which are the open circuit phase voltage waveforms induced by PM flux of a small 1 kW, slotted trapezoidal AFPM machine at half and rated speed respectively. It can be observed, firstly, that at rated speed the frequency and the magnitude of the

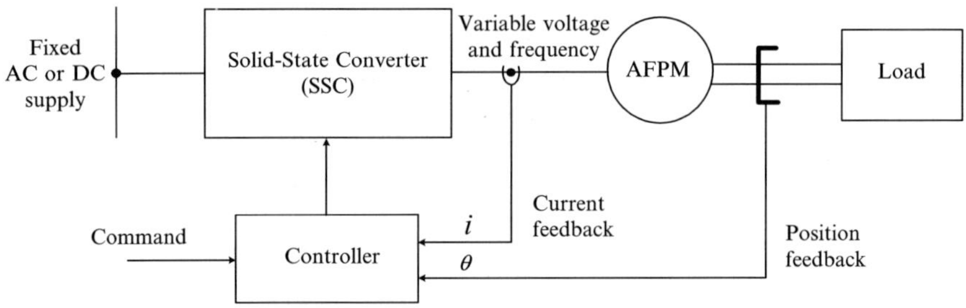

Fig. 7.1. Converter-fed AFPM machine drive.

EMF are double what they are at half rated speed. Secondly, and very important, it can be seen that the shape of the EMF waveform is not sinusoidal, but trapezoidal with a flat-topped voltage of about 120 electrical degrees. It is during this period of 120 electrical degrees that the phase is excited by means of the solid-state converter. This period is therefore called the *conduction period*, as current is flowing in that phase during that period as shown in Fig. 7.2c. Note that the time of conduction or the time of occurrence of the flat-topped voltage is directly related to the position of the rotor; the time of conduction can thus exactly be determined by means of a simple, low resolution position sensor.

During the conduction period one can consider the induced back-EMF phase voltage as a d.c. quantity due to the flat voltage waveform. With a three-phase machine the phase voltages are 120 electrical degrees out of phase, which implies that the conduction periods of two phases are always overlapping (see Fig. 7.2) and that two phases are always active. In the case of AFPM machine drives with 180 electrical degree conduction periods, e.g. [35], then all three phases are active. A different converter should be used in this case. In this section only those trapezoidal AFPM machine drives that have 120 electrical degree conduction periods are considered as they are commonly used.

7.1.1 Voltage Equations

With two phases of the Y-connected three-phase machine always active (the third phase is switched off by means of the solid-state converter), the power circuit of the trapezoidal AFPM brushless machine can be represented by the electric circuit of Fig. 7.3a. Here, phases A and B are active and phase C is switched off (shown in dotted lines). In this circuit R_1 and L_s are respectively the phase resistance and phase synchronous inductance. The synchronous inductances in the d and q-axis are equal, i.e. $L_{sd} = L_{sq} = L_s$. The voltages e_{fA}, e_{fB} and e_{fC} are the EMFs per phase. As phases A and B are in their conduction periods, both e_{fA} and e_{fB} are at their flat-topped values. Hence, $e_{fA} = E_f^{(tr)}$ and $e_{fB} = -E_f^{(tr)}$ where $E_f^{(tr)}$ is the flat-topped value of the

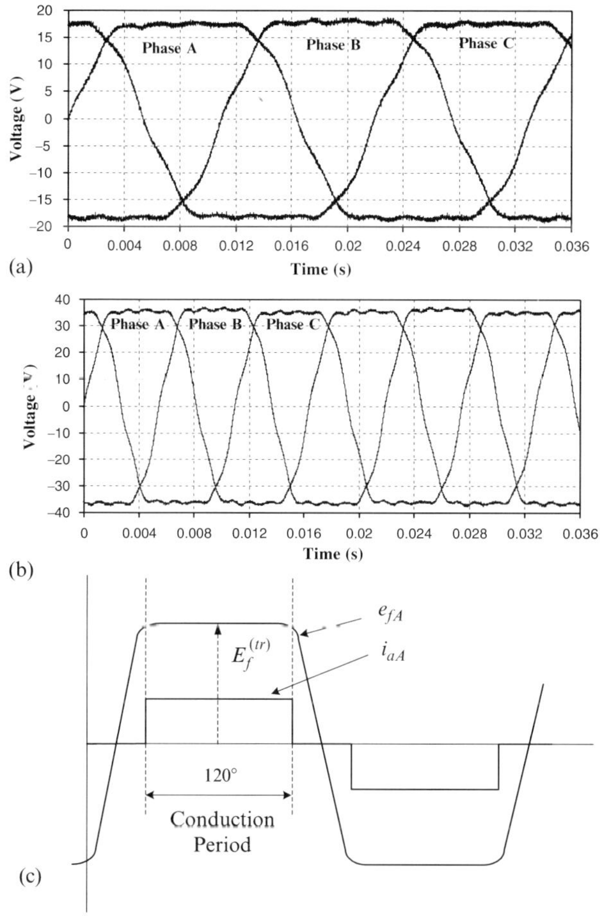

Fig. 7.2. Open-circuit induced phase voltage waveforms of a 1-kW trapezoidal AFPM machine obtained from measurements: (a) at half speed (30 Hz); (b) at rated speed (60 Hz); (c) theoretical EMF and phase current waveforms.

trapezoidal voltage waveform (Fig. 7.2c). Combining the resistances, inductances and induced voltages of Fig. 7.3a results in the equivalent circuit of Fig. 7.3b in which E_{fL-L} is taken as a d.c. voltage equal to $E_{fL-L} = 2E_f^{(tr)}$, $R_p = 2R_1$, $L_p = 2L_s$, $i_a = i_{aA} = -i_{aB}$ and $v_p = v_{1L-L}$. The dynamic voltage equation of this circuit is

$$
\begin{aligned}
v_p &= 2R_1 i_{aA} + 2L_s \frac{di_{aA}}{dt} + 2E_f^{(tr)} \\
&= R_p i_a + L_p \frac{di_a}{dt} + E_{fL-L} \qquad (7.1)
\end{aligned}
$$

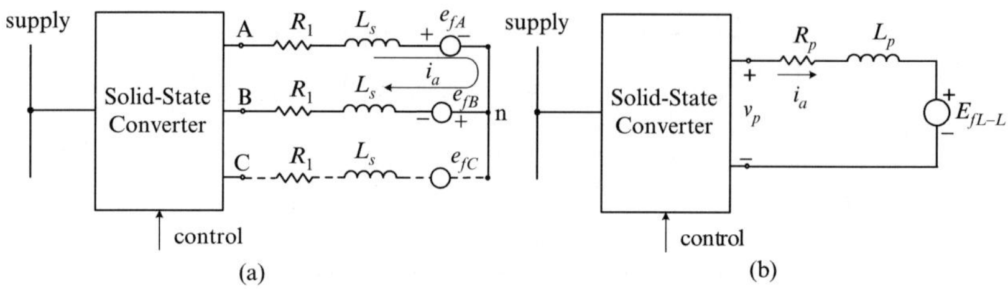

Fig. 7.3. Trapezoidal AFPM machine connected to the solid-state converter with two phases active: (a) electrical circuit; (b) combined equivalent circuit for the analysis.

Applying a d.c. voltage $v_p = V_p$ across terminals A and B by means of the solid-state converter will imply that ideally in the steady state, a d.c. current $i_a = I_a^{(sq)}$ will flow through the two phase windings according to the following steady-state voltage equation derived from eqn (7.1), i.e.

$$V_p = R_p I_a^{(sq)} + E_{fL-L} \tag{7.2}$$

The induced voltage E_{fL-L} is given by eqn (2.125), the electromagnetic power of the machine is $P_{elm} = E_{fL-L} I_a^{(sq)}$ and the developed torque is given by eqn (2.127). Note that $i_a = I_a^{(sq)}$ and E_{fL-L} are instant quantities that can change with time, similar to a brush-type d.c. machine.

7.1.2 Solid-State Converter

The complete power circuit connected to the three-phase trapezoidal AFPM brushless machine is shown in Fig. 7.4. In this circuit the fixed three-phase a.c. supply is first rectified by means of a three-phase diode rectifier to obtain a fixed d.c. voltage. The d.c. voltage is then inverted to a three-phase a.c. trapezoidal voltage. With the diode rectifier no power can flow back to the fixed a.c. supply if the AFPM machine is operating in generator mode. This means that any generated power from the machine must be dissipated at the d.c. link of the converter. This is done by switching a braking resistance at the d.c. link as shown in Fig. 7.4. If it is required that the generated power must flow back to the fixed a.c. supply, then use must be made of a so-called active rectifier. In the rest of this section only the switching and control of the d.c. to a.c. inverter of Fig. 7.4 is studied, assuming a fixed d.c.-link voltage.

The three-phase inverter of Fig. 7.4 has six power electronic transistor switches. Nowadays Insulated-Gate Bipolar Transistors (IGBTs) or MOS field effect transistors (MOSFETs) are very popular for use as switches for the inverter. The power transistor switches are switched on and off at a relatively high switching frequency (typically from 1 to 20 kHz or even higher). Either the top transistor of an inverter-leg is switched on and the bottom transistor

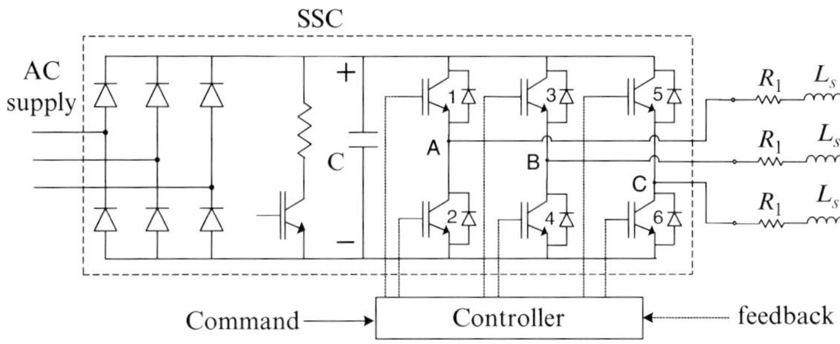

Fig. 7.4. Solid-state a.c.–a.c. converter and equivalent three-phase AFPM machine connection.

is switched off, or vice versa, or both transistors are switched off. It was explained in the previous sections that only two phases of the converter-fed trapezoidal AFPM machine drive are active at any time. The third phase is switched off by switching off the top and bottom transistors of the inverter-leg connected to that phase winding. The three-phase inverter of Fig. 7.4 also has six anti-parallel diodes to make current flow possible when a switch is turned off (this is due to inductive stored energy) and also to make the generator (braking) operation of the drive possible.

The circuit diagrams of Figs 7.3 and 7.4 show that the trapezoidal AFPM machine drive can be analysed by considering only the two active phases of the machine and the two active switching legs of the inverter. The circuit then reduces to the general circuit of Fig. 7.5.

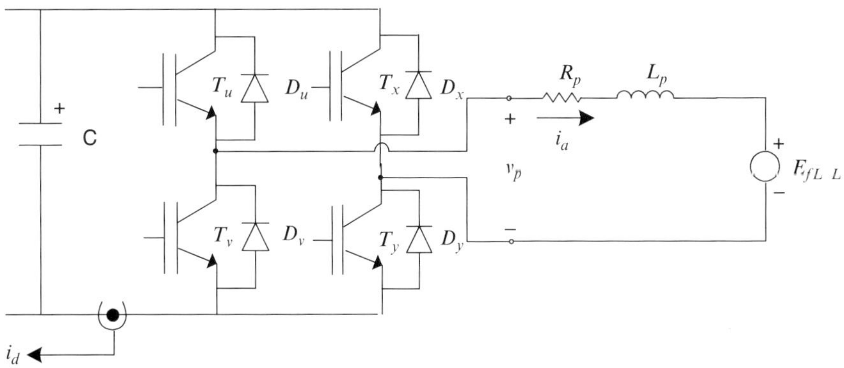

Fig. 7.5. Equivalent circuit of converter-fed trapezoidal AFPM machine (two phases active).

One method of controlling the power semiconductor switches of the inverter shown in Fig. 7.5 is explained as follows. Consider in the forward speed direction that the voltage E_{fL-L} and the average current $i_{a(avg)} = I_a^{(sq)}$ are positive, then the average supply voltage $v_{p(avg)}$ also has to be positive according to the steady-state equation, eqn (7.2), i.e. $v_{p(avg)} = V_p > E_{fL-L}$. A positive average supply voltage can be obtained by switching on transistor T_y and pulse-width-switching (PWM-switching) transistors T_u and T_v of Fig. 7.5 at duty cycles D and $(1 - D)$ respectively, such that in the steady-state

$$DV_d = v_{p(avg)} \tag{7.3}$$

where

$$D = \frac{t_{on}}{t_s} = t_{on} f_s \tag{7.4}$$

and t_{on} is the switching-on time, t_s is the switching period and f_s is the switching frequency of the transistors. This switching and the switching waveform of the supply voltage v_p across the terminals of the two phases of the machine are shown in Fig. 7.6(a). As is shown $v_{p(avg)}$ is the switching period average of the voltage v_p. Note that in Fig. 7.6 the drive is not in the steady-state as the instant average current $i_{a(avg)}$ and the average voltage $v_{p(avg)}$ change with time. Also note that the drive system only reacts on the average of the supply voltage and not on the high frequency components thereof, as the switching frequency is relatively high and the time constant of the mechanical system is relatively large.

Let us consider next the flow of current in the converter and the machine in motoring mode. When transistors T_u and T_y of Fig. 7.5 are switched on, current flows from the d.c. supply through T_u and through the machine winding and T_y, and then back to the supply. If T_v and T_y are switched on, the current then free-wheels through diode D_v and through the machine and T_y. In this case no current is flowing from or to the d.c. supply. An example of the phase winding and d.c.-link current waveforms in motoring mode is shown in Fig. 7.6a. As mentioned before $i_{a(avg)}$ is the instant average of the phase current i_a.

In generating (braking) mode with E_{fL-L} positive, the duty cycle D of eqn (7.3) is adjusted in such a way that $v_{p(avg)} < E_{fL-L}$ to allow $i_{a(avg)}$ to become negative as shown in Fig. 7.6b. In this case D_y is always conducting and the current either flows through T_v when switched on, or through D_u to the supply when switched off. An example of the d.c.-link current waveform in generating mode is shown in Fig. 7.6b.

When operating the machine in the reverse speed direction, in which case E_{fL-L} is negative with respect to the polarity given in Fig. 7.5, transistor T_v is switched on during the conduction cycle and transistors T_x and T_y are PWM-switched. In this case the supply voltage $v_{p(avg)}$ is negative and the current can be controlled again to be positive or negative in the same way as in the forward speed mode.

The advantage of the above switching method is that only one inverter-leg is switched at a high frequency at any time, keeping the switching losses of the inverter low. Which inverter-leg of Fig. 7.5 should be selected to be PWM-switched is simply a logic decision that depends on the sign (positive or negative) of the supply voltage $v_{p(avg)}$. Note that $v_{p(avg)}$ is controlled directly by the controller of the drive (Fig. 7.1).

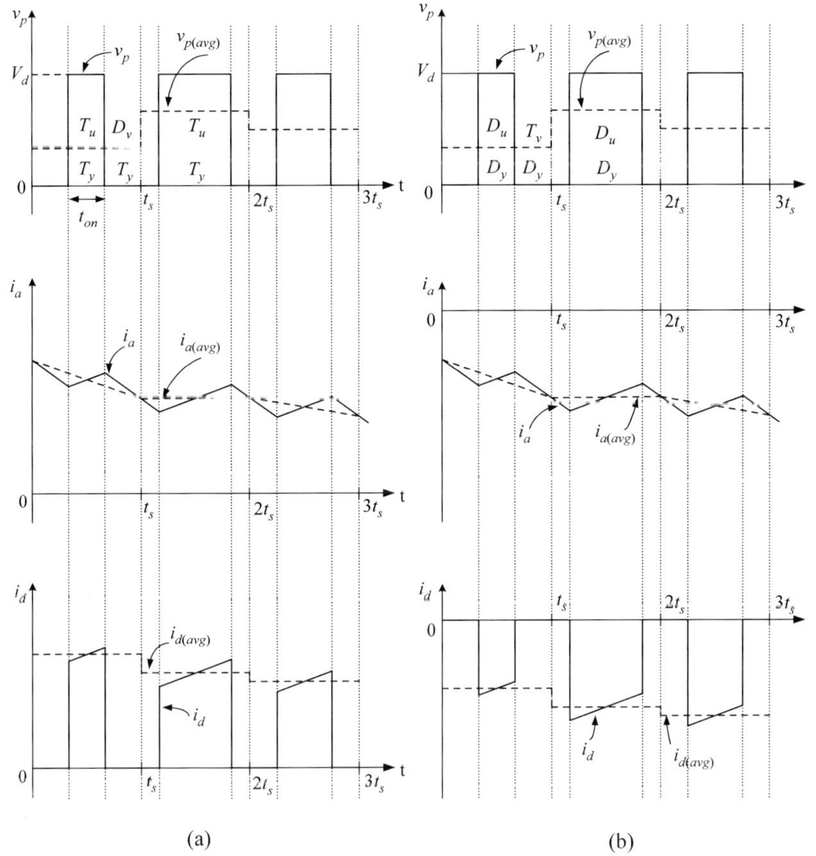

Fig. 7.6. Voltage and current waveforms of a converter-fed trapezoidal AFPM machine: (a) motoring mode; (b) generating mode (see Fig. 7.5).

7.1.3 Current Control

For good speed control of any drive a *fast torque control* is necessary. According to eqn (2.127) the torque of the trapezoidal AFPM machine drive can be controlled directly by controlling the stator current i_a and hence the instant average current $i_{a(avg)}$ explained in Fig. 7.6. Close examination of the currents of Fig. 7.6 reveals that the required information regarding the current $i_{a(avg)}$ is also present in the average d.c.-link current $i_{d(avg)}$, i.e. if the sample-measurement of current i_d is at the right instant. The current of all the phases of the machine can be controlled by just controlling the d.c.-link current $i_{d(avg)}$. Thus, only one current sensor is necessary. One aspect of this method of current sensing, however, must be kept in mind. If the machine is operating, e.g. in forward motoring mode or reverse motoring mode, then in both cases the current $i_{d(avg)}$ will be positive. This creates a problem from a control point of view as the reference current $i_{a(avg)}{}^*$ will be, say, positive for forward motoring mode, but negative for reverse motoring mode. However, this problem can easily be solved in the controller by always multiplying the d.c.-link current signal with the sign (positive or negative) of the controlled supply voltage $v_{p(avg)}{}^*$ signal.

A block diagram of the single sensor current controller is shown in Fig. 7.7. The current controller acts on the comparison between the desired current $i_{a(avg)}{}^*$ and the measured d.c.-link current $i_{d(avg)}$, and then controls the supply output voltage by controlling the signal $v_{p(avg)}{}^*$. The PWM generator outputs the PWM signal in terms of a logic 1 or 0 to the programmable logic device (PLD). The sign (positive or negative) of $v_{p(avg)}{}^*$ is also an logic input (1 or 0) to the logic device. Three logic position signals s_a, s_b and s_c are also inputs to the PLD. These three signals are received from three Hall-effect sensors mounted in the machine. The signals supply information about which two phases must be active according to the rotor position. The outputs of the PLD directly control the switching of the six power semiconductor switches.

Two aspects in the block diagram of Fig. 7.7 have not been referred to yet. The first is that a speed voltage signal is added to the output of the current regulator. This is to decouple the current regulator from the effects of speed variation. The gain, K_i, represents the gain of the solid-state inverter. The second aspect is that the d.c.-link current is *sampled* at certain times to get the average value $i_{d(avg)}$ as shown in Fig. 7.6.

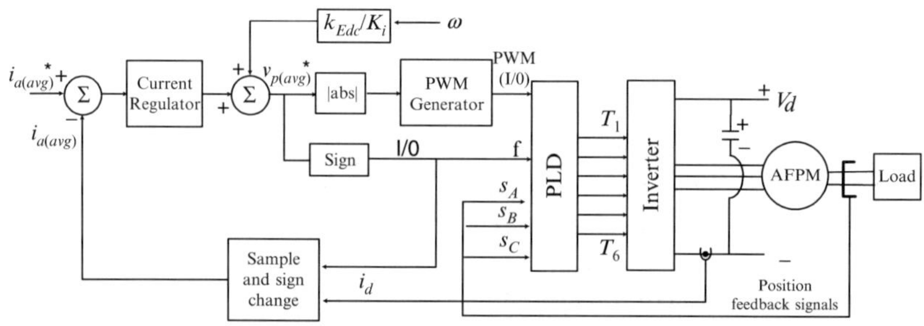

Fig. 7.7. Block diagram of single sensor current controller for trapezoidal AFPM.

To determine the parameters of the current regulator the transfer function of the current control system must be obtained. According to Fig. 7.3b and eqn (7.1), the dynamic voltage equation of the machine in terms of average components is

$$v_{p(avg)} = R_p i_{a(avg)} + L_p \frac{d}{dt} i_{a(avg)} + E_{fL-L} \tag{7.5}$$

where $v_p = v_{p(avg)} + v_r$, $i_a = i_{a(avg)} + i_r$, and v_r and i_r are the ripple voltage and ripple current respectively (see Fig. 7.6). In Laplace transform eqn (7.5) becomes

$$v_{p(avg)}(s) = (R_p + L_p s) i_{a(avg)}(s) + E_{fL-L}(s) \tag{7.6}$$

With the use of the decoupling speed voltage signal in the current control system of Fig. 7.7, the speed voltage term $E_{fL-L}(s)$ of eqn (7.6) has no effect on the current regulator's response. Hence, the control block diagram of the current control system can be represented by the block diagram shown in Fig. 7.8. In this block diagram the current regulator can be, e.g. a gain or a PI-regulator. The gain K_i represents the voltage gain of the solid-state converter (inverter) as mentioned before. The transfer function of the current control system can be determined from the block diagram of Fig. 7.8 and the parameters of the current regulator can be designed according to classical control system methods.

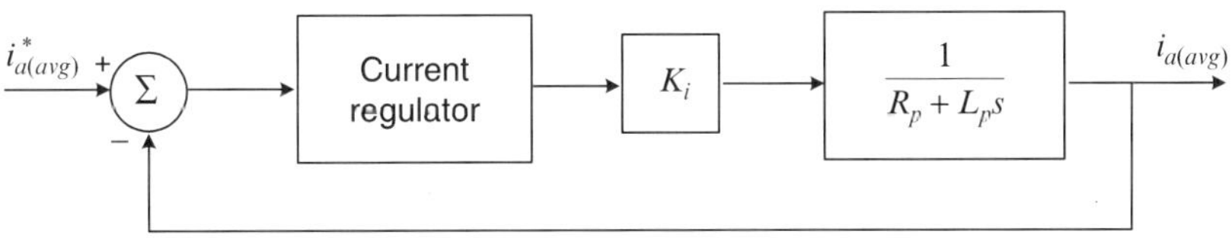

Fig. 7.8. Block diagram of decoupled current control system.

7.1.4 Speed Control

With the current control system in place, as explained in section 7.1.3, a speed control loop can be applied around the current control loop. The electrical and mechanical torque balance equations for the system are as follows:

$$T_{d(avg)}(s) = k_{1dc} i_{a(avg)}(s) \tag{7.7}$$

$$T_{d(avg)}(s) - T_L(s) = (B + Js)\omega(s) \tag{7.8}$$

where $T_L(s)$ is the load torque, B is the equivalent total damping of the system, J is the equivalent total moment of inertia of the system and ω is the rotor speed in rad/s. For a two-pole machine the angular mechanical speed is equal to the angular frequency of the stator currents. Note that eqns (7.7) and (7.8) refer to the instant average components of current, torque and speed, and not to the high frequency ripple components due to the inverter switching. From equations (7.7) and (7.8) and from Fig. 7.8, the speed control block diagram of the system can be represented as in Fig. 7.9. Hence, the transfer function of the speed system can be determined and the parameters of the speed regulator can be designed to obtain a certain speed response. A speed signal can be obtained electronically from the signals of the three Hall-effect position sensors installed in the machine.

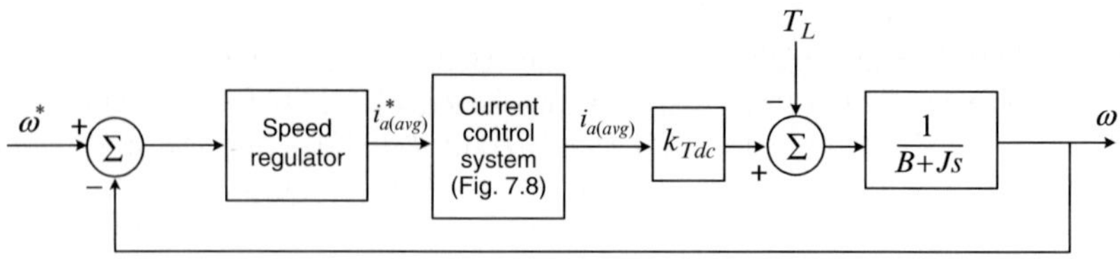

Fig. 7.9. Block diagram of speed control system.

7.1.5 High Speed Operation

Equations (2.125), (7.2) and (7.3) show that as the speed of the machine increases, the EMF E_{fL-L}, the voltage $v_{p(avg)}$ and the duty cycle D will increase till the speed is reached where D becomes unity. The speed where D becomes unity is simultaneously the speed where the output voltage of the converter (inverter) reaches a maximum. This speed is often called the *base speed* of the drive. Field or flux weakening in the high speed region of the machine to keep $E_{fL-L} = constant$ is not possible with high-energy-product surface mounted PMs. Therefore, at speeds higher than the base speed, $E_{fL-L} > V_d$, and control of the current and torque of the machine is lost. In fact, the current and torque rapidly drop to zero at speeds slightly above the base speed. This implies that the speed-range of the trapezoidal AFPM machine using surface mounted PMs is very limited. Techniques to increase the speed substantially beyond base speed are: (a) change smoothly from a 120° current conduction period, as discussed earlier in this chapter, to a 180° six-step conduction period [19, 97, 233] and (b) advance the current wave with respect to the EMF wave [136]; note that during sub-base speed operation the current wave is controlled to be in-phase with the EMF waveform as shown in Fig. 7.2c. The disadvantages of these techniques may be the following: (i) an increase in machine torque ripple, (ii) danger of converter switches being exposed to high DC-link voltages if for some reason the converter switches are switched-off at high speeds (at this condition the high EMFs are rectified by the six diodes of the inverter to a high d.c.-link voltage) and (iii) a higher resolution position sensor is necessary for the smooth advancing of the current wave as the speed increases.

A technique to obtain high speed operation that avoid the disadvantages of the above techniques and obtain true flux weakening is to increase mechanically the air gap of the AFPM machine with increase in speed; this technique is not possible in commercial RFPM machines. However, it is questionable if this technique is economically viable.

7.2 Control of Sinusoidal AFPM Machine

Sinusoidal AFPM machines are designed to have *sinusoidal or near sinusoidal EMF waveforms*. Therefore, in variable speed drive applications where a solid-state converter is used to supply the machine (Fig. 7.1), the output voltage of the converter must be sinusoidal or sinusoidal PWM modulated. Hence, the current in the machine is sinusoidal or else the current regulators of the drive will force the phase currents to be sinusoidal.

This section focuses on the control aspects of the sinusoidal AFPM machine drive. First, the modelling and equivalent circuits of the sinusoidal AFPM machine are described. Secondly, the current and speed regulators of the drive are dealt with. Finally, the hardware of the sinusoidal AFPM machine drive is considered.

7.2.1 Mathematical Model and *dq* Equivalent Circuits

The phase voltage equations of any three-phase a.c. electrical machine can be written as

$$v_{1ABC} = R_1 i_{aABC} + \frac{d\psi_{ABC}}{dt} \tag{7.9}$$

where ψ_{ABC} is the total stator magnetic flux linkage of phase A, B or C – eqn (5.20). Eqn (7.9) is expressed in the so-called ABC stationary reference frame where the circuit variables (voltage, current and flux linkage) are expressed in a reference frame that is fixed to the stationary stator. For synchronous machines, like the AFPM machine, it is very convenient to transform or refer the stationary stator variables to a reference frame that is fixed to the rotor. In this transformation the stationary ABC stator windings of the machine are replaced by fictitious $dq0$ windings that rotate with the rotor.

To explain this, let us consider the cross-section of a two-pole AFPM machine as shown in Fig. 7.10a (side view) and Fig. 7.10b (front view). Fig. 7.10a shows that the PMs are embedded into the rotor steel yoke, but the magnets in the sinusoidal AFPM machine may also be on the surface of the yoke like in a trapezoidal AFPM machine. With the magnets embedded, the AFPM machine is not a pure PM machine anymore, but a combined PM-reluctance machine because the electromagnetic torque has two components: PM synchronous torque and reluctance torque.

Fig. 7.10 explains the *ABC* and *dq* winding layout of the machine. The rotating d and $q-$axis windings are in quadrature of each other and their magnetic axes are aligned with the d and $q-$axis of the rotor. In this case the $d-$axis of the rotor is chosen to be aligned with the centre lines of the PMs and the $q-$axis between the magnets. The position of the rotor and dq-windings with respect to the stationary *ABC* windings is defined in Fig. 7.10 as the angle between the rotating $d-$axis and a defined phase $A-$axis. Note, that in Fig. 7.10b the direction of the flux (Φ) is into (or out of) the paper ($z-$direction), while the current is in the $xy-$plane.

To transfer the circuit variables from the stationary *ABC* reference frame to the $dq0$ reference frame, or vice versa, the *Park transformation*

$$[f_{dq0}] = [K_p]\,[f_{ABC}] \tag{7.10}$$

and inverse Park transformation

$$[f_{ABC}] = [K_p]^{-1}\,[f_{dq0}] \tag{7.11}$$

are used [94], where $[K_p]$ is according to eqn (5.26) and $[K_p]^{-1}$ is the inverse matrix of $[K_p]$. In the above eqns (7.10) and (7.11), f_{dqo} and f_{ABC} are the $dq0$ and ABC variables of voltage, current or flux linkage respectively. In eqn (5.26) the angle θ is the electrical angle defined in Fig. 7.10. With a constant rotor angular velocity ω the angle $\theta = \omega t$. For a balanced three-phase AFPM machine it is only necessary to consider the d and $q-$axis circuits as the zero-sequence circuit is playing no role in the performance of the machine. Applying eqn (7.10) to the voltage eqn (7.9) leads to the dq voltage equations (5.5) and (5.6) of the balanced AFPM machine (omitting the zero-sequence component). In eqns (5.5) and (5.6) ω is the electrical speed of the rotor, and ψ_d and ψ_q are the d and $q-$axis flux linkages. Let us consider the following three cases of operating conditions of the AFPM machine:

(1) The first case is when the machine is open circuit and is mechanically driven at a certain speed ω. The dq stator currents, thus, are zero and only the magnets are active. From Fig. 7.10 it is clear that the magnet flux will only link the $d-$axis winding and not the $q-$axis winding. In this case $\psi_q = 0$ and $\psi_d = \psi_f$, and from eqns (5.5) and (5.6) $v_{1d} = 0$ and $v_{1q} = \omega\psi_f$.

(2) The second case is when both the magnets and the $d-$axis winding are not active, but d.c. (steady-state) current is flowing in the $q-$axis stator winding. The dq stator windings and the rotor are at a rotating speed of ω. Clearly with no magnets and no $d-$axis current, $\psi_d = 0$. With positive d.c. $q-$axis current, i.e. $i_{aq} = \sqrt{2}I_{aq}$, then $\psi_q = \sqrt{2}L_{sq}I_{aq}$ where L_{sq} is the

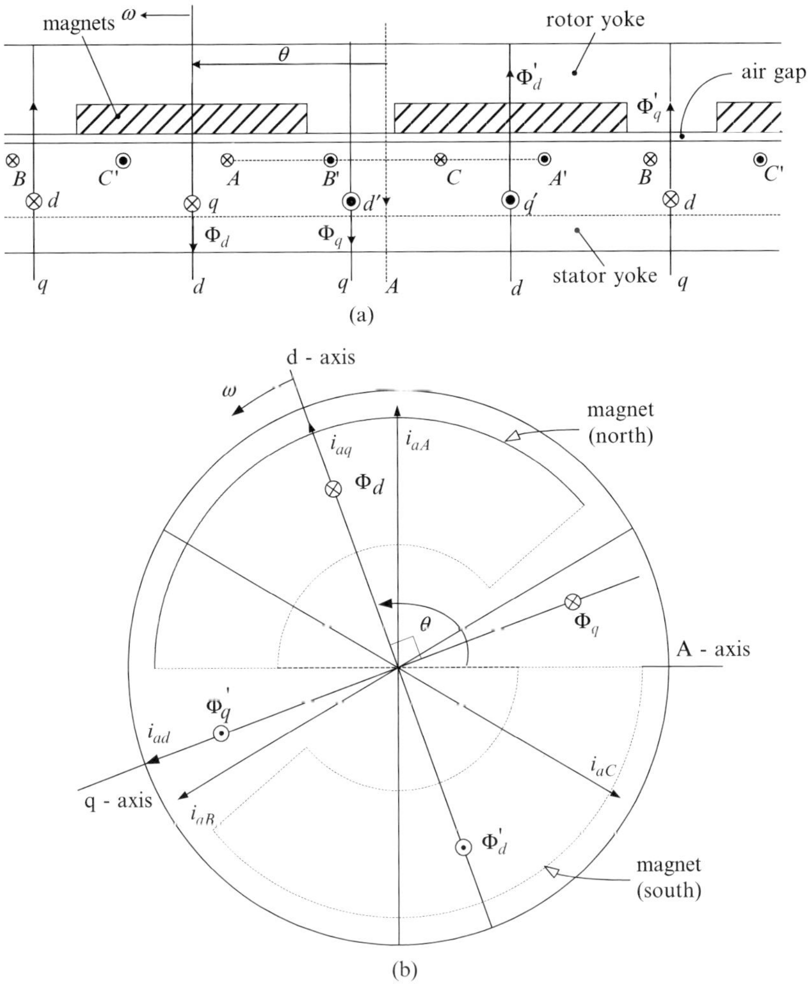

Fig. 7.10. Stationary *ABC* and rotating *dq* windings of a simple two-pole AFPM machine: (a) side view; (b) front view (PMs are shown in dotted lines).

$q-$axis winding synchronous inductance. Note that the relation between ψ_q and i_{aq} is nonlinear due to magnetic saturation; L_{sq} will thus vary with saturation. From eqns (5.9) and (5.10) it follows in this case that $v_{1q} = R_1 i_{aq}$ and $v_{1d} = -\omega L_{sq} i_{aq}$.

(3) In the third case only the $d-$axis winding is active in terms of a positive d.c. current flowing of $i_{ad} = \sqrt{2} I_{ad}$. Therefore $\psi_q = 0$ as the magnets are not active and $i_{aq} = 0$, and $\psi_d = \sqrt{2} L_{sd} I_{ad}$ where L_{sd} is the synchronous inductance of the $d-$axis winding. Again, the relation between ψ_d and i_{ad}

is nonlinear and L_{sd} will vary with saturation. In this case from eqn (5.9) and (5.10) $v_{1q} = \omega L_{sd} i_{ad}$ and $v_{1d} = R_1 i_{ad}$.

If the magnets and the dq stator windings are active, then the above three cases can be summarized by eqns (5.7) and (5.8). From eqns (5.5) to (5.10) it follows that the electromagnetic power of a three-phase machine is given by

$$P_{elm} = \frac{3}{2}(i_{aq}\omega\psi_d - i_{ad}\omega\psi_q) = \frac{3}{2}\omega(\psi_d i_{aq} - \psi_q i_{ad}) \tag{7.12}$$

The factor 3/2 has been explained in Section 5.13. The developed torque of the machine is expressed by eqn (5.13). The first term in eqn (5.13) expresses the torque component due to PM excitation, while the second term expresses the reluctance torque component. For surface mounted AFPM machines it can be assumed that $L_{sd} = L_{sq}$, hence the torque is directly proportional to i_{aq}. In the case of an embedded AFPM machine like in Fig. 7.10a it is clear that $L_{sd} < L_{sq}$, which implies that i_{ad} in eqn (5.13) must be negative in order to generate a positive reluctance torque. Fig. 7.11a shows the phasor diagrams of the motor for $i_{ad} = 0$ while Fig. 7.11b shows the phasor diagram of the motor at unity power factor.

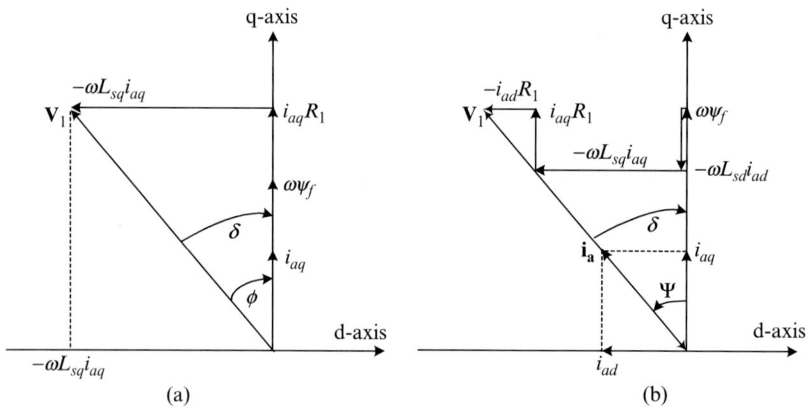

Fig. 7.11. Phasor diagrams for motoring action of overexcited AFPM machine: (a) for $i_{ad} = 0$; (b) at unity power factor. Note that δ is negative for motoring action and positive for generation action.

The inductances in the d and $q-$axis windings are expressed by eqns (5.17) and the magnet flux linkage ψ_f is further expressed as

$$\psi_f = L_{ad}\, i_f \tag{7.13}$$

where L_{ad} is the armature reaction (mutual) inductance in the d-axis and i_f is a fictitious magnetising current representing the PM excitation.

The flux linkage and mutual inductance between the $d-$axis winding and the fictitious permanent magnet (field) winding need further explanation. For the $d-$axis winding we have the following equation (ignoring cross coupling effects with the $q-$axis winding and ignoring any stator slotting effects):

$$\begin{aligned}\frac{d\psi_d}{dt} &= L_{sd}\frac{di_{ad}}{dt} + L_{fd}\frac{dI_f}{dt} \\ &= (L_l + L_{ad})\frac{di_{ad}}{dt} + L_{fd}\frac{dI_f}{dt}\end{aligned} \tag{7.14}$$

where L_{ad} is given by eqn (2.116) and L_l is a leakage inductance. I_f is the fictitious magnet (field) winding current. L_{fd} is the mutual inductance between the $d-$axis winding and the fictitious magnet (field) winding given by

$$L_{fd} = \frac{k_{wf}N_f\Phi_{df}}{I_{ad}} = \frac{k_{w1}N_1\Phi_{fd}}{I_f} \tag{7.15}$$

where $k_{wf}N_f$ and $k_{w1}N_1$ are the effective turns in series of the fictitious magnet-field and $d-$axis windings respectively. Φ_{df} and Φ_{fd} are the proportions of flux produced by the $d-$axis and magnet-field windings respectively that link with the opposite winding. Note that in this case Φ_{fd} is equal to Φ_f of eqn (2.26). From eqn (7.15) the flux linkage of the $d-$axis winding due to the field winding, ψ_f, is given by

$$\psi_f = k_{w1}N_1\Phi_{fd} = L_{fd}I_f \tag{7.16}$$

However, this flux linkage is also expressed by eqn (7.13) and is represented in the $d-$axis equivalent circuit of Fig. 7.12b. Hence, from eqns (7.13) and (7.16)

$$i_f = \frac{L_{fd}}{L_{ad}}I_f \tag{7.17}$$

Furthermore, the $d-$axis inductance L_{ad} is given by eqn (2.116) as

$$L_{ad} = \frac{k_{w1}N_1\Phi_{ad}}{I_{ad}} \tag{7.18}$$

where Φ_{ad} is the $d-$axis flux passing the air gap, linking the rotor. The total flux of the $d-$axis winding, Φ_d, is given by

$$\Phi_d = \Phi_l + \Phi_{ad} \tag{7.19}$$

where Φ_l is the leakage flux. Now, if Φ_{df} of eqn (7.15) [this is the proportion of the flux produced by the $d-$axis winding that links the field winding] is equal to $k\Phi_{ad}$, i.e. $\Phi_{df} = k\Phi_{ad}$, where $0 < k < 1$ (i.e. not all the $d-$axis flux linking the rotor necessarily links the fictitious magnet-field winding), then from eqns (7.15) and (7.18) it follows that

$$L_{ad} = \left[\frac{k_{w1}N_1}{k_{wf}N_f k}\right] L_{fd} \tag{7.20}$$

and furthermore from eqn (7.17) that

$$i_f = \left[\frac{k_{wf}N_f k}{k_{w1}N_1}\right] I_f \tag{7.21}$$

Hence, if the effective turns of the $d-$axis and field windings are the same and $k = 1$, then L_{ad} is equal to L_{fd} and i_f is equal to I_f.

Using eqns (5.7), (5.8) and (7.13) and assuming i_f is not varying with time, eqns (5.5) and (5.6) can be written as

$$v_{1d} = R_1 i_{ad} + (L_l + L_{ad})\frac{di_{ad}}{dt} - \omega(L_l + L_{aq})i_{aq} \tag{7.22}$$

$$v_{1q} = R_1 i_{aq} + (L_l + L_{aq})\frac{di_{aq}}{dt} + \omega(L_l + L_{ad})i_{ad} + \omega L_{ad} i_f \tag{7.23}$$

On the basis of eqns (7.22) and (7.23) the complete dq equivalent circuits of the sinusoidal AFPM machine can be drawn as shown in Fig. 7.12. It should be noted that the modelling according to eqns (7.22) and (7.23) is based on assumptions that

(a) there are no core losses;
(b) the magnetic circuit is linear, i.e. there is no saturation of the magnetic circuit;
(c) there is no cross magnetisation or mutual coupling between the $d-$ and $q-$axis circuits;
(d) there are no stator slotting effects (smooth stator core).

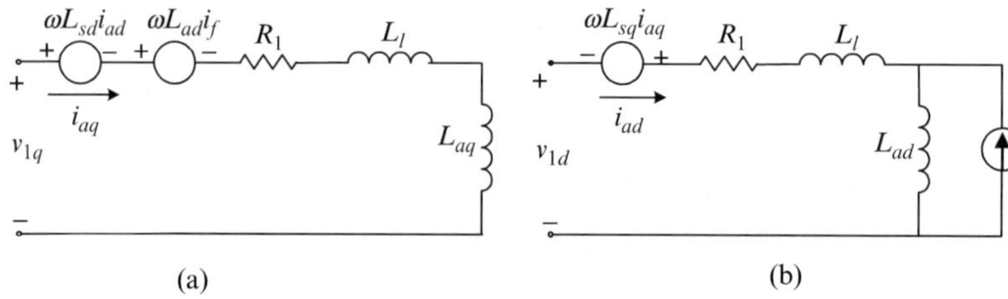

Fig. 7.12. The d and $q-$axis equivalent circuits of the sinusoidal AFPM brushless machine.

7.2.2 Current Control

The equivalent circuits of the sinusoidal AFPM machine shown in Fig. 7.12 can be used to design the current regulators for controlling the d and $q-$axis currents of the machine. Applying the principle of speed voltage decoupling, as for the trapezoidal AFPM machine (section 7.1.3), the dq current control system of the sinusoidal AFPM machine can be represented by the block diagram of Fig. 7.13. Here, the d and $q-$axis current regulators act on the errors between the desired and actual dq currents of the machine. The speed voltage terms are added (or subtracted) to the outputs of the regulators to determine the dq machine voltages. Within the machine model the speed voltages are subtracted (or added) again, hence the regulators act only on the $RL-$parts of the equivalent circuits. The block diagrams of the decoupled current controllers can therefore be simplified to those shown in Fig. 7.14. With the equivalent circuit parameters of the machine known, the transfer functions of the d and $q-$axis current controllers can be determined and the d and $q-$axis current regulators can be designed accordingly to obtain the required dq current response.

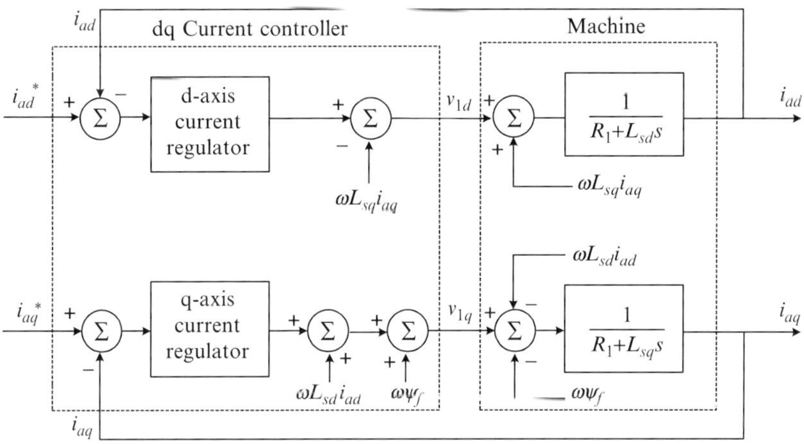

Fig. 7.13. Block diagram of d and q-axis current control system of the AFPM drive implementing speed voltage decoupling.

7.2.3 Speed Control

It is always good in terms of efficiency to control the drive at maximum torque per ampere. For the surface mounted (nonsalient) AFPM machine, where it can be taken that $L_{sd} = L_{sq}$, it is clear from eqn (5.13) that maximum torque per ampere for this machine is obtained by keeping $i_{ad} = 0$. However, for

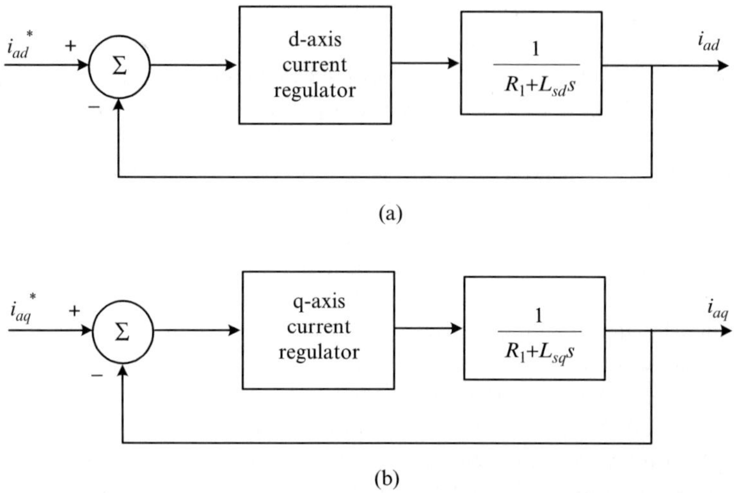

Fig. 7.14. Decoupled d and q−axis current control systems.

embedded (salient-pole) AFPM machines where $L_{sd} \neq L_{sq}$, maximum torque per ampere is obtained by making i_{ad} negative, that is by controlling the current phasor at some *current angle* Ψ as shown in Fig. 7.15 (see also Fig. 7.11b). Note from Fig. 7.15 that negative torque is obtained by keeping both i_{ad} and i_{aq} negative, or in the case of the nonsalient machine, only i_{aq} is negative. In the sub base speed region ($\omega < \omega_b$) of the drive, where there is enough voltage available so that the current regulators do not saturate, the machine is controlled to operate at maximum torque per ampere. For the nonsalient AFPM machine this implies that the current angle $\Psi = 0$, while for the salient-pole machine $\Psi = \Psi_0$ (see Fig. 7.15), which is an (average) optimum angle for all load conditions.

However, in the high speed region ($\omega > \omega_b$), the EMF voltage $\omega L_{ad} i_f = \omega \psi_f$ of eqn (7.23) (see also Fig. 7.11b) becomes larger than the possible output voltage of the inverter. The current regulators therefore start to saturate, current control is lost and the torque of the machine rapidly drops as the speed increases. To avoid this torque decrease and to ensure a wider high-speed region one has to deviate from the maximum torque per ampere control and start to inject a larger negative d−axis current to reduce the q−axis terminal voltage (Fig. 7.11b) as the speed increases. This implies that the current angle Ψ must be increased as shown in Fig. 7.15 such that always

$$2\omega\sqrt{(L_{sd}i_{ad} + L_{ad}i_f + i_{aq}R_1)^2 + (L_{sq}i_{aq} + i_{ad}R_1)^2} < V_d$$

where V_d is is the dc-link voltage of the inverter, and keeping the current amplitude $I_{am} = \sqrt{i_{ad}^2 + i_{aq}^2}$ equal or less than the rated current amplitude depending on the load. At a maximum current angle, Ψ_m, the speed range can be extended much more if the current angle is held constant at $\Psi = \Psi_m$ and the amplitude of the phase current is reduced as shown in Fig. 7.15. Note from Fig. 7.11b that as the current angle Ψ increases the apparent power of the machine decreases and the power factor improves.

The above current control strategy can be used for both salient and non-salient AFPM machines. It must be noted, however, that the speed range obtained with the above control method depends very much on the value of the stator phase inductance of the machine (see *Numerical example 7.2*). If the phase inductance is relatively small (this implies a low per unit stator phase reactance), as is the case with slotless and air-cored AFPM machines, the high speed range of the machine is very limited.

From above, thus, the current of the machine can be controlled by controlling the magnitude of the current and keeping the current angle constant at an optimum value, and advancing the current angle at higher speeds. To simplify this control method, the current angle Ψ is defined as a positive angle between 0 to 90°. The current angle is also taken as a function of the *absolute value* of the speed, $|\omega|$. Furthermore, a current i_T is defined, which controls the torque of the machine and is proportional to the current amplitude of the machine. Hence,

$$\begin{aligned} i_{ad} &= -|i_T| \sin \Psi \\ i_{aq} &= i_T \cos \Psi \quad \text{for} \quad \Psi \geq 0 \end{aligned} \tag{7.24}$$

Substituting eqn (7.24) into eqn (5.13) and noting that $L_{sd} \leq L_{sq}$, the electromagnetic torque is

$$T_d = \frac{3}{2} p \psi_f i_T \cos(\Psi) + \frac{3}{4} p (L_{sq} - L_{sd}) i_T |i_T| \sin(2\Psi) \tag{7.25}$$

It can be seen from the first term of eqn (7.25) that there is a linear relation between the PM torque component and the current i_T for a given current angle Ψ. The second term of eqn (7.25) is the reluctance component of the torque. It is found for reluctance machines [143] that the inductance difference $\Delta L = L_{sq} - L_{sd}$ is not a constant at a certain current angle, but varies inversely with load current i_T due to saturation and cross magnetization. The cross magnetization or cross coupling is the magnetic coupling between the fictitious d and q–axis windings of the machine. It is thus a fairly good approximation to assume for this component of the torque a linear relation between reluctance torque and current i_T for a given current angle Ψ [145]. Thus, eqn (7.25) can be simplified as

$$T_d \approx k_{Tf} i_T + k_{Trel} i_T \approx k_T i_T \quad \text{for} \quad \Psi = constant \tag{7.26}$$

where k_{Tf} and k_{Trel} are the PM torque and reluctance torque constants, respectively, and $k_T = k_{Tf} + k_{Trel}$. Note from eqn (7.26) that to generate a

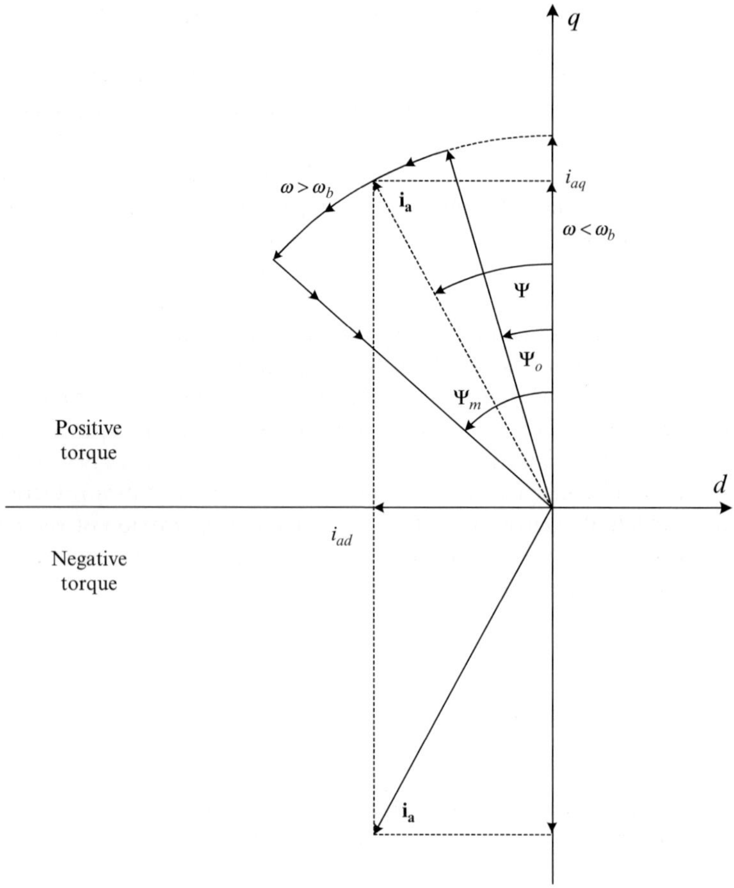

Fig. 7.15. Current phasor control of the AFPM drive in the low and high speed regions of the machine.

negative torque, i_T must be negative; Ψ is taken in the control as a positive quantity between 0 to 90°. In the case where $\Psi = 0$, then $k_{Trel} = 0$, $k_T = K_{Tf} = 3p\psi_f/2$ and $i_T = i_{aq}$.

With the current control strategy in place and with the torque as a function of current as given by eqn (7.26), a speed controller for the AFPM machine drive can be implemented around the current controller as shown in Fig. 7.16. The block diagram of Fig. 7.16 describes the basic principle of the current and speed control of the AFPM machine drive. The speed regulator acts on the error between the desired and measured speed of the system and controls the current i_T of eqn (7.26). The maximum current of the machine is limited by the limiter as shown in Fig. 7.16. The current angle Ψ is, as a positive quantity, a function of the absolute value of the speed of the machine, $|\omega|$, as shown in Fig. 7.16. In the low speed region the angle Ψ is kept constant. In the speed region above the base speed ($\omega > \omega_b$) the angle Ψ is increased with speed. In the very high speed region the angle Ψ is kept constant again. With the current and current angle Ψ known, the desired dq currents, ${i_{ad}}^*$ and ${i_{aq}}^*$, are determined by eqn (7.24). With the measured speed and measured dq currents as the other inputs, the current controller outputs the necessary dq voltages according to the block diagram of Fig. 7.13. These voltages are transformed to ABC voltages using the inverse transformation given by eqn (7.11), and are then put through the PWM generator to control the inverter and hence the terminal voltages of the machine.

To design the speed regulator of the system of Fig. 7.16 a few approximations can be made to simplify the model of the speed control drive. Firstly, the mechanical time constant of drive systems is in general much longer than the electrical time constant of the drive so that the latter can be ignored. Together with this, a good approximation is to assume that the current controllers are forcing the actual currents equal to the command (desired) currents, i.e. $i_{ad} = {i_{ad}}^*$, $i_{aq} = {i_{aq}}^*$, and $i_T = {i_T}^*$. Applying further eqn (7.26), the speed control system can be simplified to that shown in Fig. 7.17. From the simplified control system the approximate transfer function of the system can be determined and the speed regulator can be designed to obtain the required speed response.

It must be noted that the speed control method of Fig. 7.16 is only one way of controlling the speed of the PM brushless motor and is recognized as only a basic method; the method e.g. has the disadvantage that in the high speed regions under low-load conditions the drive is not necessarily operating at best (highest efficiency) operating conditions. Several other techniques and algorithms for wide-range speed control of PM motor drives have been proposed [24, 57, 137, 164, 179, 196, 280] and can be applied to the speed control of the sinusoidal AFPM brushless motor drive. The important aspect is the prevention of saturation of the dq current regulators in the high speed region of the motor.

7.2.4 Hardware of Sinusoidal AFPM Machine Drive

For the sinusoidal AFPM brushless motor the same solid-state converter as for the trapezoidal AFPM brushless motor can be used (Fig. 7.4). A more detailed diagram of the hardware of the sinusoidal AFPM brushless drive system is shown in Fig. 7.18. The fixed a.c. supply voltage is rectified by means of a diode rectifier or voltage-source controlled (active) PWM rectifier to obtain a fixed d.c. voltage. The d.c. voltage is then inverted by means of a voltage-source PWM inverter to a three-phase a.c. voltage of variable frequency and magnitude. The difference between the trapezoidal and sinusoidal drives lies in the PWM switching-control of the inverter bridge. In the case of the trapezoidal AFPM motor drive, quasi-square PWM voltages are applied, whereas

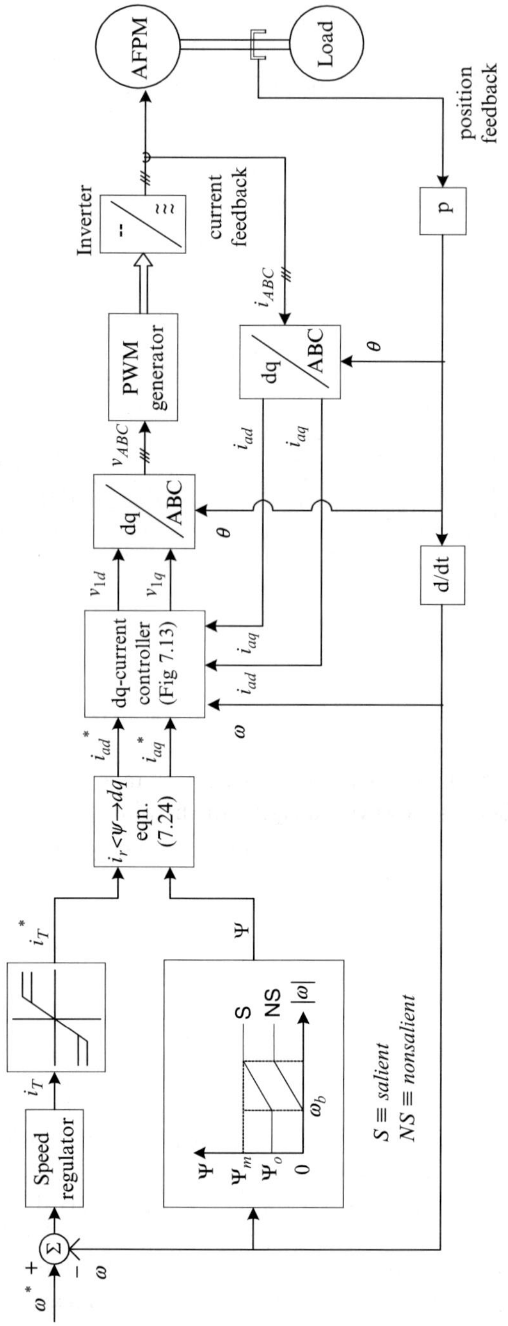

Fig. 7.16. Basic current and speed control system of the converter-fed sinusoidal AFPM drive.

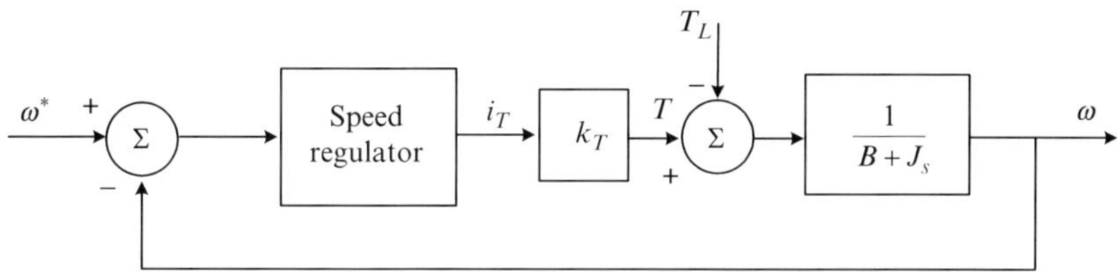

Fig. 7.17. Block diagram of simplified speed control system.

in the case of the sinusoidal AFPM motor drive, sinusoidal PWM voltages are applied and the phase currents of the stator are sinusoidally regulated as described in the previous sections.

The peak value of the fundamental frequency voltage component of the inverter-legs of Fig. 7.18 with respect to the neutral point N is clearly $\frac{1}{2}V_d$. As the fundamental frequency voltage components of the inverter-legs are 120° out of phase, the maximum *rms* line–to–line voltage of the inverter at the fundamental frequency with sinusoidal PWM is

$$V_{1L-Lmax} = \frac{\sqrt{3}}{\sqrt{2}}\frac{V_d}{2} = 0.612V_d \tag{7.27}$$

This result shows the limitation on the maximum *rms* output voltage of the inverter. This affects, amongst other things, the maximum speed at which rated torque can be delivered by the drive (see *Numerical example 7.2*).

The currents in the three phases of the AFPM machine drive are sinusoidally regulated. There is, however, a ripple in the phase currents due to the PWM switching. It can be shown that the peak–to–peak ripple in the phase current, Δi_{p-p}, is directly proportional to

$$\Delta i_{p-p} \propto \frac{V_d}{L_{ph}f_s} \tag{7.28}$$

where f_s is the switching frequency of the inverter and L_{ph} the total load inductance per phase of the inverter. The ripple current is thus inversely proportional to L_{ph} and f_s. Note that L_{ph} can be increased by the addition of an external inductor in series with the stator (armature) of the AFPM machine as shown in Fig. 7.18. Careful consideration should be given to the selection of L_{ph} and f_s. As the air gaps of AFPM machines, specifically those AFPM machines with coreless and slotless iron-cored stators, are relatively large, the internal phase inductances of these machines are relatively small. Thus to keep the ripple current in these machines within limits, the switching frequency f_s is required to be high. With power MOSFETs the switching frequency can be made high (say $f_s \leq 50$ kHz), but with IGBTs in the medium–to–large power range the switching frequency is very much limited to $f_s \leq 10$ kHz. In such a case the phase inductance must be increased. The high ripple current that can occur in most of the AFPM machine drives is often overlooked. A high ripple current has serious consequences on the current rating of the inverter and the efficiency, stator winding temperature, torque quality and acoustic noise of the AFPM brushless motor.

For rotor position feedback of the sinusoidal AFPM drive a position sensor with high resolution is necessary. Resolvers are often used for this as they are robust and can be used in harsh environments. Integrated and analog circuitry are necessary to interface the resolver with the digital controller (see Fig. 7.18). Digital interface outputs of up to 16 bit resolution are possible. Normally 12 bit resolution gives more than enough accuracy for position feedback.

All the measured information of voltage, current and rotor position are fed to the digital controller as shown in Fig. 7.18. The controller outputs PWM signals via fibre optic links to the inverter bridge. Digital signal processors (DSPs) are nowadays frequently used as drive controllers [254]. Examples of dedicated fixed point DSPs from *Texas Instruments* include TMS320F240 (20 MHz), TMS320LF2407A (40 MHz), TMS320F2812 (150 MHz). These DSPs have 12 or 16 PWM channels, 16 A/D channels (10 or 12 bit resolution) and operate at frequencies of between 20 to 150 MHz. In another recent development on drive controllers *International Rectifier* introduced their motion control chipset IRACO210 [109]. This chipset uses a high speed configurable field programmable gate array (FPGA), tightly coupled to gate driver and current sense interface chips. In this way flexible drive control algorithms are implemented in FPGA hardware, rather than in software.

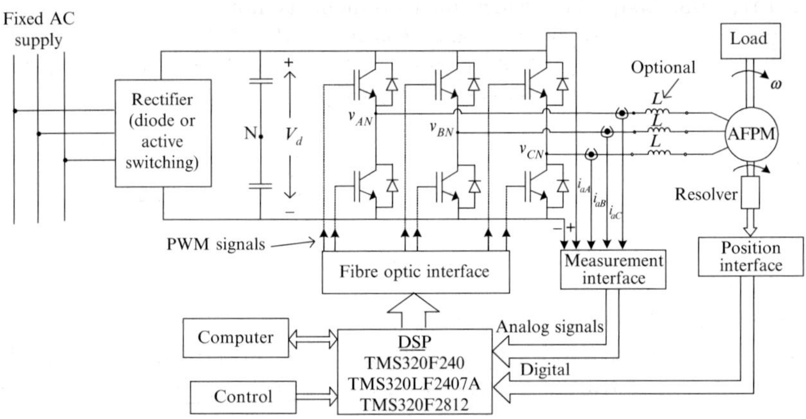

Fig. 7.18. Hardware of sinusoidal AFPM drive system.

7.3 Sensorless Position Control

It is obvious that information of the rotor position of the AFPM brushless motor is absolutely necessary for proper current and speed control. For trapezoidal AFPM motors a *low resolution position sensor* may be used, but for sinusoidal AFPM motors a *high resolution position sensor* is necessary. The use of a position sensor that is mechanically connected to the machine and electrically connected to the controller reduces the electromechanical drive system reliability and increases its cost. Moreover, in some applications insufficient space may make the use of a position sensor difficult. *Sensorless position control* techniques, in which an accurate information of the rotor position can be obtained without the need of a mechanical sensor, are superior.

Various methods and techniques have been proposed the last two decades for the sensorless position control of PM brushless motors. Most of these methods are based on the measurement of the voltage and/or current of the machine. Thus, instead of using a mechanical sensor, electrical sensors (not mounted in the machine but placed at the solid-state converter) are used to obtain the rotor position signal. In this way, the position of the rotor is determined indirectly, as measured electrical quantities together with the model of the machine are used. State observers, filters and other calculations are used to estimate the mechanical position of the rotor.

The methods of sensorless control can be classified into two types: (i) methods that are based on the back EMF estimation [23, 90, 271], and (ii) methods that make use of the magnetic saliency present in the machine [138, 167, 168, 234]. The EMF measurement is not suitable for low and zero speed operation of the motor as the induced voltage is proportional to the rotor speed. At stand still, the induced voltage is zero and there is no position-information anymore in the stator quantities. The second method, which makes use of the magnetic saliency, is more suitable for the detection of the rotor position at low speeds and standstill. With this method either test voltage pulses are applied to the machine or additional high-frequency signals superimposed on the fundamental voltages (or currents) are injected. The injection of the revolving high-frequency voltage vector is commonly used. All these methods can be used successfully for the position sensorless control of AFPM motor drives [205]. However, it is still questionable if the high-frequency injection sensorless control method can be applied to coreless type AFPM brushless motors as these machines have practically no magnetic saliency.

Fig. 7.19 explains the basic principle of the *high frequency voltage injection sensorless position control* technique. High frequency voltage signals are injected to the system by adding them to the *dq* voltage signals v_{1d} and v_{1q} of the drive. The current response of the drive system to these voltage signals is monitored by band pass filtering (BPF) the *dq* currents, so as to obtain the high frequency current components i_{adh} and i_{aqh}. These high frequency current components contain information about the rotor position as they are affected by the magnetic saliency of the machine. These currents, together with the high frequency injected signals are used by an observer or position estimator to estimate the rotor position θ_m. A low pass filter (LPF) is furthermore used to extract from the dq currents the fundamental current components i_{ad} and i_{aq}, which are necessary for the fundamental current controller of the drive (Fig. 7.13). The frequency of the injected voltage signals must be much higher than the fundamental frequency, but also much lower than the switching frequency of the inverter, and is typically between 500 Hz and 2 kHz.

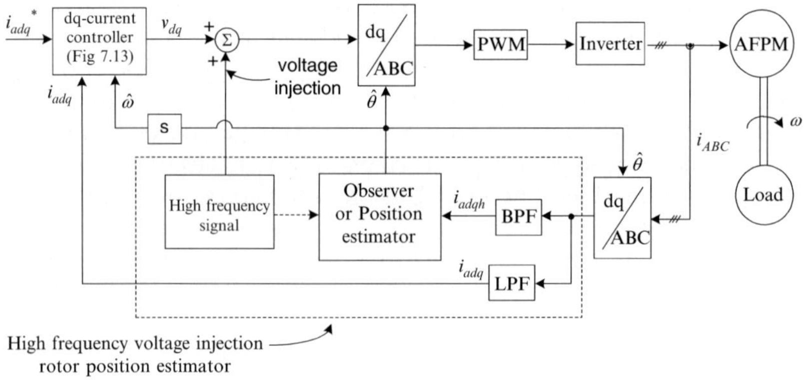

Fig. 7.19. High frequency voltage injection sensorless position control.

Numerical Example 7.1

A small, 35-W, 4300 r/min, 12 pole, Y-connected air-cored, trapezoidal AFPM brushless motor is used as part of a reaction wheel for a micro-satellite (see Fig. 7.20). Reaction or momentum wheels are used on satellites to keep the satellite steady and to orientate it. Orientation control is necessary as, e.g. the photovoltaic panels must be pointed at all times to the sun for maximum power generation. The motor has a stator phase inductance of $L_p = 3.2$ μH and a phase resistance of $R_1 = 22$ mΩ. The motor is fed by a 3-phase MOSFET inverter with $V_d = 14$ V, and is under square-wave current control as described in Section 7.1.3. Assuming the voltage drop across a MOSFET switches during switching-on as 1 V, calculate:

(a) the switching frequency f_s of the inverter;
(b) the external phase inductance required for a maximum peak–to–peak ripple current less than $\Delta i_{rmax} = 0.5$ A if $f_s = 40$ kHz.

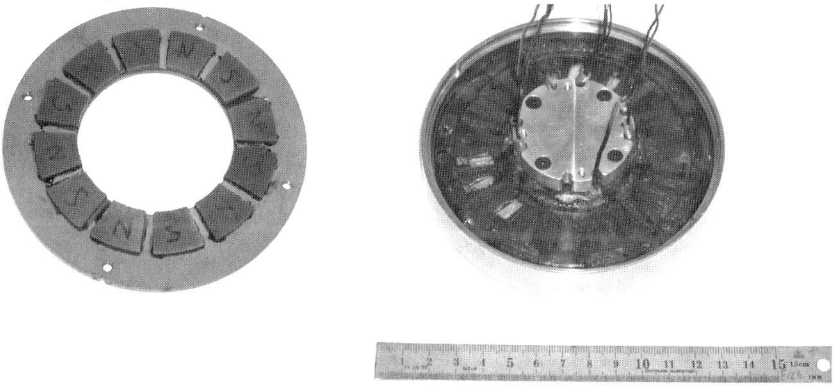

Fig. 7.20. Small coreless AFPM brushless motor. Photo courtesy of the *University of Stellenbosch*, South Africa.

Solution

First, an approximate equation for the maximum peak–to–peak phase current ripple of the motor must be derived. For the circuit shown in Fig. 7.5 during switching-on, i.e. with the transistors T_u and T_y switched on and assuming ideal transistor switches and drive in the steady-state, the voltage equation (7.1) can be written as

$$V_d = R_p I_a + R_p i_r + L_p \frac{di_r}{dt} + E_{fL-L} \tag{7.29}$$

where the phase current $i_a = I_a + i_r$, I_a is the steady-state average current and i_r is the ripple current. Putting eqn (7.3) and neglecting the voltage drop $R_p i_r$, eqn (7.29) can be brought to

$$V_d - DV_d \approx L_p \frac{di_r}{dt} \tag{7.30}$$

From eqns (7.4) and (7.30) the peak–to–peak ripple current Δi_r is

$$\Delta i_r \approx \frac{V_d(1-D)t_{on}}{L_p} = \frac{V_d(1-D)D}{L_p f_s} \tag{7.31}$$

The maximum peak–to–peak ripple current will occur at a duty cycle of $D = 0.5$, hence

$$\Delta i_{rmax} \approx \frac{V_d}{4L_p f_s} \tag{7.32}$$

(a) Switching frequency. Using eqn (7.32) and taking into account the voltage drop across the transistor switches, the switching frequency for a maximum ripple current of 0.5 A is

$$f_s \approx \frac{V_d - 2}{4L_p \Delta i_{rmax}} = \frac{14 - 2}{4. \times (2 \times 3.2 \times 10^{-6}) \times 0.5} = 937.5 \text{ kHz}$$

(b) External inductance. Eqn (7.32) gives the following additional phase inductance required to keep the ripple current less than 0.5 A at $f_s = 40$ kHz

$$L_{ph(add)} \approx \frac{V_d - 2}{4 f_s \Delta i_{rmax}} - L_s = \frac{14 - 2}{4 \times 40\,000 \times 0.5} - 3.2 \times 10^{-6} = 71.8 \ \mu\text{H} \tag{7.33}$$

These results show the low phase inductance problem of coreless AFPM brushless motors when fed from solid-state inverters. Either the switching frequency of the inverter must be high or relatively large external inductors must be added in series with the phase windings of the machine to keep the ripple current within limits. Usually, external inductances are used, as a high switching frequency causes the power losses in the inverter to be high.

Numerical Example 7.2

Consider a 75-kW, 460-V, Y-connected, 100-Hz, 1500 r/min, 8-pole AFPM motor with single PM rotor disc and double slotted stators connected in series. The equivalent circuit parameters and flux linkage of this machine are: $R_1 = 0.034\ \Omega$, $L_{sd} = L_{sq} = 2.13$ mH, $\psi_f = 0.538$ Wb. The machine is under dq current control by using amongst other things a three-phase IGBT voltage-source inverter with d.c.-link voltage $V_d = 755$ V. Find:

(a) the maximum speed and the developed power of the drive with the machine at rated current and the machine controlled at maximum torque per ampere;
(b) the per unit high speed of the drive developing the same power as in (a) with the drive at rated current, rated voltage and using an advanced current angle.

Solution

(a) Maximum speed and electromagnetic power

With a surface-mounted AFPM machine maximum torque per ampere is obtained with $i_{ad} = 0$ A. Thus, at rated stator current $i_{aq} = \sqrt{2} \times 105 = 148.5$ A. To obtain maximum speed the output voltage of the inverter must be at a maximum, i.e. from eqn (7.27) the *rms* phase voltage $V_1 = 0.612 \times 755/\sqrt{3} = 266.8$ V. Thus, the peak phase voltage, $V_{1m} = \sqrt{2} \times 266.8 = 377.3$ V or simply $1/2V_d$, and note that $V_{1m} = \sqrt{{v_{1d}}^2 + {v_{1q}}^2}$. By considering the drive in the steady state and ignoring the resistive voltage drop, from eqn (7.22), eqn (7.23) and Fig. 7.11 we have

$$v_{1d} = V_{1m} \sin\delta \approx -\omega L_{sq} i_{aq} \tag{7.34}$$

$$v_{1q} = V_{1m} \cos\delta \approx \omega \psi_f \tag{7.35}$$

or else

$$\tan\delta \approx \frac{-L_{sq} i_{aq}}{\Psi_f} = \frac{-2.13 \times 10^{-3} \times 148.5}{0.538} \quad \text{and} \quad \delta \approx -30.45^{\text{o}}$$

Using eqn (7.34) the angular frequency is

$$\omega \approx \frac{V_{1m} \sin\delta}{-L_{sq} i_{aq}} = \frac{377.2 \times \sin(-30.45)}{-2.13 \times 10^{-3} \times 148.5} = 604.5 \text{ rad/s}$$

and the mechanical speed $n = 30\omega/p\pi = 1443$ rpm. The maximum speed with maximum torque per ampere control is thus just less than the rated speed of 1500 r/min of the machine. Using eqn (7.34), an approximate equation for the steady-state electromagnetic power (7.12) can be found as follows:

$$P_{elm} = \Omega T_d = \frac{3}{2}\,\psi_f\;\omega i_{aq} \approx \frac{3}{2}\,(\frac{\psi_f V_{1m}}{-L_{sq}}) \sin\delta \tag{7.36}$$

Thus,

$$= \frac{3}{2} \times \frac{0.538 \times 377.3}{-2.13 \times 10^{-3}} \times \sin(-30.45) \approx 72.45 \text{ kW}$$

which is also just less than the rated power of 75 kW of the machine.

(b) Per unit high speed.

The speed of the AFPM brushless motor can be increased beyond base speed by increasing the current angle Ψ, as shown in Fig. 7.11(b), keeping the peak value of the sinusoidal phase current $I_{am} = constant$. Note that $I_{am} = \sqrt{{i_{ad}}^2 + {i_{aq}}^2}$.

At rated conditions $I_{am} = 148.5$ A and $V_{1m} = 377.3$ V. The same power as in (a), i.e. 72.45 kW must be obtained. Hence, it is clear from eqn (7.36) that δ will be the same as in (a), i.e. $\delta = -30.45^{\text{o}}$. With the machine considered as 100% efficient, the electromagnetic power is equal to the input power

$$P_{elm} \approx \frac{3}{2} V_{1m} I_{am} \cos\phi$$

where

$$\cos\phi \approx \frac{72\,450}{(3/2)\times 377.3\times 148.5} = 0.862 \quad \text{and} \quad \phi \approx 30.45^{\circ}\ \text{(leading)}$$

The q−axis current i_{aq} can now be calculated as

$$i_{aq} = I_{am}\cos(|\delta|+\phi) \approx 148.5\cos(30.45+30.45) = 72.2\ \text{A}$$

$$\Psi = |\delta| + \phi = 30.45 + 30.45 = 60.9^{\circ}$$

Using eqn (7.34) that is still valid, the angular frequency

$$\omega \approx \frac{V_{1m}\sin\delta}{-L_{sq}i_{aq}} = \frac{377.3\times\sin(-30.45)}{-2.13\times 10^{-3}\times 72.2} = 1243\ \text{rad/s}$$

and the mechanical speed $n = 30\omega/p\pi = 2967$ rpm. Taking the mechanical speed 1443 rpm in (a) as the base speed, the per unit speed is $2972/1443 = 2.0$ p.u. Hence, the same power can be developed by the drive at two times base speed by advancing the current angle Ψ from zero to almost 61°. It must be noted that this increase in speed is very much dependent on the q-axis synchronous inductance L_{sq} as can be seen clearly from all the equations used above. If L_{sq} becomes relatively small, the speed increase will be much less. The per unit phase reactance of the above iron-cored, slotted AFPM machine is $x_s = 0.53$ p.u. For slotless and coreless AFPM machines the per unit phase reactance is much less (typically 0.1 p.u.) and the speed increase will be little.

Numerical Example 7.3

Simulate the dq and ABC current response of the drive system of *Numerical example 7.2*. Assume that

(a) the speed is constant at 1000 r/min;
(b) the drive is operating at maximum torque per ampere;
(c) the switching frequency of the inverter is 1.5 kHz;
(d) the dq current regulators consist only of a proportional gain $K_c = 3$ V/A.

Solution

Various software, such as PSpice and Simplorer, can be used for simulating converter-fed electrical machine drives. Assuming the switching of the inverter as ideal, Matlab-Simulink has been used in this example to simulate the current response of the drive.

A block diagram of the simulation set-up is shown in Fig. 7.21. The simulation is done as "continuous", i.e. without any sampling and so forth like in a digital control system. The current control is done in the dq reference frame

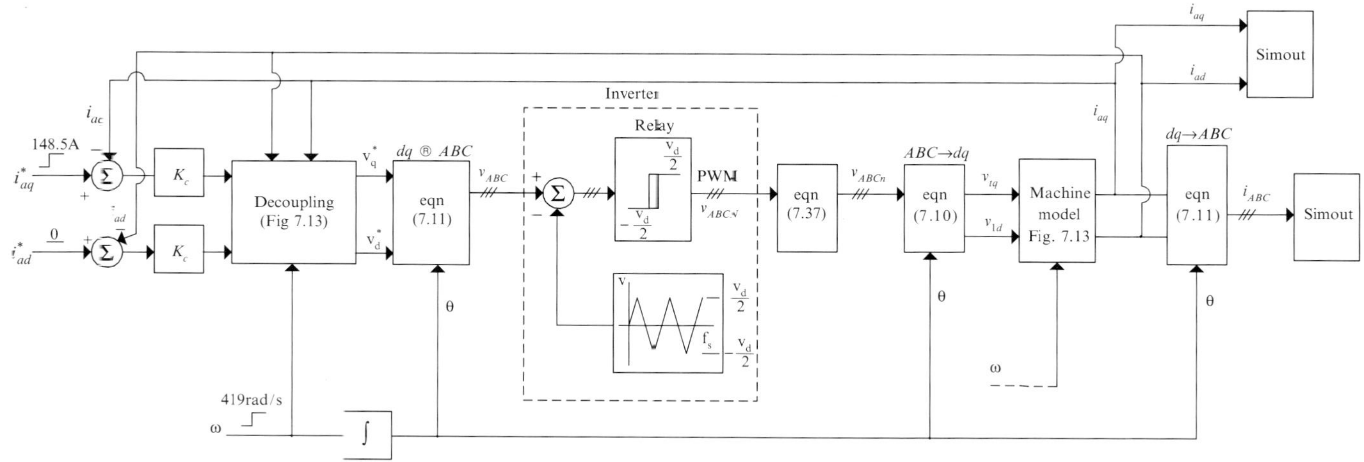

Fig. 7.21. Simulation block diagram of current controlled voltage source inverter drive system. *Numerical example 7.3*.

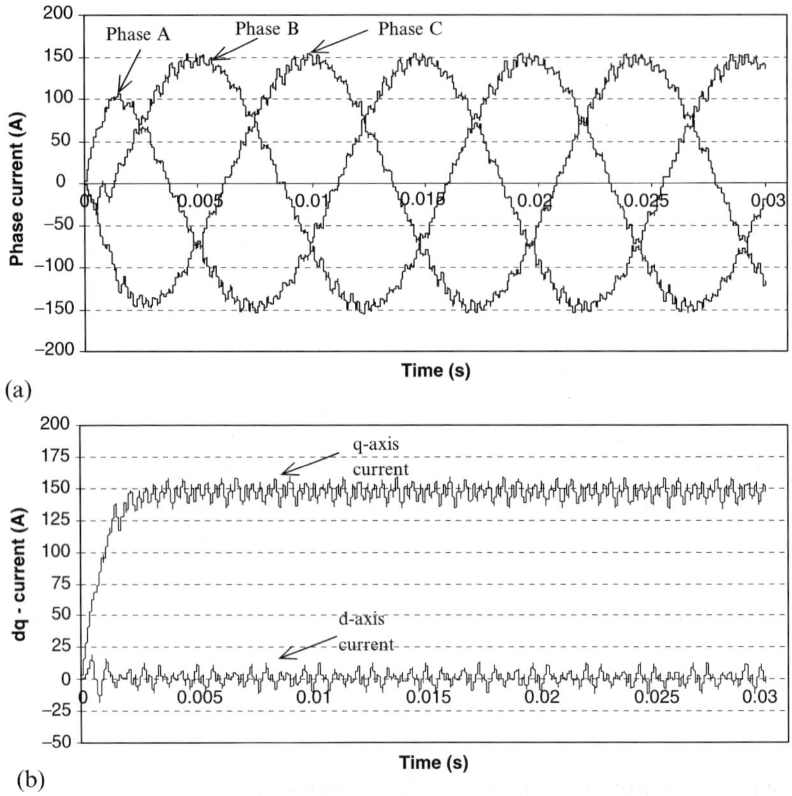

Fig. 7.22. Simulated current response for step input in the q-axis current i_{aq}: (a) phase currents i_{aA}, i_{aB} and i_{aC}; (b) dq currents i_{ad} and i_{aq}. *Numerical example 7.1.*

using the current controller of Fig. 7.13. For both the d and $q-$axis regulators of Fig. 7.13 a gain K_c is used as shown in Fig. 7.21. The decoupling is done as shown in Fig. 7.13 with the actual dq currents and speed as inputs. For simulation of the inverter, the dq voltages are transformed to ABC voltages by using eqn (7.11). To generate the PWM phase voltages, the ABC voltages are compared with a 1.5 kHz triangle wave (amplitude of $V_d/2$), by using a subtractor (sum) and comparator (relay). The relay outputs $\pm V_d/2$. The generated PWM voltages, v_{ABCN}, are with respect to the d.c.-link midpoint N of the inverter (see Fig. 7.18), and have to be converted to true phase voltages, v_{ABCn}, with respect to the floating neutral point n of the machine as follows:

$$\begin{aligned} v_{An} &= \frac{1}{3}(2v_{AN} - v_{BN} - v_{CN}) \\ v_{Bn} &= \frac{1}{3}(2v_{BN} - v_{CN} - v_{AN}) \end{aligned} \tag{7.37}$$

$$v_{Cn} = \frac{1}{3}(2v_{CN} - v_{AN} - v_{BN})$$

The next step is to either solve the machine equations in the *ABC* or in the *dq* reference frame. Using the *dq* reference frame, the true PWM phase voltages must be converted back to *dq* voltages as shown in Fig. 7.21. Using the machine model of Fig. 7.13, the i_{ad} and i_{aq} currents can be determined and fed back to be compared with the input command currents as shown. To obtain information of the actual phase currents, the i_{ad} and i_{aq} currents are transformed to *ABC* currents.

The whole block diagram of Fig. 7.21 can be set up in Matlab-Simulink using the Simulink blocks. For the machine model, eqns (7.22) and (7.23) can directly be set-up in Simulink using integrators. To simulate the inverter switching, the relay block can be used. All the other source and mathematical function blocks are available in Simulink. The simulated results can be output in various ways, in this case to the Matlab workspace.

All the machine parameters are known from *Numerical example 7.2*. The mechanical speed is taken as constant at 419 rad/s (1000 rpm). For maximum torque per ampere the *dq* command currents ${i_{ad}}^* = 0$ and ${i_{aq}}^* = 148.5$ A are given as step inputs to the system (Fig. 7.21). The *dq* and *ABC* current responses of the drive for a step input in ${i_{aq}}^*$ are shown in Fig. 7.22. The effect of the switching of the inverter on the current waveforms is clear. In this case, the $q-$axis current builds up to the rated current in less than 5 ms. This depends amongst others on the value of the proportional gain K_c. The effect of the decoupling on the current response can be self-studied by removing the decoupling signals in the simulation.

8

Cooling and Heat Transfer

8.1 Importance of Thermal Analysis

During the operation of an electrical machine, heat is generated due to power losses in electric and magnetic circuits and mechanical (rotational) losses. To ensure a long operational life for the machine, these losses must be removed as far as possible from the machine so that the temperature limitations established for the machine materials, such as insulating materials, lubricants and PMs are complied with. In addition to the consideration of the machine's operational life, a lower operating temperature reduces extra winding losses introduced by the temperature coefficient of the electric resistance — eqn (3.47).

Whereas extensive research has been devoted to the thermal studies of conventional electrical machines, AFPM machines have received very little attention [128, 245, 260]. Owing to the fact that AFPM machines possess a relatively large air gap volume and quite often have multi-gaps, the general perception is that AFPM machines have better ventilation capacity than their radial field counterparts [52, 106].

Since the external diameter increases rather slowly with the increase of output power, i.e. $D_{out} \propto \sqrt[3]{P_{out}}$ (Fig. 2.16), the existing heat dissipation capacity may be insufficient to cope with excessive heat at certain power ratings, so that more effective means of cooling have to be enforced. Thus, quantitative studies of the heat dissipation potential of AFPM machines with vastly different topologies is important.

8.2 Heat Transfer Modes

Heat transfer is a complex phenomenon presenting formidable analytical difficulties. Heat is removed from an electrical machine by a combination of *conduction*, *radiation* and *convection* processes to the ambient air and surroundings.

8.2.1 Conduction

When a temperature gradient exists in a solid body, such as in the copper, steel, PMs or the insulation of an electrical machine, heat is transferred from the high-temperature region of temperature ϑ_{hot} to the low-temperature region of temperature ϑ_{cold} according to *Fourier's law*, which is given as:

$$\Delta P_c = -kA\frac{\partial \vartheta}{\partial x} = \frac{kA}{l}\left(\vartheta_{hot} - \vartheta_{cold}\right) \tag{8.1}$$

where ΔP_c is the rate of heat conduction, A is the area of the flow path, l is the length of the flow path and k is the thermal conductivity of the material. The latter is experimentally determined and is relatively insensitive to temperature changes. Thermal properties of typical materials used for AFPM machines are given in Table 8.1, where c_p is the specific heat of material at constant pressure.

Table 8.1. Selected thermal properties of materials

Material	Grade	ρ	c_p	k
(20°C)		kg/m^3	J/(kg °C)	W/(m °C)
Air	-	1.177	1005	0.0267
Water	-	1000	4184	0.63
Mica	-	3000	813	0.33
Epoxy resin	-	1400	1700	0.5
Copper	-	8950	380	360
Aluminum	Pure	2700	903	237
	Alloy (cast)	2790	883	168
Steel	1% Carbon	7850	450	52
	Silicon	7700	490	20-30
Permanent magnet	Sintered NdFeB	7600 to 7700	420	9

8.2.2 Radiation

The net *radiant energy* interchange between two surfaces with a temperature difference is a function of the absolute temperature, the emissivity and the geometry of each surface. If heat is transferred by radiation between two gray surfaces of finite size, A_1 and A_2, and temperature, ϑ_1 and ϑ_2 (in Celsius degree), the rate of heat transfer, ΔP_r, may be written as

$$\Delta P_r = \sigma\frac{(\vartheta_1 + 273)^4 - (\vartheta_2 + 273)^4}{\frac{1-\varepsilon_1}{\varepsilon_1 A_1} + \frac{1}{A_1 F_{12}} + \frac{1-\varepsilon_2}{\varepsilon_2 A_2}} \tag{8.2}$$

where σ is the *Stefan-Boltzmann* constant, F_{12} is the shape factor which takes into account the relative orientation of the two surfaces and ε_1 and ε_2 are their respective emissivities which depend on the surfaces and their treatment. An ideal surface or body that absorbs and emits energy at maximum rate is called *black surface* or *body*. Real surfaces or bodies are normally approximated as gray surfaces or bodies where $0 < \varepsilon < 1$. Some selected emissivities related to AFPM machines are given in Table 8.2.

Table 8.2. Selected emissivities relevant to AFPM machines

Material	Surface condition	Emissivity, ε
Copper	Polished	0.025
Epoxy	Black	0.87
	White	0.85
Mild steel	-	0.2-0.3
Cast iron	Oxidized	0.57
Stainless steel	-	0.2-0.7
Permanent magnet, NdFeB	Uncoated	0.9

8.2.3 Convection

Convection is the term describing heat transfer from a surface to a moving fluid. The rate of convective heat transfer, ΔP_v, is given according to *Newton's law of cooling* as:

$$\Delta P_v = hA(\vartheta_{hot} - \vartheta_{cold}) \tag{8.3}$$

where h is the convection heat transfer coefficient, which is a rather complex function of the surface finish and orientation, fluid properties, velocity and temperature, and is usually experimentally determined. The coefficient h increases with the velocity of the cooling medium relative to the cooled surface. For a surface with forced ventilation, the following empirical relation may be used [159], i.e.

$$h_f = h_n(1 + c_h\sqrt{v}) \tag{8.4}$$

where h_f and h_n are the coefficients of heat transfer for the forced and natural convection respectively, v is the linear velocity of cooling medium and $c_h \approx 0.5$ to 1.3 is an empirical coefficient.

Some important formulae for evaluating convective heat transfer coefficients of AFPM machines are discussed in the following sections.

Convection Heat Transfer in Disc Systems

The rotating disc system plays a major role in the cooling and ventilation of the AFPM machine. Accurately determining the convection heat transfer coefficients needs thorough theoretical and experimental investigation because of the complexity of flow regimes.

In this section the convection heat transfer coefficients in different parts of the AFPM machine are evaluated, exploiting a number of existing models.

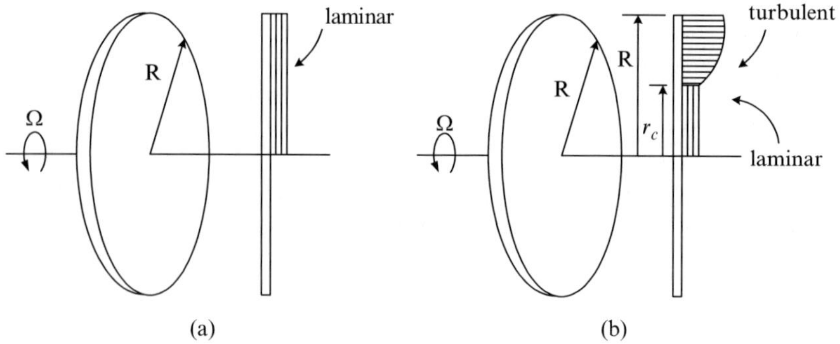

Fig. 8.1. Free rotating disc: (a) in laminar flow, (b) transition from laminar to turbulent flow.

Free rotating disc

The average heat transfer coefficient at the outside surface of a rotating disc may be evaluated using the formula developed for a free rotating disc [270], i.e.

$$\bar{h} = \frac{k}{R}\overline{Nu} \tag{8.5}$$

where R is the radius of the disc and the average Nusselt number $\overline{Nu}$ is given according to the different flow conditions as follows:

(i) For combined effects of free convection and rotation in laminar flow (Fig. 8.1a) [270]

$$\overline{Nu} = \frac{2}{5}(Re^2 + Gr)^{\frac{1}{4}} \tag{8.6}$$

$$G_r = \frac{\beta g R^3 \pi^{3/2} \Delta\vartheta}{\nu^2} \tag{8.7}$$

where Re is the Reynolds number according to eqn (2.72), β is the coefficient of thermal expansion, ν is the kinematic viscosity of the fluid (m^2/s) and $\Delta\vartheta$

is the temperature difference between the disc surface and surrounding air.

(ii) For a combination of laminar and turbulent flow with the transition at a radius r_c (Fig. 8.1b) [270]

$$\overline{Nu} = 0.015Re^{\frac{4}{5}} - 100(\frac{r_c}{R})^2 \tag{8.8}$$

where

$$r_c = (2.5 \times 10^5 \nu/\Omega)^{1/2} \tag{8.9}$$

The angular speed $\Omega = 2\pi n$ when n is the rotational speed in rev/s. It is instructive to compare the heat transfer capabilities between a rotation disc and a stationary disc. If we consider a steel disc, which has a diameter of 0.4 m and rotates at 1260 rpm., the convection heat transfer coefficient may be calculated as 41 W/(m^2 °C), which is about ten times that of the same disc at standstill. Alternatively, one can say that the effective heat dissipation area of the same disc can be increased by a factor of 10 when the disc rotates at the specific speed.

Rotor radial peripheral edge

The heat transfer correlations for the radial periphery of the rotor disc are similar to those of a rotating cylinder in air. In this case the average heat transfer coefficient is given as

$$\bar{h}_p = (k/D_{out})\overline{Nu} \tag{8.10}$$

where D_{out} is the outer diameter of the rotor disc, the average Nusselt number is given by

$$\overline{Nu} = 0.133\ Re_D^{2/3} Pr^{1/3} \tag{8.11}$$

and Reynolds number at the disc periphery is

$$Re_D = \Omega D_{out}^2/\nu \tag{8.12}$$

Note that a uniform temperature distribution in the cylinder is normally assumed when eqn (8.10) is used. Since $\bar{h}_p$ is proportional to the angular speed Ω, it may be concluded that the rotor periphery plays an increasingly important role in the heat dissipation as Ω increases.

Rotor-stator system

As seen in Fig. 8.2, an AFPM machine consists of a number of rotating and stationary discs. The heat transfer relations between a rotating and a stationary disc are of paramount importance in the thermal calculations. Due to centrifugal effects, there is a forced flow between the two discs, which increases the local heat transfer rate compared with that of a free disc. The relative increase will depend on the gap ratio, $G = g/R$, where g is the clearance between the rotor and the stator and R is the radius of the disc, the mass flow rate and the rotational speed of the system [209].

Having radial channels and thick impellers, an air-cooled AFPM machine may be regarded as a poorly designed fan from a fluid flow perspective. Its tangential velocity component is much larger than the radial component. Thus, the heat transfer rate near the rotating disc shows more dependence on the rotational Reynolds number, Re_r, given by eqn (2.72).

Owen [207] provided an approximate solution for the flow between a rotating and a stationary disc, which relates the average Nusselt number to the moment coefficient of the stator-side rotor face, C_{mo}, by the following equation:

$$\begin{cases} \overline{Nu} = Re_r C_{mo}/\pi \\ \\ C_{mo} Re_r^{1/5} = 0.333\lambda_T \end{cases} \tag{8.13}$$

where λ_T is a *turbulence parameter* given as a function of volumetric flow rate, Q, as follows

$$\lambda_T = \frac{Q}{\nu R} Re_r^{-\frac{4}{5}} \tag{8.14}$$

By replacing λ_T in eqn (8.13) with eqn (8.14), the average Nusselt number becomes

$$\overline{Nu} = 0.333 \, \frac{Q}{\pi \nu R} \tag{8.15}$$

As discussed in [208], it has been shown that for a small gap ratio ($G < 0.1$) the flow in the air-gap space between the rotor and stator can be treated as a boundary layer. Whilst it is not absolutely true that the convective heat transfer coefficient from the stator to the air flow is close to that of the air flow to a rotating disc, the same heat transfer coefficient may be assumed in the thermal circuit simulation.

8.3 Cooling of AFPM Machines

Depending on the size of the machine and the type of enclosures, different arrangements for cooling may be used. From a cooling perspective, AFPM machines may be classified into two categories as follows:

- *machines with self-ventilation*, in which cooling air is generated by a rotating disc, PM channels or other fan-alike devices incorporated with the rotating part of the machine, and

- *machines with external ventilation*, in which the cooling medium is circulated with the aid of external devices, for an example, a fan or a pump.

8.3.1 AFPM Machines With Self-Ventilation

The majority of AFPM machines are air-cooled. Compared with conventional electrical machines, a particularly advantageous feature of disc-type AFPM machines from a cooling perspective is that they possess inherent self-ventilation capability. Fig. 8.2 shows the layout and active components of a typical AFPM machine. A close examination of the machine structure reveals that an air stream will be drawn through the air inlet holes into the machine and then forced outwards into the radial channel as the rotor discs rotate. The PMs in fact act as impeller blades. The fluid behaviour of the AFPM machine is much like that of a centrifugal fan or compressor.

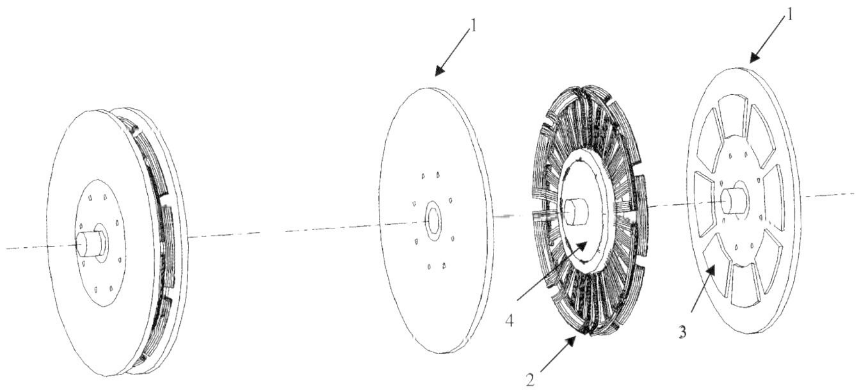

Fig. 8.2. Exploded view of an AFPM machine: 1 — rotor disc, 2 — stator winding, 3 — PM, and 4 — epoxy core.

The Ideal Radial Channel

According to the theory of an ideal impeller, a number of assumptions have to be made to establish the one-dimensional model of the ideal radial channel [78, 232]:

(a) there are no tangential components in the flow through the channel;
(b) the velocity variation across the width or depth of the channel is zero;
(c) the inlet flow is radial, which means that air enters the impeller without pre-whirl;

(d) the pressure across the blades can be replaced by tangential forces acting on the fluid;
(e) the flow is treated as incompressible and frictionless.

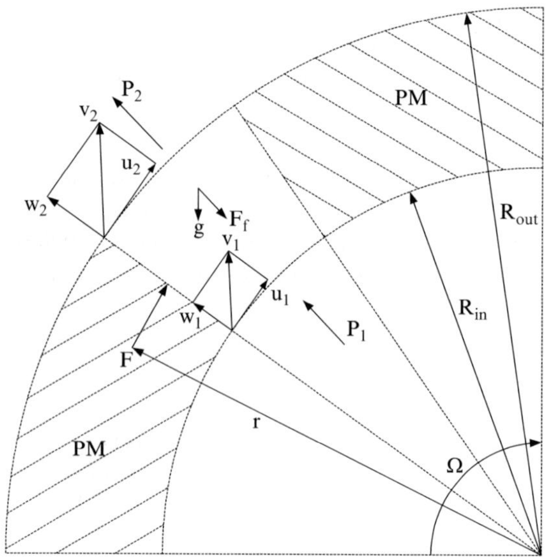

Fig. 8.3. Velocity triangles for a PM channel.

Figure 8.3 shows a radial channel with the velocity triangles drawn at the inlet and the outlet. It can be observed that the pressures at the inlet p_1 and the outlet p_2, and friction F_{fr} make no contribution to the sum of the momentum, $\sum M_0$. If gravity is ignored, the general representation of conservation of momentum takes the following form [266]:

$$\sum M_0 = \frac{\partial}{\partial t}[\int_{cv} (\mathbf{r} \times \mathbf{v})\rho dV] + \int_{cs} (\mathbf{r} \times \mathbf{v})\rho(\mathbf{v} \cdot \mathbf{n})dA \tag{8.16}$$

where $\mathbf{r}$ is the position vector from 0 to the elemental control volume dV and $\mathbf{v}$ is the velocity of the element.

For steady-state, one-dimensional air flowing between the entrance and exit of the channel, eqn (8.16) may be simplified as:

$$\sum M_0 = T_0 = (\mathbf{R}_{out} \times \mathbf{u_2})\dot{m}_2 - (\mathbf{R}_{in} \times \mathbf{u_1})\dot{m}_1 \tag{8.17}$$

where $\dot{m}_2 = \dot{m}_1 = \rho Q$, $u_1 = \Omega R_{in}$ and $u_2 = \Omega R_{out}$. The input shaft power P_{in} is then given by:

$$P_{in} = T_0 \Omega = \rho Q \Omega^2 (R_{out}^2 - R_{in}^2) \tag{8.18}$$

Re-arranging the above equation gives:

$$\frac{P_{in}}{Q} = \rho \Omega^2 (R_{out}^2 - R_{in}^2) \tag{8.19}$$

Based on the principle of conservation of energy, the input shaft power may be given as:

$$P_{in} = \dot{m}(\frac{p_2 - p_1}{\rho} + \frac{w_2^2 - w_1^2}{2} + z_2 - z_1 + U_2 - U_1) \tag{8.20}$$

If the potential $(z_2 - z_1)$ and internal energy $(U_2 - U_1)$ (friction) are ignored, eqn (8.20) may be written in the same units as eqn (8.19) as:

$$\frac{P_{in}}{Q} = (p_2 - p_1) + \rho \frac{w_2^2 - w_1^2}{2} \tag{8.21}$$

If equations (8.19) and (8.21) are equated and noting that $w_1 = Q/A_1$ and $w_2 = Q/A_2$, where A_1 and A_2 are the cross-section areas of the inlet and outlet of the channel respectively, the pressure difference Δp between the entrance and exit of the radial channel (shown in Fig. 8.3) may be expressed as:

$$\Delta p = p_2 - p_1 = \rho \Omega^2 (R_{out}^2 - R_{in}^2) - \frac{\rho}{2}(\frac{1}{A_2^2} - \frac{1}{A_1^2})Q^2 \tag{8.22}$$

Eqn (8.22) may be termed the *ideal equation* describing the air flow through the radial channel.

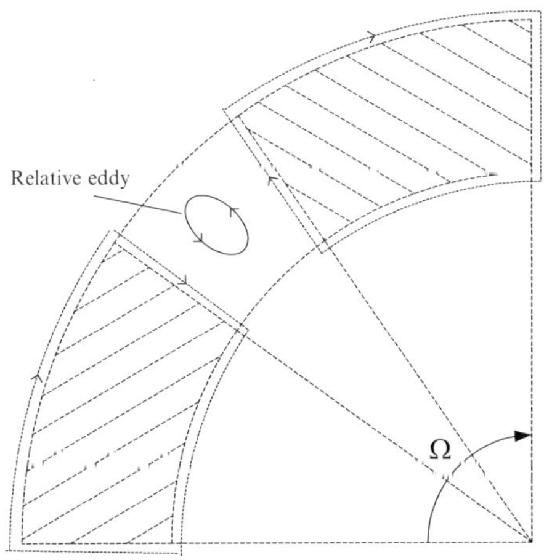

Fig. 8.4. The relative eddy in the PM channel.

The Actual Radial Channel

The actual characteristics of a hydraulic machine differ from the ideal case owing to two reasons: (i) the uneven spatial distribution of velocities in the blade passages, and (ii) the leakage and recirculation of flow and hydraulic losses such as friction and shock losses. These are completely different issues [78] and shall be dealt with separately.

Slip factor

As a result of the unbalanced velocity distribution of the leading and trailing edges of a PM channel and the rotation effects [232], there exists, according to Stodola [198, 232], a *relative eddy* within the blade passage shown in Fig. 8.4. This results in the reduction of the tangential velocity components and is called *slip*, which is usually accounted for using a *slip factor*. For approximately radial blades, the *Stanitz* slip factor k_s ($80^{\circ} < \beta_2 < 90^{\circ}$) is

$$k_s = 1 - 0.63\pi/n_b \tag{8.23}$$

where β_2 is the blade angle at exit and n_b is the number of the blades. When applying a slip factor, the pressure relation (8.22) becomes

$$\Delta p = \rho\Omega^2(k_s R_{out}^2 - R_{in}^2) + \frac{\rho}{2}(\frac{1}{A_1^2} - \frac{1}{A_2^2})Q^2 \tag{8.24}$$

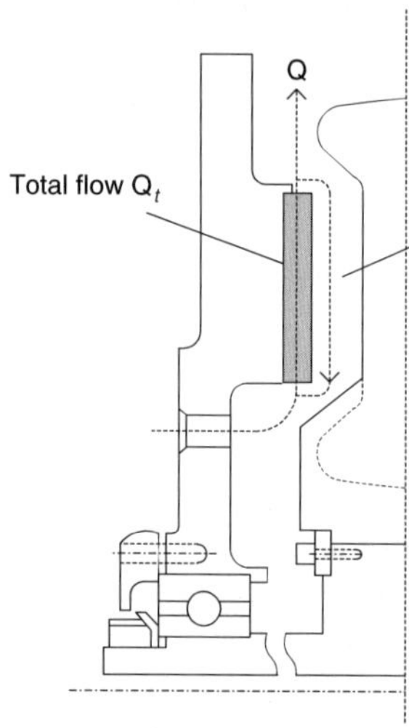

Fig. 8.5. Leakage flow in an AFPM machine (not to scale).

Shock, leakage and friction

Energy losses due to friction, separation of the boundary layer (shock loss) and leakage should also be considered in the flow analysis. As illustrated in Fig. 8.5, if the total volumetric flow rate through the PM channel is Q_t, the pressure difference between the PM exit and the entrance will cause a *leakage* or recirculation of a volume of fluid Q_l, thus reducing the flow rate at outlet to $Q = Q_t - Q_l$. The Q_l is a function of mass flow rate and discharge and leakage path resistances. The leakage flow reaches its maximum when the main outlet flow is shut.

These losses can be accounted for by introducing a pressure loss term Δp_l in eqn (8.24) as follows [232]

$$\Delta p = \rho \Omega^2 (k_s R_{out}^2 - R_{in}^2) + \frac{\rho}{2}(\frac{1}{A_1^2} - \frac{1}{A_2^2})Q^2 - \Delta p_l \tag{8.25}$$

System losses

As the air passes through the AFPM machine, the system pressure loss due to friction must be taken into account. The sum of these losses is given by:

$$\Delta p_{fr} = \frac{\rho Q^2}{2} \sum_{i=1}^{n} \frac{k_i}{A_i^2} \tag{8.26}$$

where k_i and A_i are the loss coefficient and the cross section area of the flow path i respectively.

There are a number of sections through which the air flows in the AFPM machine (see Fig. 8.6). They are:

(1) entry into the rotor air inlet holes;
(2) passage through rotation short tube;
(3) bending of air from pipe to annulus (90°);
(4) round bend air passage (90°);
(5) contraction of air in the tapering passage;
(6) bending of air through the 90° elbow;
(7) entry to the permanent magnet channels;
(8) expansion of air through the tapering enlargement;
(9) abrupt expansion of the air on the exit of the channel;
(10) expansion as the air leaves the opening of the parallel rotor discs.

The loss coefficients associated with each section in eqn (8.26) are given in [61, 188, 198]. When the section is not circular, use is made of the hydraulic diameter to characterise the cross section. The hydraulic diameter is defined as $D_h = 4A/\wp$ where A is the cross-sectional area of the flow path and $\wp$ is the wetted perimeter.

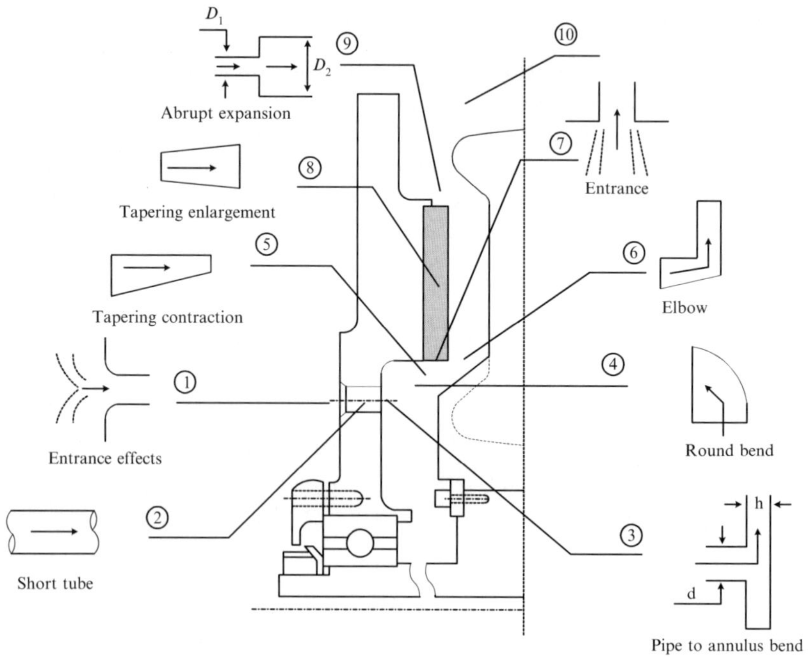

Fig. 8.6. System losses of an AFPM machine.

The loss coefficient for a pipe is given by $\lambda L/d$ where λ is a friction factor obtained as a function of Reynolds number Re and surface roughness from a *Moody diagram* [190]. To facilitate numeric calculations, the Moody diagram may be represented by [61]:

$$\begin{cases} \lambda &= 8\{(8/Re)^{12} + (X+Y)^{-\frac{3}{2}}\}^{\frac{1}{12}} \\ X &= \{2.457\ \ln\{(7/Re)^{0.9} + 0.27\gamma/D\}^{-1}\}^{16} \\ Y &= \{37530/Re\}^{16} \end{cases} \tag{8.27}$$

where $Re = \frac{\rho D_h Q}{\mu A}$ and where γ is the equivalent sand grain roughness [61].

Characteristics

It is now possible to relate the theoretical prediction obtained from the ideal flow model to the actual characteristic by accounting for the various losses discussed above.

Assuming that the AFPM machine (shown in Fig. 8.1) operates at a constant speed of 1200 rpm, the ideal developed pressure characteristic for a radial channel is a function described by eqn (8.22) as shown in Fig. 8.6. After introducing the slip factor, the resultant curve is shown as a dotted line as eqn (8.24). It was not possible to obtain a suitable correlation in the literature [248] for the pressure loss due to shock and leakage as was the case for the slip.

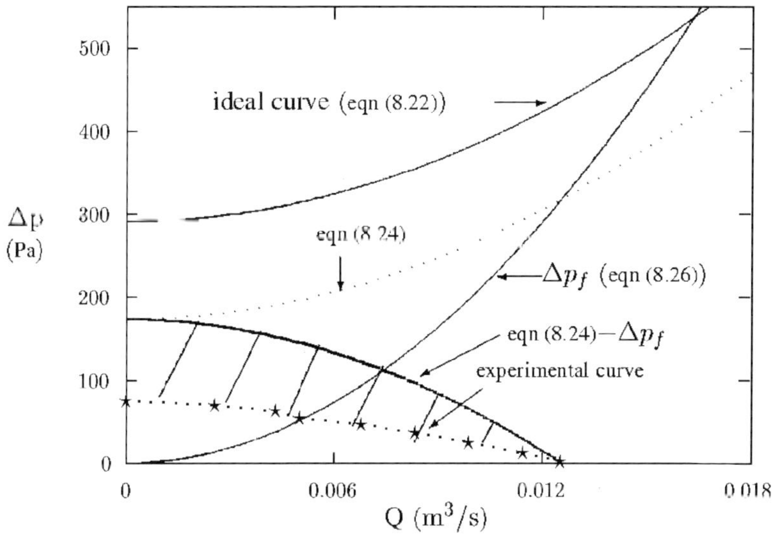

Fig. 8.7. Losses and characteristic curves at 1200 rpm.

The calculated characteristic curve without considering shock and leakage losses, i.e. eqn (8.24)–Δp_{fr} shown in Fig. 8.7, is significantly higher than the experimental one. The shaded area in Fig. 8.7 represents the shock and leakage losses. It can be seen that at low flow rates the shock and leakage losses are greater but tend to zero at the maximum flow rate. This has been discussed and experimentally validated in [260].

The derived characteristics describes the pressure relations of a single rotating PM disc facing a stator. For a double-sided AFPM machine with two identical coaxial rotating discs (Fig. 8.2) operating on the same stator, the characteristic curve presented in Fig. 8.7 represents only half of the AFPM machine. The characteristic curve of the whole machine may be obtained by adding flow rate at the same pressure, which is similar to two identical fans in parallel.

Flow and Pressure Measurements

Due to the nature and complexity of thermofluid analysis, the form of the system characteristics curve can at best be established by test. Depending on the machine topologies and size, the measurements may be taken either at the air inlet or outlet. The AFPM machine under test is normally driven by another motor. Fig. 8.8 shows the experimental arrangements of the flow measurements at the machine outlet, in which a discharge duct is set up to provide good conditions for observing the flow. Along one side of the duct, several tapping points were made for measuring the static pressure with a pressure gauge (manometer). Near the outlet of the duct, provision was made for measuring velocity using a hot-wire anemometer probe.

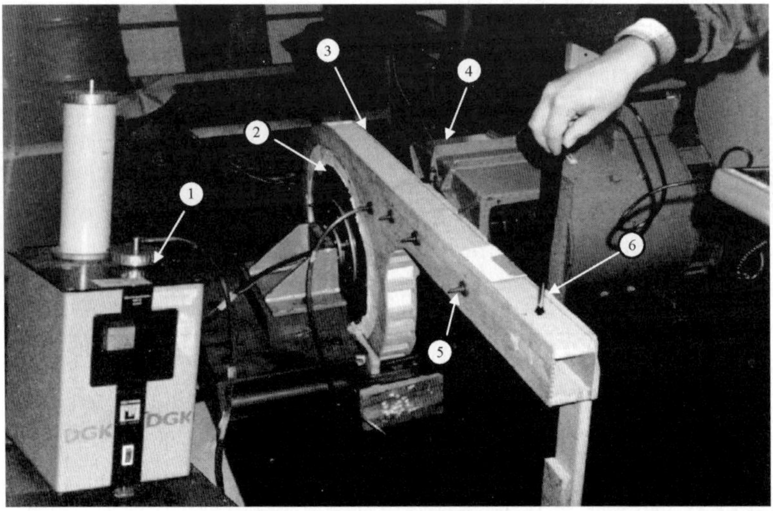

Fig. 8.8. The experimental set up. 1 — manometer, 2 — AFPM machine, 3 — discharge duct, 4 — prime mover (drive machine), 5 — pressure tapping point, 6 — wind speed probe. Photo courtesy of the *University of Stellenbosch*, South Africa.

To vary the flow rate, the test duct was fitted at its outer end with an obstruction. The test was started with no obstruction at the end of the discharge duct. The only resistance was then the duct friction, which was small and could be readily computed out of results. As the end of the duct was obstructed progressively, the flow was reduced and the static pressure increased to a maximum at zero volumetric flow rate. The static pressure difference Δp is measured as a function of volumetric flow rate $Q = A \times v$ for different motor speeds, where v is the linear speed.

The air flow-rate measurement can also be carried out by measuring inlet air pressure difference Δp, which is then used for calculating mass flow-rate $\dot{m}$ according to the following equation:

$$\dot{m} = \sqrt{2\rho \Delta p}\, A_d \tag{8.28}$$

Fig. 8.9. Flow measurement at the air intake of AFPM machine. 1 — Prime mover (drive machine), 2 — inlet duct, 3 — bell mouth, 4 — manometer, and 5 — pressure tapping point. Photo courtesy of the *University of Stellenbosch*, South Africa.

where A_d is the cross section area of the inlet duct. Fig. 8.9 shows the setup of the flow measurement. A specially designed inlet duct with a bell mouth was mounted to the inlet (hub) of the AFPM machine. A few tapping points were made on the inlet duct for pressure measurements. The pressure drop through the bell mouth and inlet duct may often be assumed negligible.

8.3.2 AFPM Machines With External Ventilation

For medium to large power AFPM machines, the loss per unit heat dissipation area increases almost linearly with the power ratings. Thus the forced-cooling with the aid of external devices may be necessary. Some common techniques are described as follows.

External Fans

Large AFPM machines may require a substantial amount of air flow per unit time in order to bring out the heat generated in the stator windings. Depending on the operation conditions obtained on site, either an air-blast or a suction fan may be used as shown in Fig. 8.10. In both cases, intake and/or discharge ducts are needed to direct and condition the air flow. Since the inlet air temperature, for a given volumetric flow rate, has a significant effect on the machine temperature, this cooling arrangement can also help prevent recirculation of hot air should the machine operate in a confined space (e.g. small machine room). For high speed AFPM machines, a shaft-integral fan may be

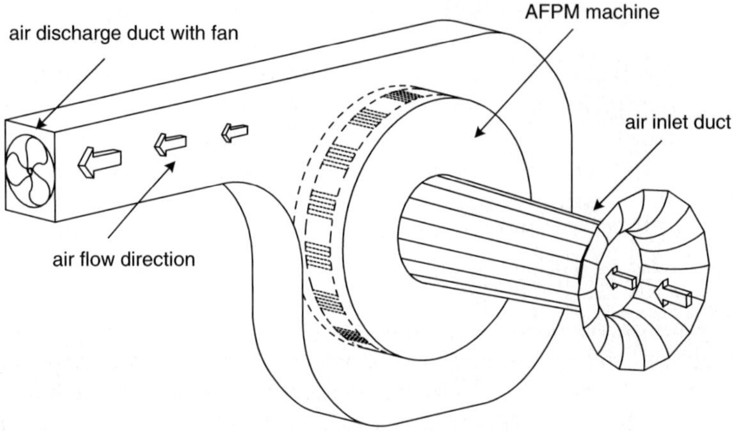

Fig. 8.10. AFPM machine with external air cooling.

a good option. Fig. 8.11 shows the assembly of a large power AFPM machine developed in the Department of Electrical and Electronic Engineering at the University of Stellenbosch, South Africa, in which the rotor hub part serves as both cooling fan and supporting structure for the rotor discs. It can be seen that the "blades" of the hub are not curved as the machine may operate in both directions of rotation.

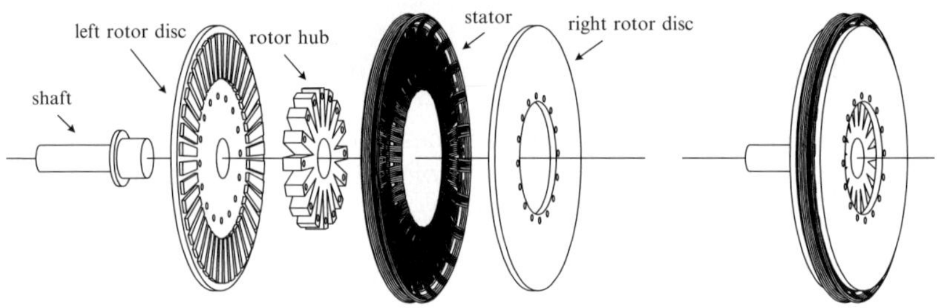

Fig. 8.11. Configuration of the AFPM machine with shaft-integral fan.

Heat Pipes

The concept of a passive two-phase heat transfer device capable of transferring large amount of heat with a minimal temperature drop was introduced by R.S.

Gaugler of the General Motors Corporation in 1942. This device received little attention until 1964, when Grover and his colleagues at Los Alamos National Laboratory, NM, U.S.A., published the results of an independent investigation and first used the term *heat pipe*. Since then, heat pipes have been employed in many applications ranging from temperature control of the permafrost layer under the Alaska pipeline to the thermal control of optical surfaces in spacecraft.

A typical *heat pipe* consists of a sealed container with wicking material. The container is evacuated and filled with just enough liquid to fully saturate the wick. A heat pipe has three different regions, namely (i) *an evaporator* or heat addition region of the container, (ii) *a condenser* or heat rejection region, and (iii) *an adiabatic* or isothermal region. If the evaporator region is exposed to a high temperature, heat is added and the working fluid in the wicking structure is heated until it evaporates. The high temperature and the corresponding high pressure in this region cause the vapour to flow to the cooler condenser region, where the vapour condenses, dissipating its latent heat of vaporisation. The capillary forces existing in the wicking structure then pump the liquid back to the evaporator. The wick structure thus ensures that the heat pipe can transfer heat whether the heat source is below the cooled end or above the cooled end.

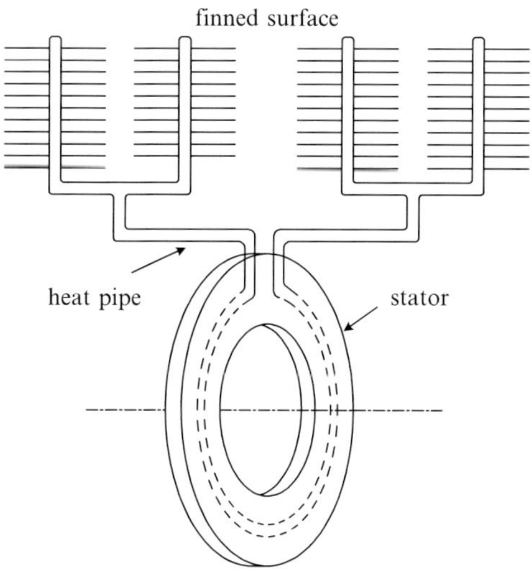

Fig. 8.12. AFPM machine cooled by heat pipes.

A heat pipe presents an alternative means for removing the heat from the AFPM machine. The heat pipe in an AFPM machine may be configured as shown in Fig. 8.12. Heat is transferred into the atmosphere through the finned surface. The finned surface is cooled by air moving over the fins. The heat loss removed by the heat pipe, ΔP_{hp}, is given by [235]:

$$\Delta P_{hp} = \frac{\vartheta_{hot} - \vartheta_{cold}}{\frac{1}{h_{hot}A_{hot}} + \frac{1}{h_{cold}A_{cold}} + \frac{1}{\eta_{fin}h_{fin}A_{fin}}} \tag{8.29}$$

where ϑ_{hot} is the average temperature of the elements that surround the heat pipe in the stator, ϑ_{cold} is the average temperature of the air cooling the finned surface, h_{hot} is the convective heat transfer coefficient on the inside wall of the heat pipe in the stator, A_{hot} is the exposed area of the heat pipe in the stator, h_{cold} is the convection heat transfer coefficient on the inside wall of the heat pipe in the finned area, A_{cold} is the exposed area of the heat pipe at the finned surface, η_{fin} is the efficiency of the finned surface, h_{fin} is the convection heat transfer coefficient on the surface of the fins and A_{fin} is the total exposed area of the finned surface.

Direct Water Cooling

Depending on the conditions at the site of operation, it is often necessary to use forced water circulation to cool the stator windings directly, especially for large power AFPM machines. An external water pump is required for forcing water circulation. A longitudinal section of a water cooled double-disc motor has been shown in Fig. 2.7. For AFPM machines with an internal iron core stator, an ideal location of placing cooling channel is around the outer periphery of the stator disc, where the effective heat transfer area is the largest. For coreless winding AFPM machines, the winding coils may have a rhomboidal shape so that the space between the two active sides of each coil may be utilised for placing a cooling water duct [45].

The heat removed through cooling pipes can be calculated by using eqn (8.29). However, the heat transfer coefficients, h_{hot} and h_{cold}, are calculated using the following relationships:

(i) for laminar flow, i.e. $\mathrm{Re}_d = \frac{\rho v d}{\mu} < 2000$, where v is the flow velocity and d is the diameter of the water pipe, the Nusselt number may be obtained using following empirical relation [124]

$$Nu_d = 1.86(Re_d Pr)^{\frac{1}{3}}(\frac{d}{L_p})^{\frac{1}{3}}(\frac{\mu}{\mu_w})^{0.14} \tag{8.30}$$

where L_p is the length of the water pipe, μ and μ_w are the dynamic viscosity of water at inlet and wall temperature respectively.

(ii) for turbulent flow, i.e. $\mathrm{Re}_d = \frac{\rho v d}{\mu} > 2000$, the Nusselt number may be calculated as [134]

$$Nu_d = 0.023{Re_d}^{0.8}Pr^n \tag{8.31}$$

where

$$n = \begin{cases} 0.4 & \text{for heating of the water,} \\ 0.3 & \text{for cooling of the water.} \end{cases}$$

8.4 Lumped Parameter Thermal Model

Lumped-parameter circuits, consisting of a network of thermal resistances, thermal capacitances, nodal temperatures and heat sources, have been used extensively to represent the complex distributed thermal parameters of electrical machines [84, 159, 245].

8.4.1 Thermal Equivalent Circuit

A *thermal equivalent circuit* is essentially an analogy of an electrical circuit, in which the heat (analogous to *current*) flowing in each path of the circuit is given by a temperature difference (analogous to *voltage*) divided by a thermal resistance (analogous to *electrical resistance*). For conduction, the thermal resistance depends on the thermal conductivity of the material, k, and the length, l, and cross-sectional area, A_d, of the heat flow path and may be expressed as

$$R_d = \frac{l}{A_d k} \tag{8.32}$$

Thermal resistances for convection is defined as:

$$R_c = \frac{1}{A_c h} \tag{8.33}$$

where A_c is the surface area of convective heat transfer between two regions and h is the convection coefficient.

The thermal resistance for radiation between two surfaces is

$$R_r = \frac{\frac{1-\varepsilon_1}{\varepsilon_1 A_1} + \frac{1}{A_1 F_{12}} + \frac{1-\varepsilon_2}{\varepsilon_2 A_2}}{\sigma[(\vartheta_1 + 273) + (\vartheta_2 + 273)][(\vartheta_1 + 273)^2 + (\vartheta_2 + 273)^2]} \tag{8.34}$$

It is seen that the radiation thermal resistance in eqn (8.34) depends on the difference of the third power of the temperature, the surface spectral property ε and the surface orientation taken into account by a form factor F.

The thermal circuit in the steady state consists of thermal resistances and heat sources connected between motor component nodes. For transient analysis, the thermal capacitances are used additionally to account for the change of internal energy in the various parts of the machine with time. The heat capacitance is defined as:

$$C = \rho V c_v = m c_v \tag{8.35}$$

where c_v is the heat capacity of the material, ρ is the density, and V and m are the volume and mass of the material respectively. Fig. 8.13a shows a sectional view of an AFPM machine with a coreless stator. It can be observed that the AFPM stator is symmetrical from a heat transfer perspective and each half of the machine from the centre line mirrors the other half. It is therefore reasonable to model only half of the machine as shown in Fig. 8.13b.

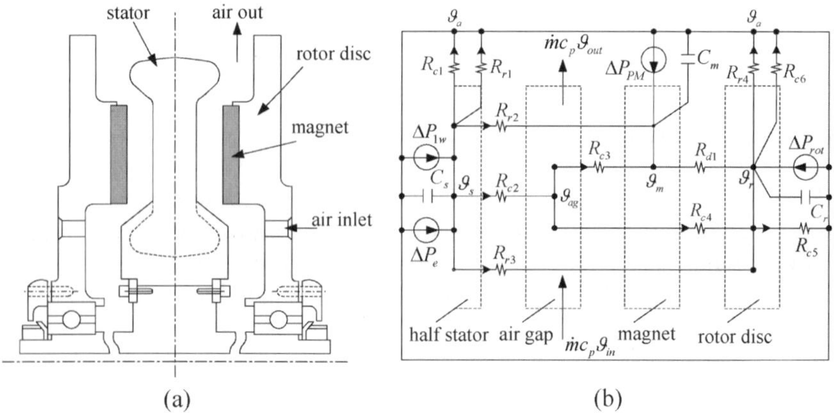

Fig. 8.13. The thermal resistance circuit of an AFPM brushless machine with coreless stator.

The heat source terms ΔP_{1w}, ΔP_e, ΔP_{PM} and ΔP_{rot} stand for winding losses (2.49), eddy current losses in one half of the stator winding (2.68), losses in PMs (2.61) and rotational losses (2.70) per one rotor disc respectively. C_s, C_m and C_r are the thermal capacitances of stator, PMs and rotor steel disc respectively. The heat resistances used in the circuit are described in Table 8.3.

In the lumped parameter thermal circuit analysis, it is often assumed that the temperature gradients within certain part of the machine is negligible. This assumption can only be made if the internal resistance to heat transfer is small compared with the external resistance [124]. The *Biot number*, B_i, is often used for determining the validity of this assumption. In the case that internal conduction resistance is compared with external convection resistance, B_i is defined as:

$$B_i = \frac{\bar{h}_c L}{k_s} \tag{8.36}$$

where k_s is the thermal conductivity of the solid material, L is the characteristic length of a solid body and $\bar{h}_c$ is the convection heat transfer coefficient. The criterion $B_i < 0.1$ ensures that the internal temperature will not differ

Table 8.3. Definition of thermal resistances

Symbols	Definition
R_{c1}	Convection resistance from stator end-winding to open air
R_{c2}	Convection resistance from stator to air-gap
R_{c3}	Convection resistance from air-gap to permanent magnets
R_{c4}	Convection resistance from air-gap to rotor disc plate
R_{c5}	Convection resistance from rotor disc to open air
R_{c6}	Convection resistance from rotor radial periphery to open air
R_{r1}	Radiation resistance from stator end-winding to environment
R_{r2}	Radiation resistance from stator to permanent magnets
R_{r3}	Radiation resistance from stator to rotor disc
R_{r4}	Radiation resistance from rotor radial periphery to environment
R_{d1}	Conduction resistance from PMs to rotor disc

from that at the surface of a solid body by more than 5% [190], and thus is frequently used.

8.4.2 Conservation of Energy

If conservation of energy is applied, the rate of internal energy change in each part of a machine (also called *control volume*) may be written as follows:

$$\frac{\Delta U}{\Delta t} = C\frac{\Delta\vartheta}{\Delta t} = \Delta P_{in} - \Delta P_{out} + \dot{m}_{in}i_{in} - \dot{m}_{out}i_{out} \tag{8.37}$$

where U is the internal energy, $\dot{m}$ is the mass flow rate and i is the *enthalpy* and C is the thermal capacitance of a control volume.

For steady-state conditions, $\frac{\Delta U}{\Delta t} = 0$ and therefore,

$$0 = \Delta P_{in} - \Delta P_{out} + \dot{m}_{in}i_{in} - \dot{m}_{out}i_{out} \tag{8.38}$$

These equations are applied to each part (the stator, air gap, PM and rotor disc) of the AFPM machine to obtain a set of equations with the temperatures of the parts being the only unknowns. This set of equations is rather complex but is readily solved using for example, the *Gauss-Seidel* iteration. It should be noted that the term $(\dot{m}_{in}i_{in} - \dot{m}_{out}i_{out})$ in the above equations represents internal heat removed due to the air flow pumped through the machine. This air flow is of paramount importance to the cooling of the machine. The determination of the air gap temperature in the thermal equivalent circuit is only possible if the mass flow rate through the air gap can be somehow predicted, hence the necessity of using the fluid-flow model described in section 8.3.1.

8.5 Machine Duties

Depending on the load conditions, there are mainly three types of duties or operation modes for all the electrical machines, i.e. continuous duty, short-time duty and intermittent duty.

8.5.1 Continuous Duty

When an electrical machine operates for a long period so that the temperatures of various parts of the machine reach steady-state at a given ambient temperature, this operation mode is called *continuous duty.* Due to the different physical properties, the final stablised temperatures in various parts of the machine may vary greatly. The machine can be continuously operated for an infinitely long time without exceeding the temperature limits specified for each component of the machine. When only the solid parts of the machine are considered, the air flow term in the eqn (8.37) is ignored. The temperature rise versus time relationship in a control volume of the machine may be derived based on the theory of solid body heating as [159]:

$$\vartheta_c = \Delta P\ R(1 - e^{-\frac{t}{\tau}}) + \vartheta_o\ e^{-\frac{t}{\tau}} \tag{8.39}$$

where R is the thermal resistance, $\tau = RC$ is the thermal time constant, C is the thermal capacitance , ΔP is the heat loss flow, and ϑ_o is the initial temperature rise of the control volume . In a case where $\vartheta_o = 0$, the above equation is simply:

$$\vartheta_c = \Delta P\ R(1 - e^{-\frac{t}{\tau}}) \tag{8.40}$$

According to the properties of an exponential function, the final temperature rise $\vartheta_{cf} = \Delta P \times R$. Fig. 8.14a shows the typical transient temperature response of an AFPM machine operated continuously.

8.5.2 Short-Time Duty

Short-time duty means that machine operates only within a short specified period followed by a long period of rest or no-load condition. The operation time is so short that the machine does not attain its steady-state temperature, and the machine practically returns to its cold state after a long period of rest. Given the operation time t_s, the temperature rise of the machine can be found by using eqn (8.40) as

$$\vartheta_s = \Delta P\ R(1 - e^{-\frac{t_s}{\tau}}) = \vartheta_{cf}\ (1 - e^{-\frac{t_s}{\tau}}) \tag{8.41}$$

Compared with continuous duty, it is obvious that $\vartheta_s < \vartheta_{cf}$. This implies that the permissible load for the same machine in short time duty can be $1/(1 - e^{-\frac{t_s}{\tau}})$ times greater than that in continuous duty. Fig. 8.14b shows the typical temperature rise curve for short time duty machines.

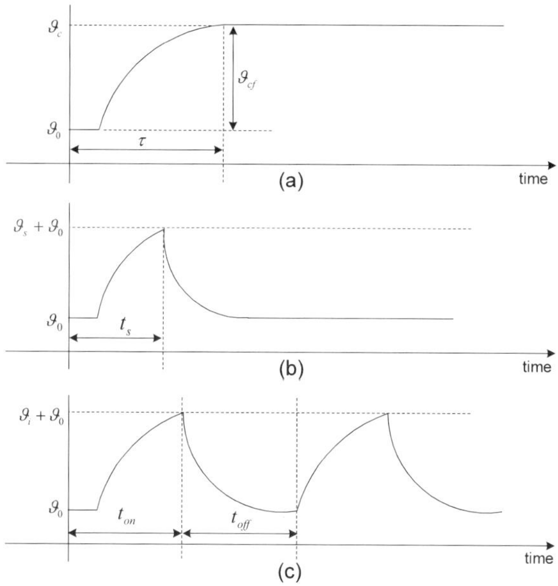

Fig. 8.14. Characteristic temperature curves of AFPM machines for different operation modes: (a) continuous duty, (b) short-time duty, (c) intermittent duty.

8.5.3 Intermittent Duty

The *intermittent duty* is characterised by short-time operations alternated with short-time pause intervals. Suppose a machine operates for a short period of t_{on} and then stops for a period of t_{off}, the cycle t_{cy} is then $t_{cy} = t_{on} + t_{off}$. The duty cycle d_{cy} may be defined as:

$$d_{cy} = \frac{t_{on}}{t_{on} + t_{off}} \tag{8.42}$$

The temperature rise of the machine ϑ_i during the t_{on} period can be calculated by using eqn (8.39) provided that thermal time constant and the steady-state temperature rise for continuous operation duty are known. During the t_{off} period, the machine loss $\Delta P = 0$ and the machine's temperature decreases according to an exponential function, i.e. $\vartheta_i e^{-t_{off}/\tau}$ before the second cycle sets in.

After many subsequent cycles, the machine's temperature variation becomes uniform and tends to be within a certain limited range (see Fig. 8.14c).

Under the same load and cooling condition, the maximum stablised temperature of a machine in intermittent operation duty is smaller than that in continuous duty. Hence, similar to short-time operation duty, a machine operated in intermittent duty has overload capacity.

Numerical Example 8.1

A self-cooled 8-pole, 16-kW AFPM generator with an ironless stator as shown in Fig. 8.2 will be considered. The outer diameter is $D_{out} = 0.4$ m and the inner diameter is $D_{in} = 0.23$ m. The magnet width–to–pole pitch ratio $\alpha_i = 0.8$ and thickness of a rotor disc $d = 0.014$ m. The measured flow characteristic curves are shown in Fig. 8.15. At rated speed 1260 rpm, the total losses are 1569 W, of which (i) rotational losses $\Delta P_{rot} = 106$ W; (ii) eddy current losses in the stator $\Delta P_e = 23$ W; (iii) stator winding losses (rated) $\Delta P_{1w} = 1440$ W. Find:

(a) Convective heat transfer coefficients of the disc system

(b) Steady-state temperatures at different parts of the machine.

Solution

(a) Convective heat transfer coefficients of the disc system

The dynamic viscosity, density and thermal conductivity of air are assumed to be $\mu = 1.8467 \times 10^{-5}$ Pa s, $\rho = 1.177$ kg/m^3 and $k = 0.02624$ W/(m °C) respectively in the following convective coefficients calculations.

Convection coefficient: outside rotor disc surface

At rated speed, the Reynolds number according to eqn (2.72) is

$$Re = \rho \frac{\Omega D_{out}^2}{4\mu} = 1.177 \times \frac{2\pi \times 1260/60 \times 0.4^2}{4 \times 1.846 \times 10^{-5}} = 336384.7$$

According to eqn (8.9) the transition between laminar and turbulent flow takes place at

$$r_c = \sqrt{\frac{2.5 \times 10^5 \nu}{\Omega}} = \sqrt{\frac{2.5 \times 10^5 \times 1.569 \times 10^{-5}}{2\pi \times 1260/60}} = 0.172 < \frac{D_{out}}{2} \text{ m}$$

The average Nusselt number of the disc according to eqn (8.8) is

$$\overline{Nu} = 0.015 \times Re^{\frac{4}{5}} - 100 \times (\frac{2r_c}{D_{out}})^2$$

$$= 0.015 \times 336384.7^{\frac{4}{5}} - 100 \times \left(\frac{2 \times 0.172}{0.4}\right)^2 = 321.9$$

The average heat transfer coefficient at the outer surface of the disc according to eqn (8.5) is

$$\overline{h}_{fr} = \frac{k}{D_{out}/2} \times \overline{Nu} = \frac{0.02624}{0.4/2} \times 321.9 = 42.2 \text{ W/(m}^2 \text{ °C)}$$

Convection coefficient: rotor disc peripheral edge

The Prandtl number is taken as $Pr = 0.7$ (Atmospheric pressure, 25 °C). The Reynolds number at disc periphery according to eqn (8.12) is

$$Re_D = \Omega \frac{D_{out}^2}{\nu} = \frac{2\pi \times 1260}{60} \frac{0.4^2}{1.569 \times 10^{-5}} = 1345538.8$$

The average Nusselt number according to eqn (8.11) is

$$\overline{Nu} = 0.133 \times Re_D{}^{\frac{2}{3}} \times Pr^{\frac{1}{3}} = 0.133 \times 1345538.8^{\frac{2}{3}} \times 0.7^{\frac{1}{3}} = 1439.3$$

The average heat transfer coefficient around the radial periphery according to eqn (8.10) is

$$\overline{h}_p = \frac{k}{D_{out}} \times \overline{Nu} = \frac{0.02624}{0.4} \times 1439.3 = 94.4 \text{ W/(m}^2 \text{ °C)}$$

Convection coefficient: rotor-stator system

The volumetric flow rate of the machine at rated speed can be looked up from Fig. 8.15. Assuming equal flow on both sides of stator, the equivalent flow rate is taken as $Q = 0.013$ m^3/s. The average Nusselt number according to eqn (8.15) is

$$\overline{Nu} = 0.333 \frac{Q}{\pi\nu(D_{out}/2)} = 0.333 \times \frac{0.013}{\pi 1.566 \times 10^{-5} \times (0.4/2)} = 440$$

The average heat transfer coefficient between two discs according to eqn (8.5)is

$$\overline{h}_{rs} = \frac{2k}{D_{out}} \times \overline{Nu} = \frac{2 \times 0.02624}{0.4} \times 440 = 57.7 \text{ W/(m}^2 \text{ °C)}$$

(b) Steady-state temperatures at different parts of the machine

If the radiation from the rotor disc to ambient, the convection from the stator overhang to air flow, and the conduction resistance between magnets and rotor are ignored, the generic thermal equivalent circuit given in Fig. 8.13 is simplified as shown in Fig. 8.16.

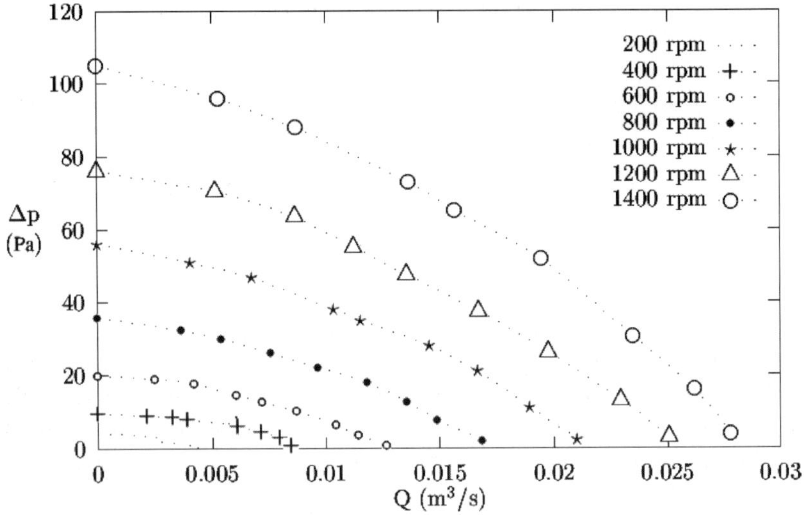

Fig. 8.15. Measured characteristic curves of the AFPM machine. *Numerical example 8.1.*

<u>Control volume 1 (*half of the stator*)</u>

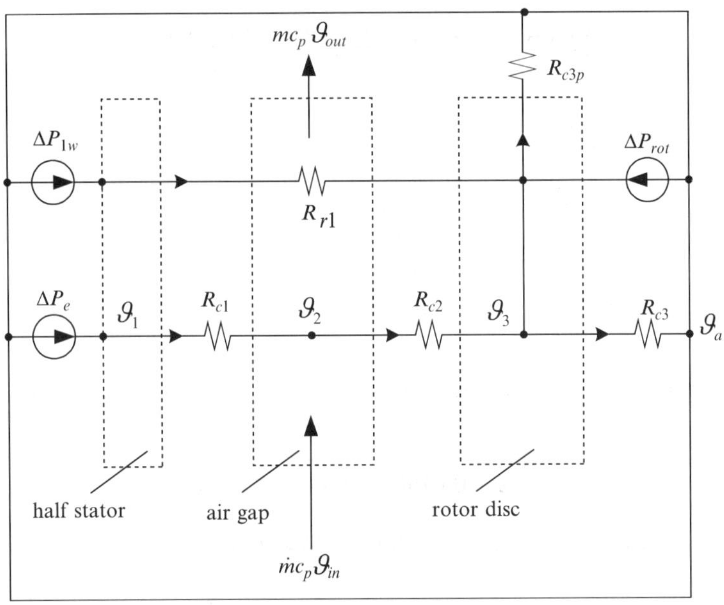

Fig. 8.16. Simplified thermal equivalent circuit. *Numerical example 8.1.*

The convection heat transfer resistance between the stator and air flow in the air gap is

$$R_{c1} = \frac{1}{\bar{h}_{rs} \times \frac{\pi}{4}(D_{out}^2 - D_{in}^2)} = \frac{1}{57.7 \times \frac{\pi}{4}(0.4^2 - 0.23^2)} = 0.223\ {}^{\circ}\mathrm{C/W}$$

The radiation heat transfer resistance between the stator and rotor discs is

$$R_{r1} = \frac{\frac{1-\varepsilon_1}{\varepsilon_1 A_1} + \frac{1}{A_1 F_{12}} + \frac{1-\varepsilon_2}{\varepsilon_2 A_2}}{\sigma[(\vartheta_1 + 273) + (\vartheta_2 + 273)][(\vartheta_1 + 273)^2 + (\vartheta_2 + 273)^2]}$$

in which the areas of both discs can be taken as the same, i.e. $A_1 = A_2 = \frac{\pi}{4}(D_{out}^2 - D_{in}^2) = 0.084\ \mathrm{m}^2$, the shape factor $F_{12} = 1$, the *Stefan-Boltzmann* constant $\sigma = 5.67 \times 10^{-8}\ \mathrm{W/(m^2}\ K^4)$, the emissivity of epoxy encapsulated stator $\varepsilon_1 = 0.85$. Since part of the rotor disc is covered with PMs, the emissivity of the rotor disc is defined based on the proportion of different materials, i.e.

$$\varepsilon_2 = \varepsilon_{fe}\alpha_i + \varepsilon_{pm}(1 - \alpha_i) = 0.3 \times 0.8 + 0.9 \times (1 - 0.8) = 0.42$$

Apparently R_{r1} is a function of ϑ_1 and ϑ_2, an iterative approach has to be used to solve R_{r1} for different temperatures.
According to conservation of energy, the steady-state energy equation for *control volume 1* can be written as

$$\frac{1}{2}(\Delta P_{1w} + \Delta P_e) - \frac{\vartheta_1 - \vartheta_2}{R_{c1}} - \frac{\vartheta_1 - \vartheta_3}{R_{r1}} = 0 \tag{8.43}$$

Control volume 2 (*air gap*)

The convection heat transfer resistance from air gap to the rotor disc may be taken as the same as that from stator to air gap, i.e. $R_{c2} = R_{c1}$. The mass flow rate $\dot{m} = \rho Q = 1.177 \times 0.013 = 0.0153$ kg. Assume that the air temperature at the machine inlet is ambient temperature, i.e. $\vartheta_{in} = \vartheta_a$ and the air gap average temperature $\vartheta_2 = \frac{1}{2}(\vartheta_{out} + \vartheta_{in})$. The heat dissipated due to the air flow is

$$\dot{m}_{out}i_{out} - \dot{m}_{in}i_{in} = \dot{m}\ c_p\ (\vartheta_{out} - \vartheta_{in}) = 2\dot{m}\ c_p\ (\vartheta_2 - \vartheta a)$$

$$= 2 \times 0.0153 \times 1005 \times (\vartheta_2 - 20)$$

The steady-state energy equation for *control volume 2* is

$$\frac{\vartheta_1 - \vartheta_2}{R_{c1}} - \frac{\vartheta_2 - \vartheta_3}{R_{c2}} - 2 \times 0.0153 \times 1005 \times (\vartheta_2 - 20) = 0 \tag{8.44}$$

Control volume 3 (*rotor disc*)

The convection heat transfer resistance at the outside surface of the disc is

$$R_{c3} = \frac{4}{\bar{h}_{fr}\pi {D_{out}}^2} = \frac{4}{42.4 \times \pi \times 0.4^2} = 0.1877\ ^{\circ}\mathrm{C/W}$$

The convection heat transfer resistance at the periphery of the disc is

$$R_{c3p} = \frac{1}{\bar{h}_p\ \pi\ D_{out}\ d} = \frac{1}{94.4 \times \pi \times 0.4 \times 0.014} = 0.602\ ^{\circ}\mathrm{C/W}$$

The steady-state energy equation for *control volume 3* is

$$\frac{\vartheta_2 - \vartheta_3}{R_{c2}} + \frac{\vartheta_1 - \vartheta_3}{R_{r1}} + \frac{1}{2}\Delta P_{rot} - \frac{\vartheta_3 - \vartheta_a}{R_{c3}} - \frac{\vartheta_3 - \vartheta_a}{R_{c3p}} = 0 \tag{8.45}$$

Having established the energy equations (8.43),(8.44) and (8.45) for each part of the machine, the steady-state temperatures can be found by solving these equations. Due to the temperature dependency of R_{r1}, a simple computer program using *Gauss Seidel* iteration has been created to find the solutions of the equations. The results are given in Table 8.4.

Table 8.4. Predicted temperature rises

Machine parts	Temperature rise, °C
Stator winding, ϑ_1	114.9
Air-gap flow, ϑ_2	21.35
Rotor disc, ϑ_3	18.32

Numerical Example 8.2

A totally enclosed AFPM brushless machine has a power loss of ΔP_{1W} =2500 W in the stator winding at continuous duty. The machine's outer and inner diameters are $D_{out} = 0.72$ m and $D_{in} = 0.5$ m respectively. To remove the heat from the stator, use is made of heat pipes for direct cooling of the motor as shown in Fig. 8.12. The convection heat transfer coefficients on the inside wall of heat pipes in the stator and in the finned area are assumed to be $h = 1000$ W/(m^2 °C). The average convective heat transfer coefficient on the fin surface is taken as $h_{fin} = 50$ W/(m^2 °C). The finned surface has an overall area of $A_{fin} = 1.8$ m^2 and an efficiency of $\eta_{fin} = 92\%$. The length of heat pipe embedded in the finned surface is $l_{fin} = 1.5$ m. Find:

(a) Steady-state temperature of the stator winding if the heat pipe with a diameter $D_{hp} = 9$ mm is placed along the average radius of the stator

(b) Steady-state temperatures of the stator winding if the heat pipe is replaced by a $d = 9$ mm water cooling pipe, in which water (with an average temperature of 60°C) flows at $v_w = 0.5$ m/s.

Solution

(a) Steady-state temperature of the stator winding if the heat pipe with a diameter $D_{hp} = 9$ mm is placed along the average radius of the stator

Assuming the outside wall of the heat pipe is in full contact with the stator winding and finned surface, the exposed area of the heat pipe in the stator will be

$$A_{hot} = \pi D_{hp} \frac{\pi(D_{out} + D_{in})}{2} = 0.009\pi \times \frac{\pi(0.72 + 0.5)}{2} = 0.0542 \text{ m}^2$$

The exposed area of the heat pipe embedded in the finned surface will be

$$A_{cold} = \pi D_{hp} l_{fin} = 0.009\pi \times 1.5 = 0.0424 \text{ m}^2$$

Assuming the temperature of cooling air over the finned area to be $\vartheta_{cold} =$ 30°C, the steady-state temperature of the stator may be obtained by using eqn (8.29)

$$\vartheta_{stator} = \vartheta_{cold} + \Delta P_{hp}\left(\frac{1}{h_{hot}A_{hot}} + \frac{1}{h_{cold}A_{cold}} + \frac{1}{\eta_{fin}h_{fin}A_{fin}}\right)$$

$$= 30 + 2500\left(\frac{1}{1000 \times 0.0542} + \frac{1}{1000 \times 0.0424} + \frac{1}{0.92 \times 50 \times 1.8}\right) = 165.3 \text{ °C}$$

(b) Steady-state temperatures of the stator winding if the heat pipe is replaced by a water cooling pipe of diameter $d = 9$ mm, in which water flows at 0.5 m/s.

The Reynolds number is first calculated to determine the flow regime. The properties of water at 60°C are $\rho = 983.3$ kg/m^3, $c_p = 4179$ J/(kg °C), $\mu = 4.7 \times 10^{-4}$ Pa s, $k = 0.654$ W/(m °C), $Pr = \mu c_p/k = 4.7 \times 10^{-4} \times 4179/0.654 = 3$, $Re_d = \rho v_w d/(2\mu) = 983.3 \times 0.5 \times 0.009/4.7 \times 10^{-4} = 9414.6 > 2000$, so that the flow is turbulent. Thus, the Nusselt number for the heating of the water is

$$Nu_{dh} = 0.023 Re_d^{0.8} Pr^{0.4} = 0.023 \times 9414.6^{0.8} \times 3^{0.4} = 53.9$$

while the Nusselt number for the cooling of the water is

$$Nu_{dh} = 0.023 Re_d^{0.8} Pr^{0.3} = 0.023 \times 9414.6^{0.8} \times 3^{0.3} = 48.3$$

The convection heat transfer coefficients of the water pipe inside the stator h_{hot} and in the finned area h_{cold} are calculated as

$$h_{hot} = \frac{k\ Nu_{dh}}{d} = \frac{0.654 \times 53.9}{0.009} = 3916.7\ W/(m^2\ {}^oC)$$

$$h_{cold} = \frac{k\ Nu_{dc}}{d} = \frac{0.654 \times 48.3}{0.009} = 3509.8\ W/(m^2\ {}^oC)$$

The steady-state temperature in the stator can be calculated as in (a) with the exception that the heat transfer coefficients are different

$$\vartheta_{stator} = \vartheta_{cold} + \Delta P_{hp}(\frac{1}{h_{hot}A_{hot}} + \frac{1}{h_{cold}A_{cold}} + \frac{1}{\eta_{fin}h_{fin}A_{fin}})$$

$$= 30 + 2500 \times (\frac{1}{3916.7 \times 0.0542} + \frac{1}{3509.8 \times 0.0424} + \frac{1}{0.92 \times 50 \times 1.8}) = 88.8\,{}^oC$$

9

Applications

9.1 Power Generation

Brushless PM electrical machines are the primary generators for *distributed generation systems*. They are compact, high efficient and reliable self-excited generators. The distributed generation is any electric power production technology that is integrated within a distribution system. Distributed generation technologies are categorized as *renewable* and *nonrenewable*. Renewable technologies include solar, photovoltaic, thermal, wind, geothermal and ocean as sources of energy. Nonrenewable technologies include internal combustion engines, combined cycles, combustion turbines, microturbines and fuel cells.

AFPM brushless generators can be used both as high speed and low speed generators. Their advantages are high power density, modular construction, high efficiency and easy integration with other mechanical components like turbine rotors or flywheels. The output power is usually rectified and then inverted to match the utility grid frequency or only rectified. A method for active output voltage regulation in an AFPM automotive alternator by means of mechanical flux weakening is proposed in [34].

9.1.1 High Speed Generators

Although the minimization of windage losses of high speed generators requires rotors with small diameters, a multidisc design has the advantages of modular and compact construction, low synchronous reactance, good voltage regulation and very high efficiency in the case of coreless stators.

A multidisc AFPM *high speed generator* is usually driven by a gas microturbine. The turbine rotor and PM rotors are mounted on the same shaft. Such a generator has compact construction, low mass and very high efficiency.

According to *TurboGenset Company*, U.K., a 100-kW, 60 000 rpm multidisc generator has an outer diameter of 180 mm, length of 300 mm and weighs only 12 kg (Fig. 9.1)[1]. The high frequency output is rectified to d.c.

[1] ULEV-TAP Newsletter, No. 2, 2000, www.ulev-tap.org

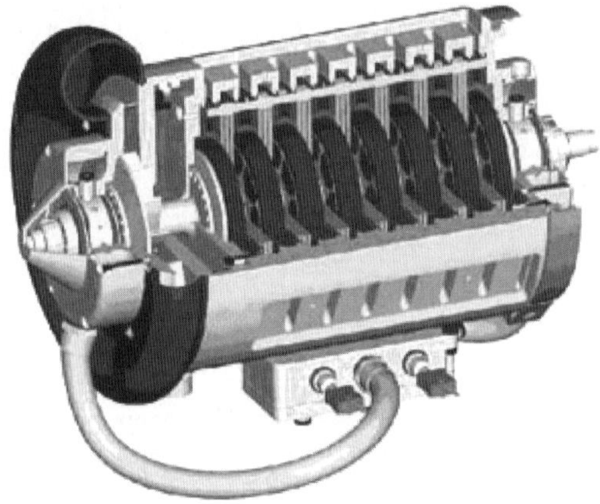

Fig. 9.1. 100-kW, 8-disc AFPM synchronous generator. Courtesy of *TurboGenset*, London, U.K.

and then inverted to 50, 60 or 400 Hz a.c. (Fig. 2.20). The generator is totally air cooled.

Armed forces are interested in applications of microturbine driven PM synchronous generators to battery chargers for soldiers. The low energy density of portable bagpack batteries imposes a major constraint and challenge to future infantry operations. With recent advancement in PM brushless machine technologies, a lightweight *miniature generator set* can minimize the

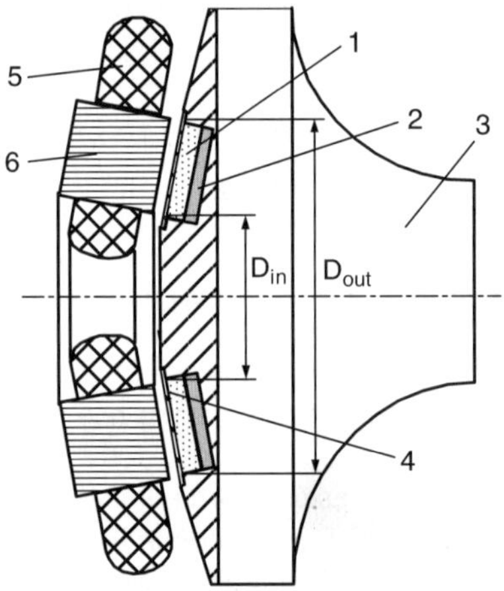

Fig. 9.2. High speed AFPM synchronous generator integrated with the rotor of a microturbine. 1 — SmCo PM, 2 — backing steel ring, 3 — rotor of microturbine, 4 — nonmagnetic retaining ring, 5 — stator winding, 6 — stator core.

mass of heavy batteries and charge them during field operations. A miniature generator can deliver electric power for a long period of time limited only by the fuel availability. Easy and rapid refueling can be done in the field by using, e.g. kerosene. Fig. 9.2 shows a tiny microturbine integrated with a miniature AFPM synchronous generator. At speed 150 000 to 250 000 rpm and outer diameter of PM ring $D_{out} \approx 50$ mm, the generator can deliver about 1 kW electric power.

Table 9.1. Specifications of five phase Kestrel AFPM synchronous generators

Type of generator	KestreL 600	KestreL 800	KestreL 2000
Rated power, W	400 at 12.5m/s	800 at 11.5m/s	2000 at 10.5m/s
Maximum power, W	600 at 14.5m/s	850 at 12m/s	2200 at 11m/s
Number of poles $2p$	48	48	200
Generation steps	2	2	4
Rated wind speed, m/s	12.5	11.5	10,5
Cut in wind speed, m/s	2.5	2.5	2.2
Rated rotational speed, rpm	1100	1010	925
Rotor diameter, m	1.2	2.1	3.6
Number of blades	6	3	3
Type of blade	Basic aerofoil	Full aorofoil	
Tower top weight, kg	23	35	80
Lateral thrust, N	250	300	450
Speed control	Self stall	Dynamic tail	
Standard rectified d.c. voltage, V	12, 24, 36, 48		48
Protection	IP55		

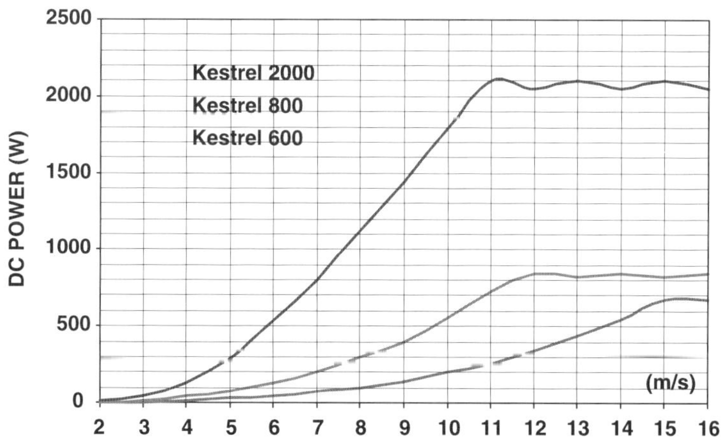

Fig. 9.3. Output power versus wind speed of Kestrel AFPM synchronous generators. Courtesy of *Kestrel Wind Turbines*, South Africa.

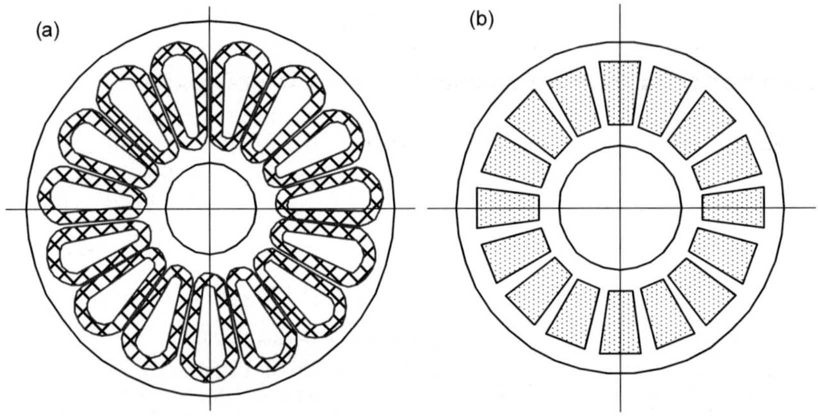

Fig. 9.4. Stator and rotor of a medium power AFPM generator with inner coreless stator and twin external rotor: (a) stator coils, (b) rotor disc with NdFeB PMs.

9.1.2 Low Speed Generators

A *low speed AFPM generator* is usually driven by a *wind turbine*. With wind power rapidly becoming one of the most desirable alternative energy sources world-wide, AFPM generators offer the ultimate low cost solution as compared with e.g. solar panels. Table 9.1 shows specifications of five-phase AFPM synchronous generators manufactured by *Kestrel Wind Turbines (Pty) Ltd*, Gauteng, South Africa. All three types of Kestrel generators use laminated cores. The KestreL 2000 is a double-sided AFPM generator with twin rotors. The KestreL 600 generator rated at 400 W is in fact capable of delivering power in excess of 600 W. Controlled by a shunt regulator the maximum charging current is 50 A into a 12 V d.c. battery bank. At wind speeds of 52.2 km/h, which translates into 14.5 m/s (28.2 knots) and rotational speed of 1100 rpm, the power output is 600 W (Fig. 9.3).

Figure 9.4 shows a medium-power AFPM generator with twin external rotor and inner coreless stator for a 5-m blade diameter wind turbine. The stator system consists of 15 non-overlapping concentrated coils, one coil per phase. The thickness of coil is 12 mm and diameter of wire 0.6 mm. Each coil is individually rectified to d.c., to reduce the commutation torque ripple effect and provide better control over the voltage. The rotor has $2p = 16$ poles. Performance characteristics are shown in Fig. 9.5.

Figure 9.6 shows the construction detail of a 2 kW 500 rpm AFPM air-cored wind turbine-driven generator with concentrated coil topology developed by Durham University, UK [27]. The stator is located between two rotor discs and contains 12 circular bobbin armature coils while each rotor disc has 16 circular NdFeB magnets. The rotor discs are mechanically joined by an outer location ring and through bolts. The magnets are secured and positioned

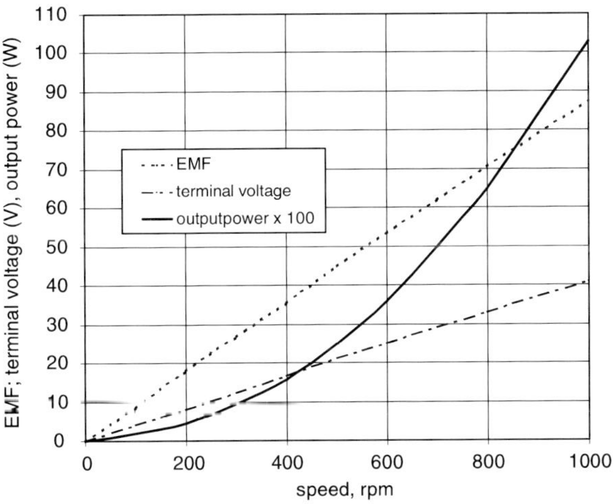

Fig. 9.5. Steady-state performance characteristics of a 10-kW AFPM generator shown in Fig. 9.4.

Table 9.2. Specifications of the AFPM wind generator [27]

Rated speed, rpm	500
Rated power, W	2000
Number of phases	3
Number of magnets	16
Number of armature coils	12
Frequency, Hz	66.67
Magnet type	NdFeB
Magnet diameter, mm	65
Magnet thickness, mm	8
Pole pitch at mean diameter, mm	70
Pole pitch, deg	22.5
Rotor thickness, mm	10.0
Generator overall diameter, mm	495
Generator axial length, mm	67
Stator thickness, mm	20
Pitch of armature coil, deg	30
Outer diameter of armature coil, mm	75.0
Inner diameter of armature coil, mm	30.0
Coil height, mm	14
Wire diameter, mm	1.0
Number of turns/coil	276
Running clearance, mm	3

in a retainer ring. The stator is supported from the inside using a double-shaft arrangement. The generator is of an overall diameter 495 mm and axial length 67 mm. The main generator parameters are given in Table 9.2. The advantages associated with this type of generator are as follows [27]:

- circular coils offer the shortest turns and lowest resistance;
- simple manufacture and low production costs;
- easy assembly as no magnetic forces between rotor and stator;
- no cogging torque and thus presents little mechanical resistance to turbine starting.

A prototype generator has been manufactured and integrated with a vertical axis wind turbine as shown in Fig. 9.7.

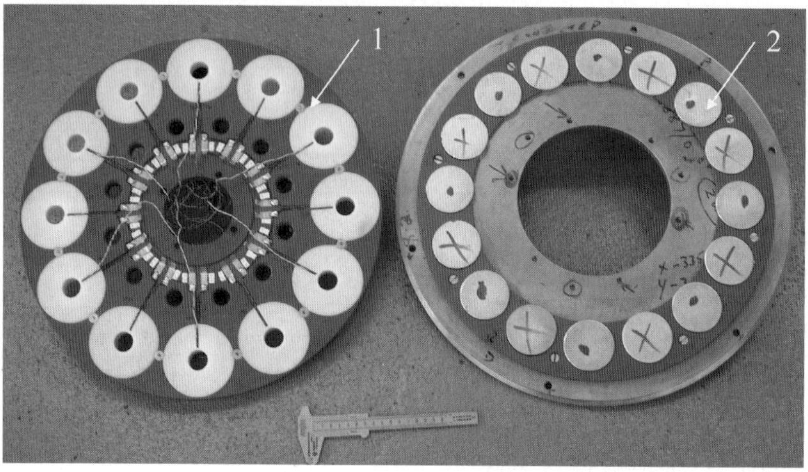

Fig. 9.6. The stator and rotor layout of the 2 kW 500 rpm air-cored AFPM wind generator with concentrated coil topology developed by Durham University: 1 — bobbin wound air-cored stator, 2 — rotor disc. Courtesy of *Durham University*, UK.

9.2 Electric Vehicles

Electric vehicles are divided into three general categories: *hybrid electric vehicles* (HEVs), *battery electric vehicles* (EVs) and *fuel cell electric vehicles* (FCEVs). The switch from gasoline to HEVs and EVs/FCEVs would reduce the total primary energy consumed for personal transportation. Traction electric motors for HEVs and EVs/FCEVs should meet the following requirements:

Fig. 9.7. Generator integrated with a vertical axis wind turbine. Courtesy of *Durham University* and *Carbon Concepts*, UK.

- power rating: high instant power, high power density;
- torque–speed characteristics: high torque at low speed for starting and climbing, high speed at low torque for cruising, wide speed range including constant torque region and constant power region, fast torque response;
- high efficiency over wide speed and torque ranges;
- high reliability and robustness under various operating conditions, e.g. at high and low temperature, rain, snow, vibration, etc.;
- low cost.

9.2.1 Hybrid Electric Vehicles

HEVs are now at the forefront of transportation technology development. HEVs combine the internal combustion engine of a conventional vehicle with the electric motor of an EV, resulting in twice the fuel economy of conventional vehicles. The electric motor/generator is usually located between the combustion engine and gear box. One end of the rotor shaft of the electric motor/generator is bolted to the combustion engine crankshaft, while the opposite end can be bolted to the flywheel or gearbox via the clutch (Fig. 9.8). The electric motor serves a number of functions, i.e.:

- assisting in vehicle propulsion when needed, allowing the use of a smaller internal combustion engine;

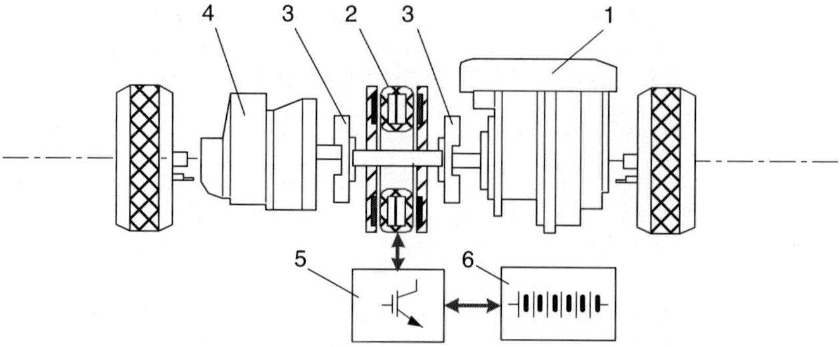

Fig. 9.8. Hybrid electric gasoline car: 1 - gasoline combustion engine, 2 - integrated motor-generator, 3 - cranking clutch, 4 - gearbox, 5 - inverter, 6 - battery.

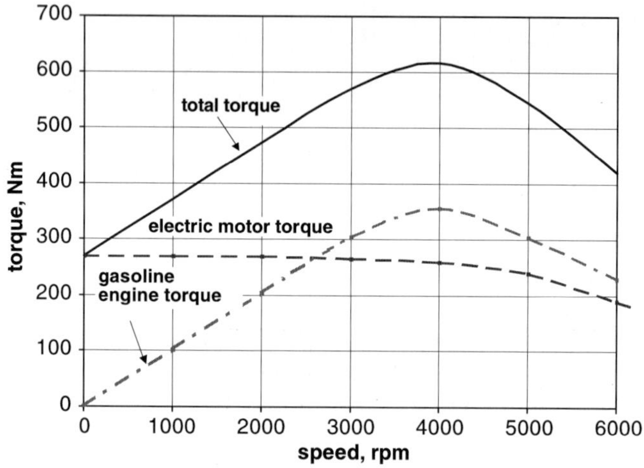

Fig. 9.9. Torque-speed characteristics of an electric motor and gasoline engine. The electric motor assists the gasoline engine at low speeds.

- operating as a generator, allowing excess energy (during braking) to be used to recharge the battery;
- replacing the conventional alternator, providing energy that ultimately feeds the conventional low voltage, e.g. 12 V electrical system;
- starting the internal combustion engine very quickly and quietly. This allows the internal combustion engine to be turned off when not needed, without any delay in restarting on demand;
- damping crankshaft speed variations, leading to smoother idle.

In a hybrid electric gasoline car the electric motor assists the gasoline engine in the low speed range by utilizing the high torque of the electric motor, as shown in Fig. 9.9. Currently manufactured hybrid electric gasoline cars

(Fig. 9.8) are equipped either with cage induction motors or PM brushless motors. A PM brushless motor can increase the overall torque by more than 50%. In most applications, the rated power of electric motors is from 10 to 75 kW. Because of limited space between the combustion engine and gear box as well as the need to increase the flywheel effect, electric motors for HEVs are short and have large diameters. AFPM brushless machines are pancake-type high torque density motors and fit perfectly the HEV requirements. An AFPM brushless machine can be liquid cooled and integrated with a power electronics converter.

In the HEV shown in Fig. 9.10 the energy is stored in a flywheel integrated with the AFPM machine. The vehicle may operate without an electrochemical battery. If the electromagnetic conversion process is included within the flywheel, there is no external shaft end, because the power flow to and from the machine is via electric connections to the stationary armature [3]. Thus, complex vacuum seals are avoided.

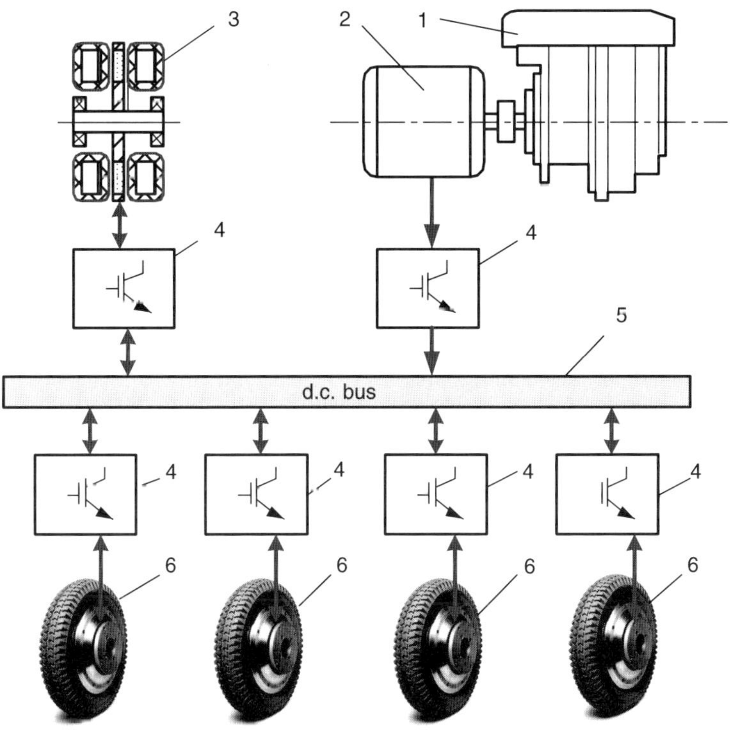

Fig. 9.10. HEV with flywheel energy storage: 1 — gasoline combustion engine, 2 — brushless generator, 3 — integrated flywheel–motor–generator, 4 – solid state converter, 5 — d.c. bus, 6 — electrical motorized wheel.

The moment of inertia and tensile stress are respectively

- for a rotating disc

$$J = \frac{1}{2} m R_{out}^2, \qquad \sigma = \frac{\rho \Omega^2 R_{out}^2}{3} \tag{9.1}$$

- for a rotating ring

$$J = \frac{1}{2} m (R_{in}^2 + R_{out}^2), \qquad \sigma = \frac{\rho \Omega^2}{3} (R_{in}^2 + R_{in} R_{out} + R_{out}^2) \tag{9.2}$$

where ρ is the specific mass density of the rotating body, m is the mass, $\Omega = 2\pi n$ is the angular speed, R_{out} is the outer radius and R_{in} is the inner radius. The kinetic energy is

- for a rotating disc

$$E_k = \frac{1}{2} J \Omega^2 = \pi^3 \rho d R_{out}^4 n^2 \quad \text{J} \tag{9.3}$$

- for a rotating ring

$$E_k = \pi^3 \rho d (R_{out}^4 - R_{in}^4) n^2 \quad \text{J} \tag{9.4}$$

where d is the thickness of the flywheel (disc or ring).

Combining together pairs of eqns (9.1), (9.3) and (9.2), (9.4), the stored energy density in a flywheel can be expressed as

- for a rotating disc

$$e_k = \frac{E_k}{m} = \frac{3}{4} \frac{\sigma}{\rho} \quad \text{J/kg} \tag{9.5}$$

- for a rotating ring

$$e_k = \frac{E_k}{m} = \frac{3}{4} \frac{R_{in}^2 + R_{out}^2}{R_{in}^2 + R_{in} R_{out} + R_{out}^2} \frac{\sigma}{\rho} \quad \text{J/kg} \tag{9.6}$$

For a very high strength steel, e.g. alloy steel (3% CrMoC) with $\sigma \approx 1000$ MPa and $\rho = 7800$ kg/m^3 [3], the stored energy density of a disc-type flywheel is $e_k = 96.1$ kJ/m^3. If we need to store a kinetic energy $E_k = 500$ kJ, the mass of the disc will be 5.2 kg.

9.2.2 Battery Electric Vehicles

Battery EVs are vehicles that use secondary batteries (rechargeable batteries or storage batteries) as their only source of energy. An EV power train can convert energy stored in a battery into vehicle motion and it can also reverse direction and convert the kinetic energy of the vehicle back into the storage battery through regenerative braking.

AFPM brushless motors are used in EVs as in-wheel motors (Figs 6.8 and 9.11) [215, 225]. The pancake shape of an AFPM motor allows for designing a compact motorized wheel.

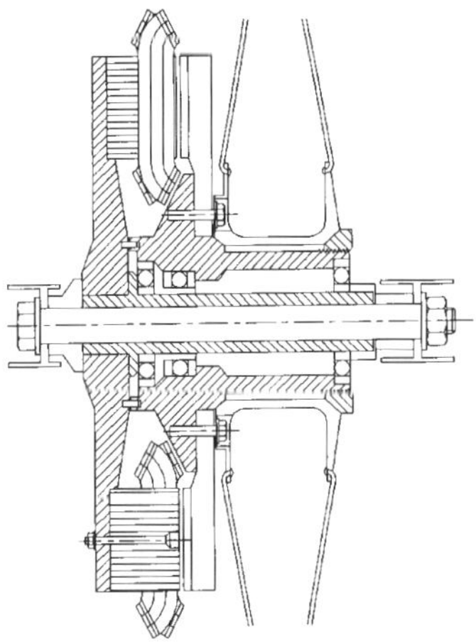

Fig. 9.11. Single-sided AFPM brushless motor fitted to spoked wheel [215].

With the aid of the AFPM brushless motor the differential mechanism can be replaced by an electronic differential system [101]. The configuration shown in Fig. 9.12a illustrates the use of a pair of electric motors mounted on the chassis to drive a pair of wheels through drive shafts, which incorporate constant velocity joints. In the configuration shown in Fig. 9.12b, the motors forming the electronic differential are mounted directly in the wheels of the vehicle. The electromechanical drive system is considerably simplified, when the motor is mounted in the wheel, because the drive shafts and constant velocity joints are now no longer needed. However, the resultant "unsprung" wheel mass of the vehicle is increased by the mass of the motor. Wheel motors of this direct drive configuration also suffer overloading because the speed of the rotor is lower than would be the case with a geared arrangement. This leads to an increase of the active materials volume required in the motor.

The disadvantages of conventional wheel motors can be overcome by using the arrangement shown in Fig. 9.13. The two stators are directly attached to the vehicle body whilst the PM rotor is free to move in radial directions. It can be observed that in this case the wheel and disc rotor form the unsprung mass, whilst the stators of the motor become sprung mass supported on the chassis [101].

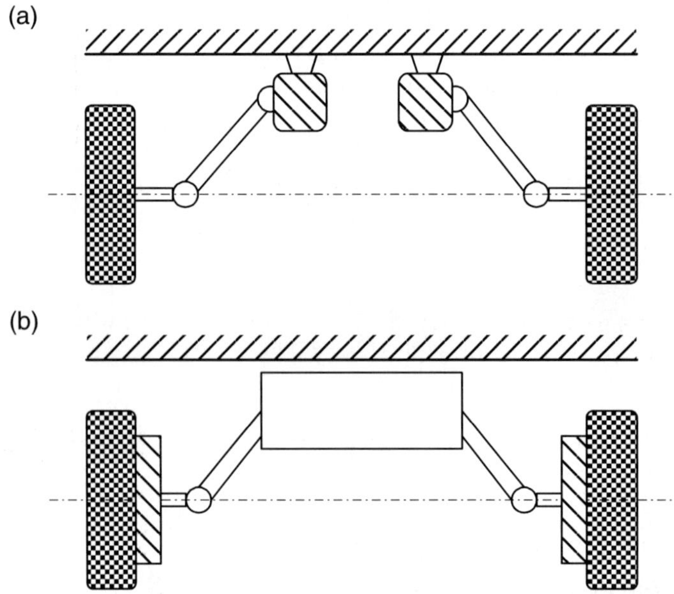

Fig. 9.12. Alternative forms of "electronic differential" drive schemes: (a) onboard motor, (b) wheel mounted motor.

Given the advantages of using wheel motor in electric vehicle drive system such as more onboard space, better vehicle control and modular drive train topology, there is a growing interest in automotive industry to develop wheel motors for future electric vehicles. Figure 9.14 shows the AFPM wheel motor developed at General Motors Advanced Technology Center, Torrance, CA, U.S.A. The developed wheel hub motor has a single stator located between two PM rotor discs. This topology features higher torque density design as it utilizes copper from both surfaces of the stator for torque production [66, 199, 223]. To remove heat from inside stator winding, provision is made of an aluminum ring with internal liquid coolant passages along end winding of the outer stator winding [P149]. The design specification of the GM's wheel motor is summarised in Table 9.3. To demonstrate the performance and robustness of the AFPM wheel motor technology, two wheel motors were fitted on GM's S-10 show truck and underwent several thousand miles road tests.

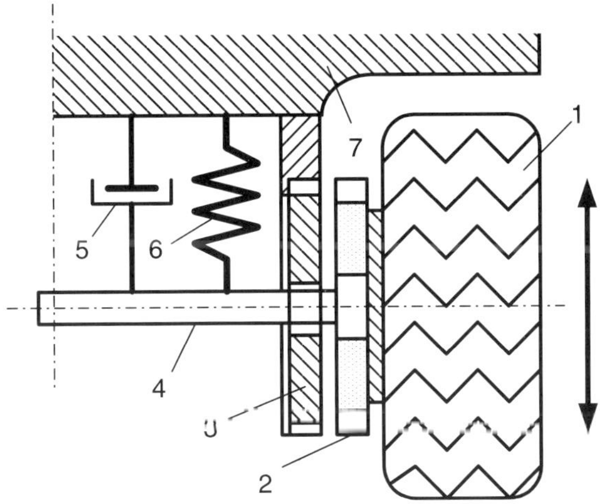

Fig. 9.13. Schematic representation of the disc-type machine with lower unsprung mass: 1 — wheel, 2 — disc rotor, 3 — stator, 4 — shaft, 5 — damper, 6 — spring, 7 — chassis.

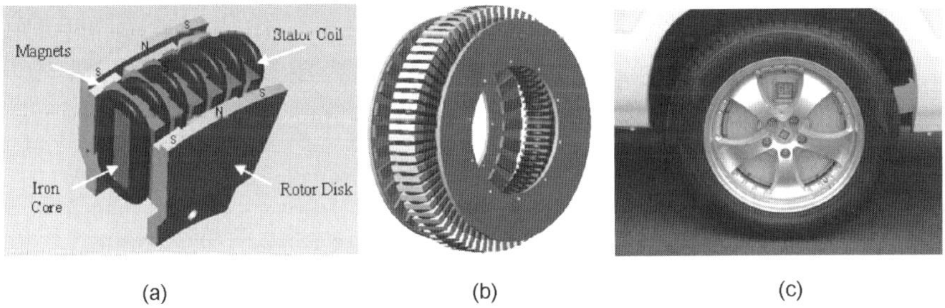

Fig. 9.14. GM's AFPM wheel motor: (a) Section of the AFPM motor showing details of slotted stator, (b) Basic layout of the AFPM motor showing complete stator located between two PM rotor discs, (c) AFPM wheel motor mounted on GM S-10 show truck. Courtesy of *General Motors Corporation*, U.S.A.

Table 9.3. GM's wheel motor design specification [66]

Number of phases	3
Number of poles	24
Peak torque, Nm	500
Stator outer diameter, m	340
Maximum axial length, m	75
Peak power, kW	25
Nominal bus voltage, V	280
Peak current, A	150
Maximum speed, RPM	1200

9.2.3 Fuel Cell Electric Vehicles

The fuel cell principle was first discovered by William Robert Grove in 1839 [272, 273]. Although fuel cells are well-proven technologies, they are presently still at a pre-commercial stage. Being widely regarded as a transforming technology, fuel cells offer the prospects of great improvements in energy efficiency for both mobile and stationary applications, and of a long-term transition to the hydrogen economy. In transport sector, the light-duty automotive applications are potentially the largest market opportunity available to fuel cell technology and have been the focus of intense development effort [76, 98]. Most major car manufacturers have on-going fuel cell vehicle programmes.

Fuel cells can be used with hydrogen or hydrocarbon fuels. The high cost of producing hydrogen, as compared to hydrocarbon fuels, has been a major barrier to its commercialisation, which would also require major infrastructural changes such as build-up of a hydrogen production and distribution infrastructure. However, with the continuing technology development in this fast-moving field and increasing possibility for introduction of renewable energy sources into fuel supply path, costs are expected to decline to an extent within the next decade.

Fig. 9.15. GM's Sequel fuel cell concept electric vehicle: (a) Front view of the AFPM wheel motor, (b) Back view of the AFPM wheel motor, (c) Sequel FCEV, (d) Rear AFPM wheel motor with electromechanical brake, rear steering actuator and integrated controlled damper. Courtesy of *General Motors Corporation*, U.S.A.

Recently developed GM Sequel FCEV is powered by an induction motor located in the front compartment and two AFPM wheel motor on the rear. Figure 9.15 shows front and back view of the AFPM wheel motor and the mechanical integration detail. This powertrain arrangement demonstrated excellent vehicle performance, driveability and safety as well as potential of axial flux motor technology. GM Sequel FCEV achieved a record-breaking range of 300 mile on a single tank of hydrogen.

9.3 Ship Propulsion

9.3.1 Large AFPM Motors

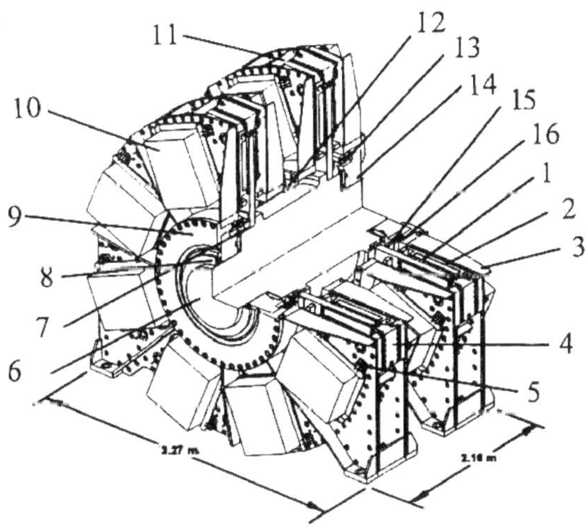

Fig. 9.16. Large power double disc AFPM brushless motor. 1 — PMs, 2 — stator assembly, 3 — housing, 4 — shock snubber, 5 — shock mount, 6 — rotor shaft, 7 — rotor disc clamp, 8 — shaft seal assembly, 9 — bearing retainer, 10 — stator segment, 11 — centre frame housing, 12 — spacer housing, 13 — rotor disc, 14 — bearing assembly, 15 — rotor seal runner, 16 — rotor seal assembly. Courtesy of *Kaman Aerospace*, EDC, Hudson, MA, U.S.A.

Stators of large AFPM brushless motors with disc type rotors usually have three basic parts [48]:

- aluminum cold plate;
- bolted ferromagnetic core;
- polyphase winding.

Table 9.4. Design data of large power, three phase AFPM brushless motors manufactured by *Kaman Aerospace*, EDC, Hudson, MA, U.S.A.

Quantity	PA44-5W-002	PA44-5W-001	PA57-2W-001
Number of poles, $2p$	28	28	36
Number of windings per phase		2	
Output power P_{out}, kW	336	445	746
Peak phase voltage, V	700	530	735
Rated speed, rpm	2860	5200	3600
Maximum speed, rpm	3600	6000	4000
Efficiency at rated speed	0.95	0.96	0.96
Torque at rated speed, Nm	1120	822	1980
Stall torque, Nm	1627	1288	2712
Continuous current (six step waveform), A	370	370	290
Maximum current, A	500	500	365
Peak EMF constant per phase, V/rpm	0.24	0.10	0.20
Winding resistance per phase at 500 Hz, Ω	0.044	0.022	0.030
Winding inductance per phase at 500 Hz, μH	120	60	100
Moment of inertia, kgm^2	0.9	0.9	2.065
Cooling		Water and glycol mixture	
Maximum allowable motor temperature, °C		150	
Mass, kg	195	195	340
Power density, kW/kg	1.723	2.282	2.194
Torque density, Nm/kg	5.743	4.215	5.823
Diameter of frame, m	0.648	0.648	0.787
Length of frame, m	0.224	0.224	0.259
Application	Traction, Drilling industry		General purpose

The *cold plate* is a part of the frame and transfers heat from the stator to the heat exchange surface. The slots are machined into a laminated core wound of steel ribbon in a continuous spiral in the circumferential direction. The copper winding, frequently a Litz wire, is placed in slots and then impregnated with a potting compound. The construction of a double disc AFPM motor developed by *Kaman Aerospace*, EDC, Hudson, MA, U.S.A. is shown in Fig. 9.16 [48]. Specifications of large axial flux motors manufactured by *Kaman* are given in Table 9.4.

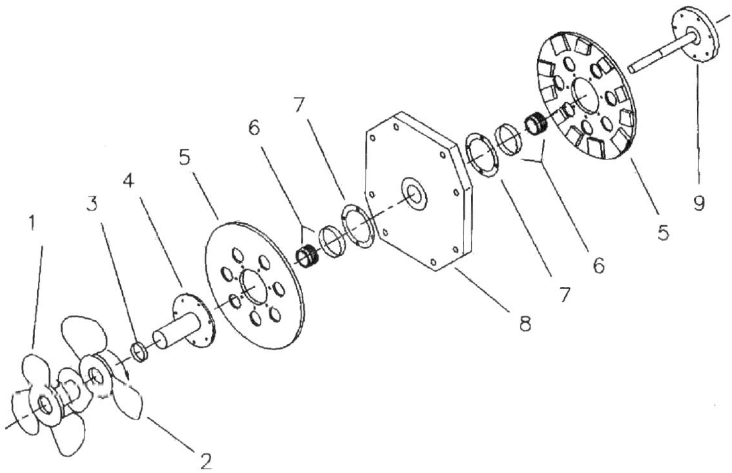

Fig. 9.17. Exploded view of the axial flux PM motor with counter-rotating rotor: 1 — main propeller, 2 — counter-rotating propeller, 3 — radial bearing, 4 — outer shaft, 5 — PM rotor, 6 — motor bearing, 7 — assembly ring, 8 — stator, 9 — inner shaft.

9.3.2 Propulsion of Unmanned Submarines

An electric *propulsion system for submarines* requires high output power, high efficiency, very salient and compact motors [64, 195]. Disc-type brushless motors can meet these requirements and run for over 100 000 h without a failure, cooled only by ambient sea water. These motors are virtually silent and operate with minimum vibration level. The output power at rated operating conditions can exceed 2.2 kW/kg and torque density 5.5 Nm/kg. The typical rotor linear speed of large marine propulsion motors is 20 to 30 m/s [195].

9.3.3 Counterrotating Rotor Marine Propulsion System

An AFPM brushless motor can be designed with the *counterrotation of two rotors* [39]. This machine topology can find applications in marine propulsion systems, which use an additional counter-rotating propeller in order to recover energy from the rotational flow of the main propeller stream. In this case the use of an AFPM motor having counter-rotating rotors allows the elimination of the motion reversal epicyclical gear.

The stator winding coils have a rectangular shape, which depends on the cross section of the toroidal core [39] (also see Fig. 2.4). Each coil has two active surfaces and each coil surface interacts with the facing PM rotor. In order to achieve the opposite motion of the two rotors, the stator winding coils have to be arranged in such a manner that counter-rotating magnetic fields are produced in the machine's annular air gaps. The stator is positioned between two rotors that consist of mild steel discs and axially magnetized NdFeB PMs. The magnets are mounted on the disc's surface from the stator sides. Each rotor has its own shaft that drives a propeller, i.e. the motor has two coaxial shafts that are separated by a radial bearing. The arrangement is shown in Fig. 9.17 [39].

Table 9.5. EMALS requirements [79]

End speed, m/s	103
Maximum peak–to–mean thrust force ratio	1.05
Cycle time, s	45
End speed variation, m/s	1.5
Maximum acceleration, m/s^2	56
Launch energy, MJ	122
Maximum thrust, MN	1.3
Maximum power, MW	133
Volume, m^3	<425
Mass, kg	<225 000

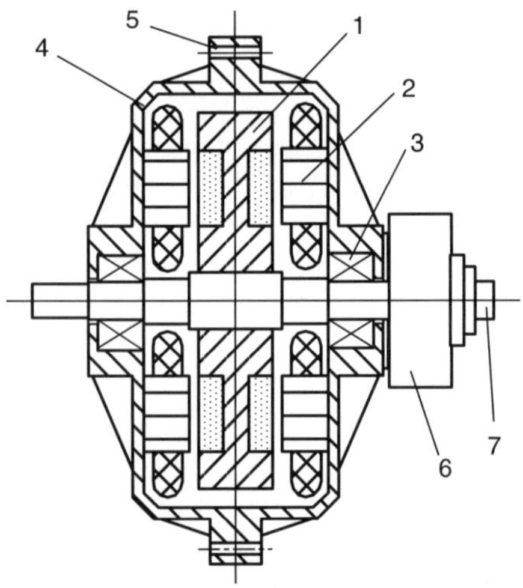

Fig. 9.18. AFPM synchronous machine for EMALS. 1 — PM rotor assembly, 2 — stator assembly, 3 — bearing, 4 — enclosure, 5 — mounting flange, 6 — brake, 7 — shaft encoder.

Table 9.6. Specifications of AFPM synchronous machine for EMALS [79]

Number of phases	6
Number of poles	$2p = 40$
Number of active stator slots	$s_1 = 240$
Type of stator	double-sided slotted
Number of stator windings	2
Generator winding	air gap side
Motor winding	bottom of slots
Maximum speed, rpm	6400
Maximum frequency, Hz	2133
Output power at maximum speed, MW	81.6
Maximum line-to-line EMF for motoring, V	1122
Maximum line-to-line peak voltage, V	1700
Maximum phase peak current, A	6400
Power losses, kW	127
Efficiency	0.893
Resistance per phase, mΩ	8.6
Inductance per phase, μH	10.4
Stator cooling	cold plate
Coolant	water and glycol mixture
Flow rate, l/min	151
Average temperature of winding, °C	84
Average temperature of stator core, °C	61
Type of PMs	sintered NdFeB
Remanence at 40°C, T	1.05
Air gap magnetic flux density, T	0.976
Tooth magnetic flux density, T	1.7
Maximum energy storage capacity, MJ	121
Energy density, kJ/kg	18.1
Mass, kg	6685

9.4 Electromagnetic Aircraft Launch System

Military aircrafts are catapulted from the decks of aircraft carriers with the aid of steam catapults. A steam catapult uses two parallel rows of slotted cylinders with pistons connected to the shuttle that tows the aircraft. The pistons under the steam pressure accelerate the shuttle until the aircraft will takeoff. Steam catapults have many drawbacks [79], i.e.:

- operation without feedback control;
- large volume (over 1100 m^3) and mass (up to 500 t);
- occupation of the "prime" real estate on the carrier and negative effect on the stability of the ship;
- low efficiency (4 to 6%);
- operational energy limit, approximately 95 MJ;
- need for frequent maintenance.

The *electromagnetic aircraft launch system* (EMALS) is free of the drawbacks listed above. The EMALS technology employs linear induction or synchronous motors fed from disc-type alternators through cycloconverters. The electrical energy obtained from the carrier power plant is stored kinetically in the rotors of AFPM synchronous generators. The energy is then released as a few-second pulses to accelerate and launch an aircraft. The cycloconverter between the AFPM generator and linear motor raises the voltage and frequency. The EMALS operates in a "real time" closed loop control [79]. The requirements are given in Table 9.5.

Specifications of the AFPM synchronous machine shown in Fig. 9.18 are given in Table 9.6 [79]. The EMALS uses four AFPM machines mounted in a torque frame and grouped in counter-rotating pairs to reduce the torque and gyroscopic effect. The rotor of the AFPM machine serves both as the kinetic energy storage when operating as a motor and a field excitation system when operating as a generator. The electric power from the on-board generators is fed to the AFPM machine via rectifier–inverters. There are two separate stator windings for motoring and generating mode. The motor winding is located at the bottom of slots to provide better thermal conductivity between the winding and enclosure.

Table 9.7. Specifications of ECI drilling systems with AFPM synchronous motors manufactured by *Tesco Corporation*, Calgary, Alberta, Canada

ECE drilling system	670	1007
Drilling unit without swivel		
Mass, kg	5,909	6,270
Operating length, m (including 2.8-m links and elevators)	4.34	4.34
Make-up Torque (typical), Nm	5,965	9,490
Breakout torque, Nm	7,592	11,389
Max. Drill Torque, Nm	5,098	7,860
Max. Speed, rpm	187	187
Power system (mechanical module)		
Approximate mass, kg	3,340	3,730
Length, m	6.4	6.4
Width, m	2.3	2.3
Power system (power module)		
Approximate mass, kg	3,545	4,320
Length, m	6.35 mm	6.30
Width, m	2.311	2.30

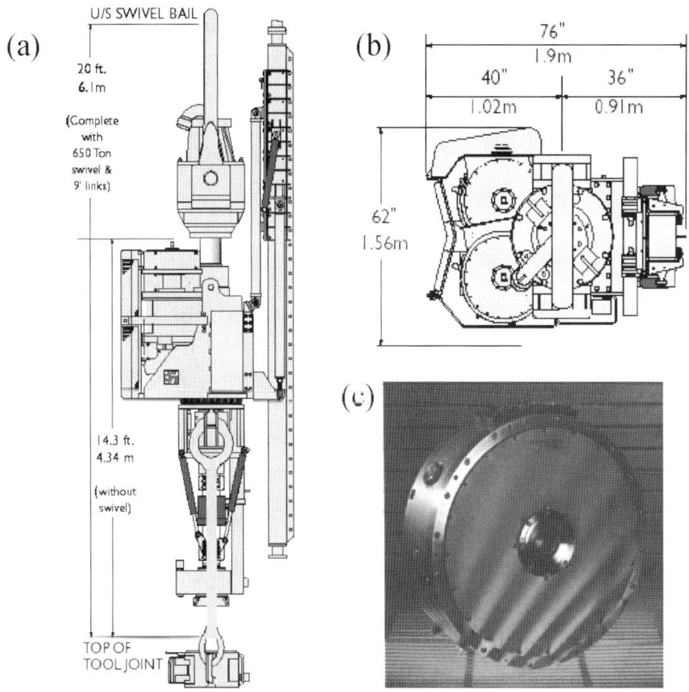

Fig. 9.19. ECI drilling system: (a) general view, (b) power unit with AFPM synchronous motors, (c) PA-44 *Kaman Aerospace* AFPM motor. Courtesy of *Tesco Corporation*, Calgary, Alberta, Canada.

9.5 Mobile Drill Rigs

Drill rigs vary from small electric motor driven tools to large oil field rigs. The basic elements of an aboveground drill rig are the power unit or motor, pump or air compressor for circulating drilling fluid to the bit and cleaning the borehole, drill head, hoisting drums and cables, derrick, mounting platform or deck and assorted equipment such as hammers for driving and removing casing, portable mud pit, racks for stacking the drill rods and samples, small tools for coupling or uncoupling and hoisting the drill string, etc.

The power unit (motor) performs the following functions:

- operates a drive weight mechanism for percussion or churn drilling or provides rotary motion to turn augers and coring equipment for rotary drilling operations;
- operates a winch for raising and lowering the drilling and sampling equipment;
- provides downward pressure for pushing boring and sampling equipment, or lifting and dropping a hammer to drive casing or sampling equipment.

For most drilling and sampling operations, the power source is the power take-off from the truck motor on which the drilling machine is mounted or from a separate engine, which is assigned or attached as an integral component of the drilling rig. It is estimated that approximately 90% of the motors are gasoline or diesel engines and 10% are compressed air or electric motors. A drive train that consists of gears or hydraulic pumps is used to convert the power supply to speed and torque for hoisting and rotating the drilling equipment. Most units have a transmission that allows 4 to 8 speeds for hoisting and drilling. In general, the hoisting capacity of the drill rig governs the depth of the borehole. A rule of thumb for selecting the power source is that the power required to hoist the drill rods should be about three times the power that is required to turn the drill string. For high elevations, the power loss is about 3% for each 300 m.

Large power AFPM brushless motors (Table 9.4) are especially suitable for portable drilling equipment because of their compact design, light weight, precise speed control, high efficiency and high reliability.

In the ECI lightweight *top drive drilling system* (Fig. 9.19) manufactured by *Tesco Corporation*, Calgary, Alberta, Canada, the traditional induction or brush type d.c. motors have been replaced with high performance liquid cooled AFPM synchronous motors. This system is compatible with most 600 V a.c. rig power systems. The product highlights include:

- *Kaman* PA44 AFPM synchronous motors that (Table 9.4) have been successfully tested to 60g triaxial loading, making them the ideal choice for rough drilling environments;
- AFPM synchronous motors that allow a high level of precision in speed and torque control not available with other motors;
- a modular design that allows drilling to continue at reduced power with one motor;
- direct connection to the rigs a.c. bus and interface with the existing power supply with minimal degradation to the rig power;
- the entire ECI system can be transported in three standard 6-m sea containers.

Specifications of the ECI drilling system are given in Table 9.7.

9.6 Oil Beam Pumps

An oil beam pump (Fig. 9.20) consists of an overground electromechanical drive and underground reciprocating piston pump. An electric motor drives one end of a heavy beam via pair of cranks with counterweights and transmission. The other end of the beam is equipped with the so called horse head with attached vertical cable. The cable is connected to a polished rod of a pump at the bottom of the well. The cranks move the beam up and down and actuates the pump. Typically, a beam pump delivers 5 to 40 liters of a crude oil-water mixture per each stroke.

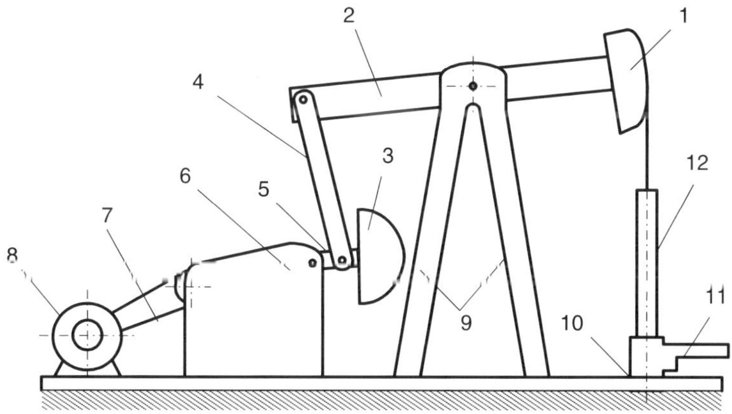

Fig. 9.20. Oil beam pump: 1 - horse head, 2 - walking beam, 3 - counterweights, 4 - Pitmans arm, 5 -cranks, 6 - transmission, 7 - sheaves and belts, 8 - electric motor or other prime mover, 9 - Samson beams, 10 - casing, 11- flow line, 12 - polishing rod.

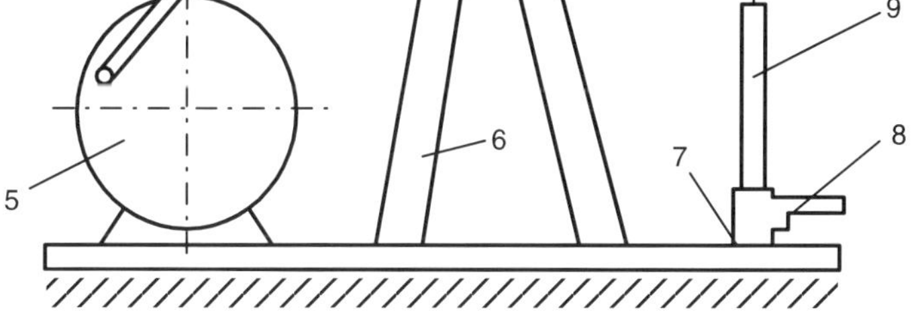

Fig. 9.21. Novel oil beam pump with AFPM brushless motor: 1 - horse head, 2 - walking beam, 3 counterweights, 4 - Pitmans arm, 5 - disc type motor 6 - Samson beams, 7 - casing, 8- flow line, 9 - polishing rod.

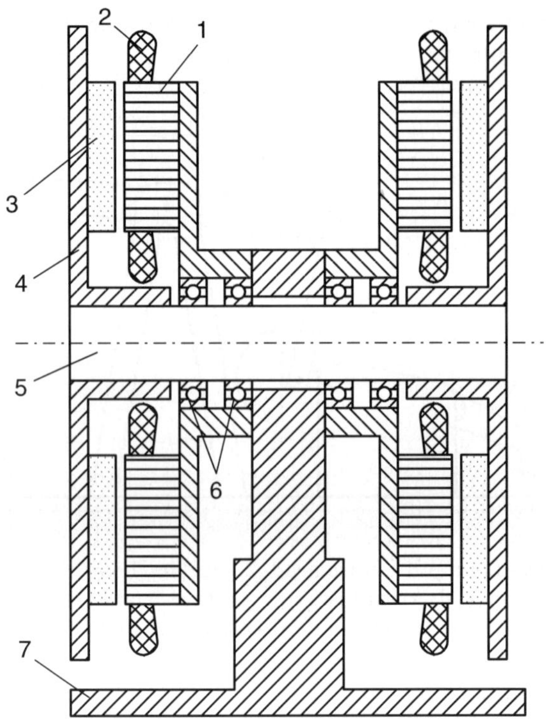

Fig. 9.22. AFPM brushless motor for oil beam pumps: 1 - stator core, 2 - stator winding, 3 - PM, 4 - rotor, 5 - shaft, 6 - bearing, 7 - base.

Table 9.8. Specifications of a prototype of AFPM brushless motor for beam pumps [130]

Rated power, kW	9.0
Rated torque, Nm	4820
Rated speed, rpm	20
Rated frequency, Hz	8
Rated voltage, V	380
Rated current, A	16
Number of rotor poles	48
Number of stator slots	144
Remanent magnetic flux density of PM, T	1.2
Thickness of PM, mm	10.0
Air gap, mm	4.0
Stator inner diameter, m	0.6
Stator outer diameter, m	0.9

The novel beam oil pump (Fig. 9.21) uses a low speed AFPM brushless motor, which replaces an induction motor, sheaves, belts and transmission [130]. There is a counterweight on the motor side of the beam that has an approximate weight as the horse head. Applications of an AFPM brushless motors simplifies the construction and maintenance of a beam pump, reduces its cost and improves the overall efficiency of the machine. Longitudinal section of the prototype of AFPM brushless motor is shown in Fig. 9.22 and specifications are given in Table 9.8 [130].

9.7 Elevators

The concept of *gearless electromechanical drive for elevators* was introduced in 1992 by *Kone Corporation* in Hyvinkää, Finland [113]. With the aid of a disc type low speed compact AFPM brushless motor (Table 9.10), the penthouse machinery room can be replaced by a space-saving direct electromechanical drive. In comparison with a low speed axial flux cage induction motor of similar diameter, the AFPM brushless motor has a doubled efficiency and a three times higher power factor.

Fig. 9.23a shows the propulsion system of the elevator while Fig. 9.23b shows how the AFPM brushless motor is installed between the guide rails of the car and the hoistway wall.

Table 9.9 contains key parameters for the comparison of different hoisting technologies [113]. The disc type AFPM brushless motor is a clear winner.

Specifications of single-sided AFPM brushless motors for gearless passenger elevators are given in Table 9.10 [113]. Laminated stators have from 96 to 120 slots with three phase short pitch winding and class F insulation. For example, the MX05 motor rated at 2.8 kW, 280 V, 18.7 Hz has the stator

Table 9.9. Comparison of hoisting technologies for 630 kg elevators [113]

Quantity	Hydraulic elevator	Warm gear elevator	Direct PM brushless motor elevator
Elevator speed, m/s	0.63	1.0	1.0
Motor shaft power, kW	11.0	5.5	3.7
Speed of motor, rpm	1500	1500	95
Motor fuses, A	50	35	16
Annual energy consumption, kWh	7200	6000	3000
Hoisting efficiency	0.3	0.4	0.6
Oil requirements, l	200	3.5	0
Mass, kg	350	430	170
Noise level, dB(A)	60..65	70...75	50...55

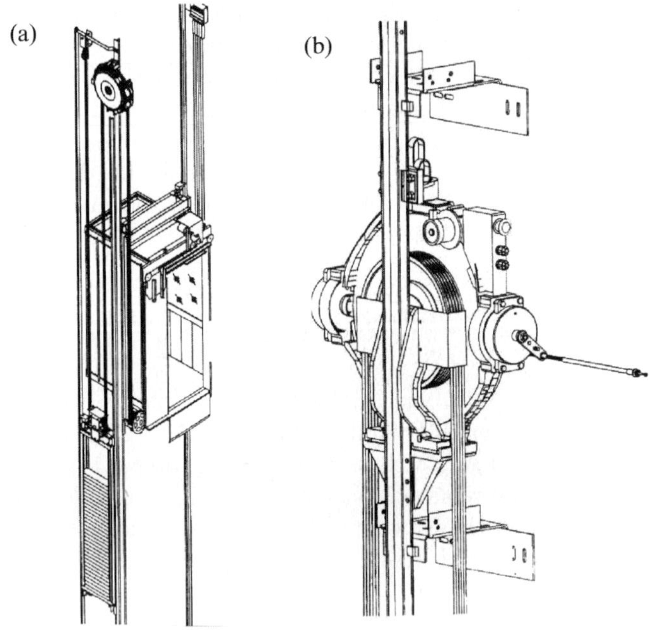

Fig. 9.23. MonoSpace^{TM} elevator: (a) elevator propulsion system; (b) Ecodisk^{TM} motor. Courtesy of *Kone*, Hyvinkää, Finland.

Table 9.10. Specifications of single-sided AFPM brushless motors for gearless elevators manufactured by *Kone*, Hyvinkää, Finland

Quantity	MX05	MX06	MX10	MX18
Rated output power, kW	2.8	3.7	6.7	46.0
Rated torque, Nm	240	360	800	1800
Rated speed, rpm	113	96	80	235
Rated current, A	7.7	10	18	138
Efficiency	0.83	0.85	0.86	0.92
Power factor	0.9	0.9	0.91	0.92
Cooling	natural	natural	natural	forced
Diameter of sheave, m	0.34	0.40	0.48	0.65
Elevator load, kg	480	630	1000	1800
Elevator speed, m/s	1	1	1	4
Location	hoistway	hoistway	hoistway	machine room

winding resistance $R_1 = 3.5\ \Omega$, stator winding reactance $X_1 = 10\ \Omega$, $2p = 20$, sheave diameter 340 mm and weighs 180 kg.

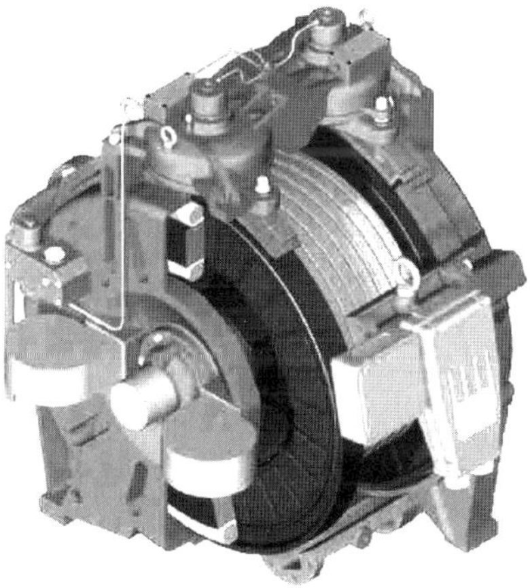

Fig. 9.24. Double-disc AFPM brushless motor for gearless elevators. Courtesy of *Kone*, Hyvinkää, Finland.

Table 9.11. Specifications of double disc AFPM brushless motors manufactured by *Kone*, Hyvinkää, Finland

Quantity	MX32	MX40	MX100
Rated output power, kW	58	92	315
Rated torque, Nm	3600	5700	14,000
Rated speed, rpm	153	153	214
Rated current, A	122	262	1060
Efficiency	0.92	0.93	0.95
Power factor	0.93	0.93	0.96
Elevator load, kg	1600	2000	4500
Elevator speed, m/s	6	8	13.5

A double disc AFPM brushless motor for gearless elevators is shown in Fig. 9.24 [113]. Table 9.11 lists specifications data of double disc AFPM brushless motors rated from 58 to 315 kW [113]. The double disc configuration offers high output torque and compensation of axial forces [P105].

Fig. 9.25. Penny-motor. Photo courtesy of *Mymotors & Actuators GmbH*, Wendelsheim, Germany.

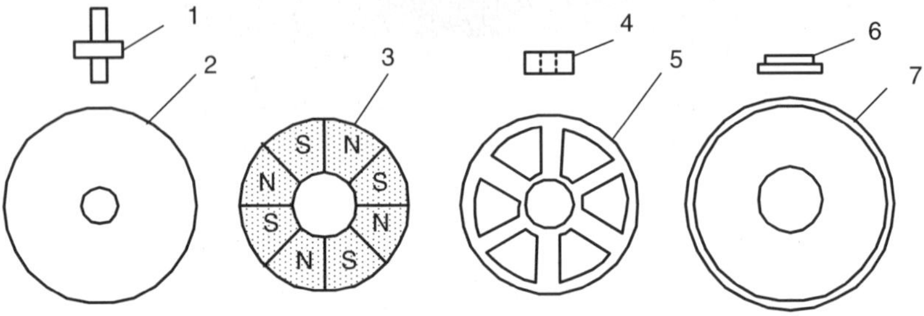

Fig. 9.26. Construction of penny-motor: 1 — shaft, 2 — soft steel yoke cover, 3 — PM ring, 4 — ball bearing, 5 — stator winding, 6 — flange, 7 — bottom steel yoke.

9.8 Miniature AFPM Brushless Motors

Ultra-flat PM micromotor, the so called *penny-motor* is shown in Figs 9.25 and 9.26 [154]. The thickness is 1.4 to 3.0 mm, outer diameter about 12 mm, torque constant up to 0.4 μNm/mA and speed up to 60,000 rpm. A 400-μm eight pole PM and a three-strand, 110-μm disc shaped, lithographically produced stator winding have been used [154]. Plastic bound NdFeB magnets are a cost effective solution. However, the maximum torque is achieved with sintered NdFeB magnets. A miniature ball bearing has a diameter of 3 mm. Penny-motors find applications in miniaturized hard disc drives, cellular phones as vibration motors, mobile scanners and consumer electronics.

Moving Magnet Technologies (MMT) two or three phase, *miniature AFPM brushless motors* (Fig. 9.27) are well adapted to high volume low cost production. In order to obtain rigidity the stator coils are overmoulded. The stator is mounted on a single-sided printed circuit board. Each part can be manufactured using standard moulding, stamping techniques and automatic coil winding with a specific design for simple and efficient assembly. Three position

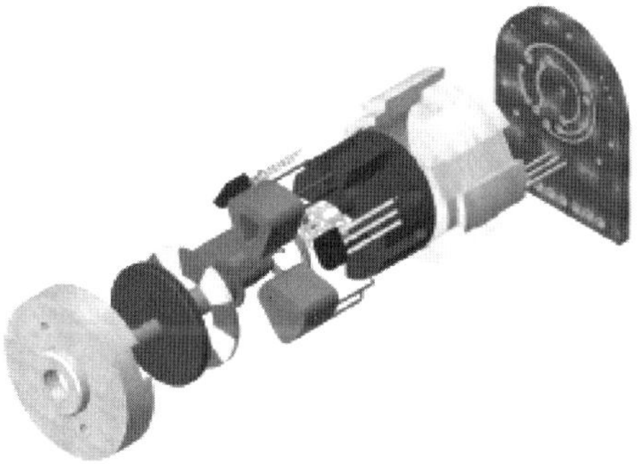

Fig. 9.27. Two-phase, four-coil, single-sided 15-mm AFPM brushless motor. Photo courtesy of *Moving Magnet Technologies, SA*, Besancon, France.

Fig. 9.28. Miniature two-phase, four-coil, 4-pole, single-sided AFPM rotary actuator. Photo courtesy of *Moving Magnet Technologies, SA*, Besancon, France.

sensing methods for the closed-loop control are used: (a) digital Hall probes located in the stator, (b) EMF signal (sensorless mode) and (c) absolute analogue position sensor (position sensor mode).

MMT AFPM *miniature rotary actuators* (Fig. 9.28) have been designed for automotive applications to provide an efficient, contactless, direct drive rotary motion on a limited stroke. Ring, disc and tile shaped PMs have been used. The main features of this family of actuators are:

- contactless actuation principle;
- constant torque independent of the angular position for a given current;
- linear torque–current characteristic;
- two directions of rotation;
- high torque density.

Additional functions, such as a magnetic return spring or analogue contactless position sensing can also be implemented. Owing to the linearity of the torque-current characteristic and its independence of position, it is possible to operate actuators in an open-loop against a spring or in a simple closed-loop mode by using a position sensor.

9.9 Vibration Motors

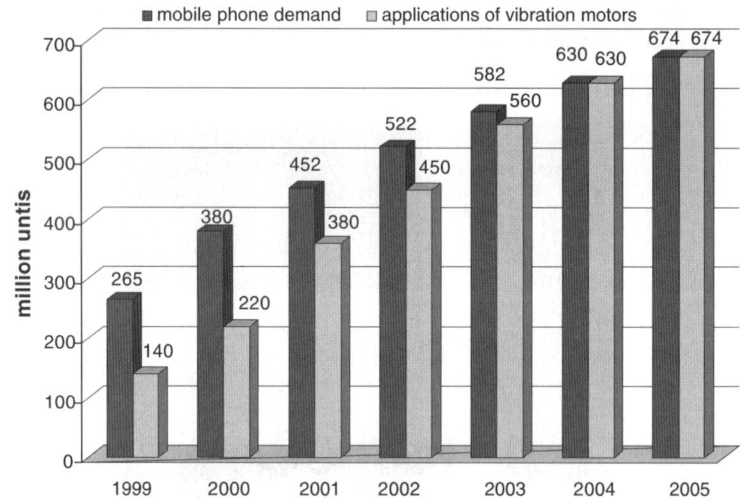

Fig. 9.29. Worldwide demand estimate for mobile phones and vibration motors.

Advances in cellular telecommunication have made *mobile phones* to be highly popular communication tool in modern society. Since a mobile phone is now a gadget as necessary as a watch or wallet, small vibration motors with diameters from 6 to 14 mm are manufactured in unbelievable large quantities (Fig. 9.29). The trends in *vibration motors* for mobile phones include reduction of mass and size, minimization of energy consumption and guaranteed stable vibration alarming in any circumstances [62].

There are two types of brushless vibration motors for mobile phones: *cylindrical* or RFPM motor and *coin type* or AFPM motor (Fig. 9.30). The *unbalanced exciting force* generated by an eccentric rotor is

$$F = m\epsilon\Omega = 2\pi m\epsilon n^2 \tag{9.7}$$

where m, ϵ and n denote the rotor mass, eccentricity and rotational speed respectively. The rotational speed is the most effective design parameter to increase the unbalanced exciting force [62].

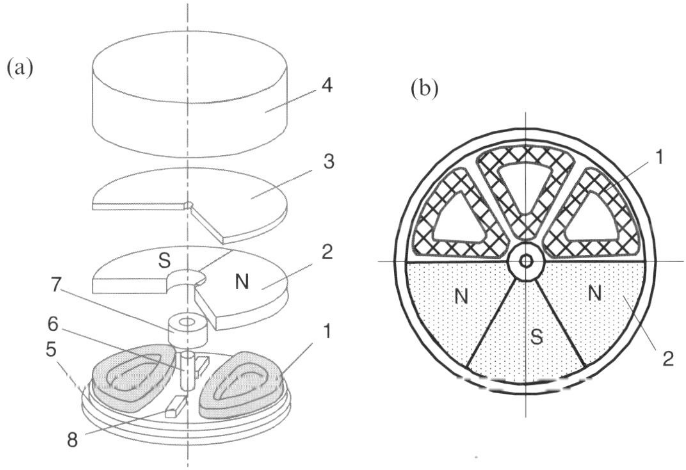

Fig. 9.30. Coin type AFPM vibration motor for mobile phones [62]: (a) two coil motor; (b) multicoil motor. 1 — phase coil, 2 — PM (mechanically unbalanced system), 3 — ferromagnetic yoke, 4 — cover, 5 — base plate, 6 — shaft, 7 — bearing, 8 — detent iron [62].

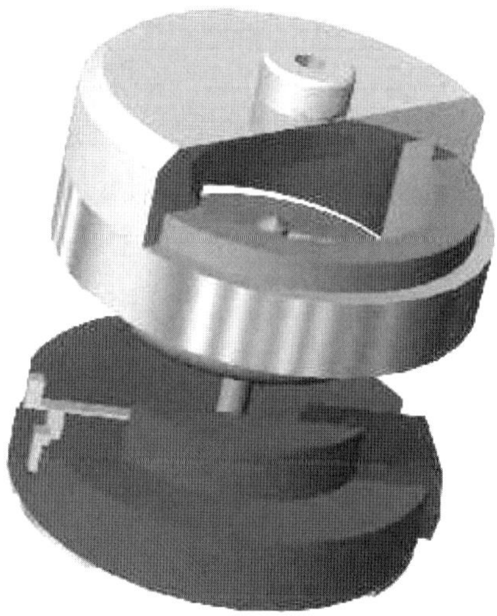

Fig. 9.31. Cost effective AFPM brushless vibration motor. Photo courtesy of *Moving Magnet Technologies, SA*, Besancon, France.

Moving Magnet Technologies single-phase AFPM brushless vibration motors provide a strong vibration feeling and silent alert with a very simple design (Fig. 9.31). Fabrication of *MMT* vibration motors is cost effective due to contactless design, scalable size and slim shape.

9.10 Computer Hard Disc Drives

The data storage capacity of a *computer hard disc drive* (HDD) is determined by the aerial recording density and number of discs. It is expected that the aerial density will soon increase from 6 Gb/cm^2 = 38.7 Gb/in^2 to 15.5 Gb/cm^2 = 100 Gb/in^2. The mass of the rotor, moment of inertia and vibration increase with the number of discs.

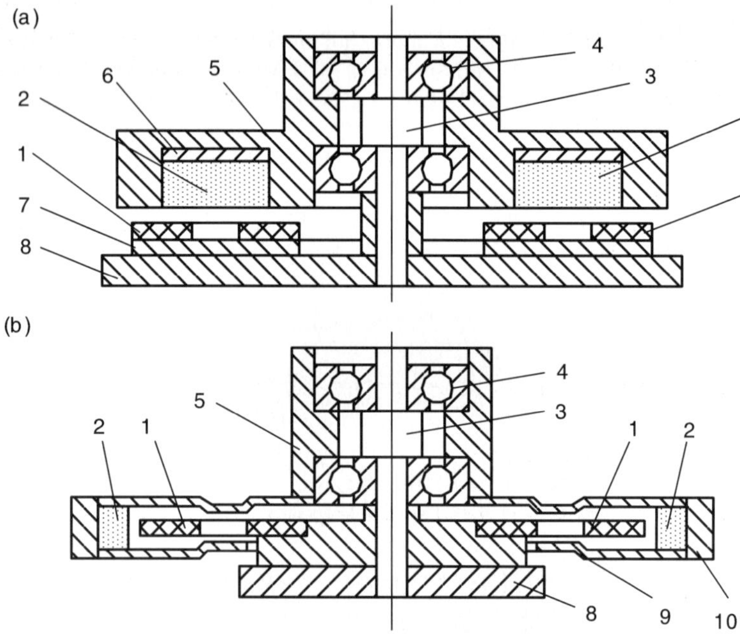

Fig. 9.32. Construction of computer HDDs with AFPM brushless motors: (a) single-sided motor; (b) double-sided motor. 1 — stator coil, 2 — PM, 3 — shaft, 4 — bearing, 5 — hub, 6 — rotor ferromagnetic yoke, 7 — stator ferromagnetic yoke, 8 — base plate, 9 — pole fingers, 10 — nonmagnetic ring.

Special design features of computer HDD motors are their high starting torque, limited current supply, low vibration and noise, physical constraints on volume and shape, protection against contamination and scaling problems. A high starting torque, 10 to 20 times the running torque is required, since the read/write head tends to stick to the disc when not moving. The starting current is limited by the computer power supply unit, which in turns, can severely limit the starting torque.

AFPM brushless motors (Fig. 9.32) can develop higher starting torque than their radial field counterparts. The additional advantage is zero cogging torque. The drawback of the single-sided AFPM HDD motor shown in Fig. 9.32a is the high normal attractive force between the stator with ferromagnetic yoke and rotor mounted PMs. In double-sided HDD AFPM motors the stator does not have any ferromagnetic core and no normal attractive force is produced at zero-current state. The stator has a three-phase winding with coreless coils fabricated with the aid of a lithographic method. In a typical design, there are two coils per phase for $2p = 8$ PM poles and three coils per phase for $2p = 12$ PM poles. To reduce the air gap and increase the magnetic flux density, the so called *pole fingers* are created in the lower part of the hub and the rotor magnetic circuit is bended by pressing it towards the stator coil centres (Fig. 9.32b). According to [140], for a HDD AFPM brushless motor with a 51-mm stator outer diameter, $N_1 = 180$ turns per phase and torque constant $k_T = 6.59 \times 10^{-3}$ Nm/A the current consumption is 0.09 A at 13 900 rpm and no load condition.

The acoustic noise of HDD brushless motors with ball bearings is usually below 30 dB(A) and projected mean time between failures MTBF = 100 000 hours. HDD spindle motors are now changing from ball bearing to fluid dynamic bearing (FDB) motors. Contact-free FDBs produce less noise and are serviceable for an extended period of time.

9.11 Ventricular Assist Devices

When electricity was new, people had high hopes that it had curative powers. For example, electropathy (electrodes between patient's hands and ailing body part), very popular from 1850 to 1900, promised to cure most diseases and conditions, including mental illness. The 21st century biomedical engineering community has resurrected magnetic fields to treat depression, e.g. magnetic seizure therapy (high frequency, powerful electromagnets) or transcranial magnetic stimulation (strong pulse magnetic fields).

At present, many medical devices use small permanent magnet (PM) electric motors as, for example, high-quality pumps, centrifuges, infusion pumps, hemodialysis machines, precision handpieces and implantable devices (ventricular assist devices, pacemakers, defibrillators, nerve stimulators, etc). Since the reliability of medical products is critical, the electric motor is considered a precision component rather than a commodity device [203]. This section focuses on very small PM brushless motors for implantable devices, in particular, motors for rotary blood pumps [102].

A left ventricular assist device (LVAD) is an electromechanical pump implanted inside the body and intended to assist a weak heart that cannot efficiently pump blood on its own. It is used by end-stage heart failure patients who are unable to receive a heart transplant due to donor availability, eligibility, or other factors.

Motor driven pumps implanted in the human body must be free of shaft seals. This problem can be solved by embedding PMs in the pump rotor placed in a special enclosure and driven directly by the stator magnetic field. In this case the nonmagnetic air gap is large and high energy PMs are required.

Electromagnetic pumps for LVADs can be classified into three categories:

- 1st generation (1G), i.e. electromagnetic pulse pumps;
- 2nd generation (2G), i.e. electromagnetic rotary pumps;
- 3rd generation (3G), i.e. electromagnetic pumps with magnetic or hydrodynamic bearings.

Electromagnetic 1G pumps were driven by electromagnets, linear oscillating motors or linear short-stroke actuators. The pump, integrated with a linear actuator, is heavy, large and noisy.

DuraHeart^R 3G LVAD developed by *Terumo Corporation*, Tokyo, Japan, combines centrifugal pump with magnetic levitation technologies (Figs 9.33 and 9.34) [102]. Magnetic levitation allows the impeller to be suspended within the blood chamber by electromagnets and position sensors. The three-phase, 8-pole, axial flux PM brushless motor with slotless stator resembles a floppy disc drive spindle motor (Fig. 9.33). NdFeB PMs are integrated with the impeller. The output power of the motor is 4.5 W, speed 2000 rpm and torque 0.0215 Nm [226].

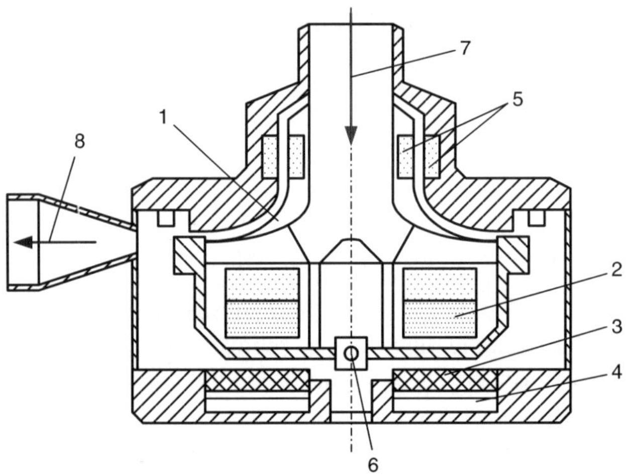

Fig. 9.33. *DuraHeart*^{R} centrifugal rotary pump with axial flux PM brushless motor and magnetic levitation bearings: 1 - impeller, 2 - PM, 3 - stator winding, 4 - stator core, 5 - magnetic suspension, 6 - ceramic pivot, 7 - inlet, 8 - outlet.

Fig. 9.34. Exploded view of *DuraHeart*^{R} centrifugal rotary pump. Courtesy of *Terumo Corporation*, Tokyo, Japan.

An axial flux slotless motor integrated with a centrifugal blood pump is shown in Figs. 9.35 and 9.36. The so-called *VentrAssist*^{TM} manufactured now by Australian company *Ventracor* is a new cardiac LVAD, which has only one moving part - a hydrodynamically suspended impeller integrated with a PM rotor. The hydrodynamic forces act on tapered edges of the four blades. The stator of the brushless electric motor is of slotless type and has only upper and lower coils. The three coil winding and four pole rotor use the second harmonic of the magnetic field wave to produce the torque. To provide redundancy, body coils and cover coils are connected in parallel so that the motor still can run even if one coil is damaged.

The housing and impeller shell are made of titanium alloy Ti-6Al-4V. Vacodym 510 HR NdFeB PMs are embedded into impeller. To reduce the reluctance for the magnetic flux, laminated silicon steel return paths (yokes) are designed.

The 2D FEM simulation of the magnetic field distribution is shown in Figs 9.37 and 9.38 [102]. The measured performance characteristics are shown in Figs 9.39a and 9.39b [226]. For output power between 3 and 7 W and speed between 2000 and 2500 rpm, the efficiency is from 45 to 48% (Fig. 9.39a). At 3 W and 2250 rpm the winding losses are 1.7 W, eddy current loses in titanium

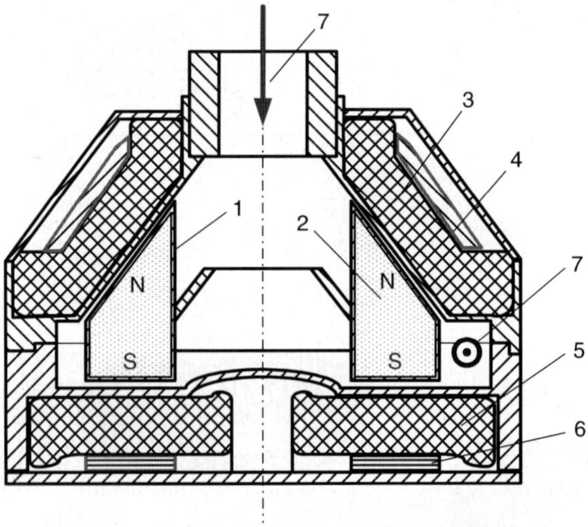

Fig. 9.35. Longitudinal section of *VentrAssist*^{TM} hydronamically levitated centrifugal blood pump: 1 - impeller, 2 - PM, 3 - body coil, 4 - body yoke, 5 - cover coil, 6 - cover yoke, 7 - inlet, 8 - outlet.

Fig. 9.36. Computer generated 3D image of *VentrAssist*^{TM} centrifugal blood pump. Photo Courtesy of *Ventracor Global Company*, Australia, www.ventracor.com.

1.0 W and losses in laminated yokes are 0.7 W [226]. At load torque 0.03 Nm the fundamental phase current is 0.72 A (Fig. 9.39b).

The device weighs 298 g and measures 60 mm in diameter, making it suitable for both children and adults.

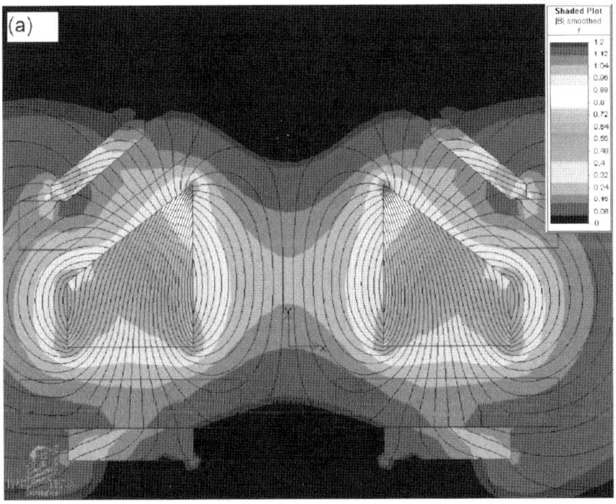

Fig. 9.37. Magnetic flux (contour plot) and magnetic flux density (shaded plot) excited by NdFeB PMs at no load [102].

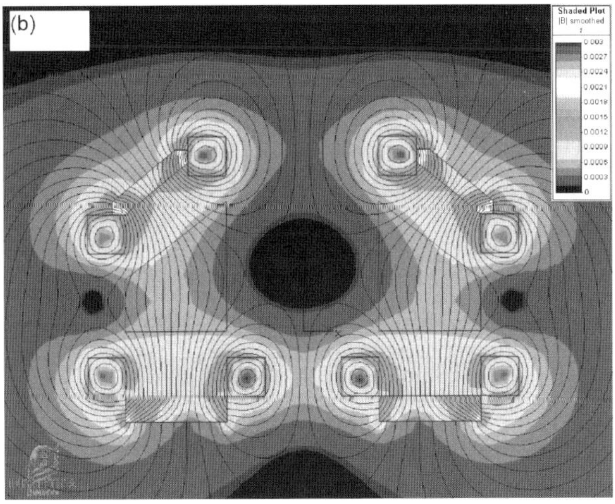

Fig. 9.38. Magnetic flux (contour plot) and magnetic flux density (shaded plot) excited by armature (stator) coils [102].

9.12 Axial Flux Machines with Superconducting Field Excitation System

Superconducting excitation system can provide stronger magnetic field in the air gap than the best rare-earth PMs. A superconducting axial flux machines is at least 30% smaller in volume than an equivalent PM machine.

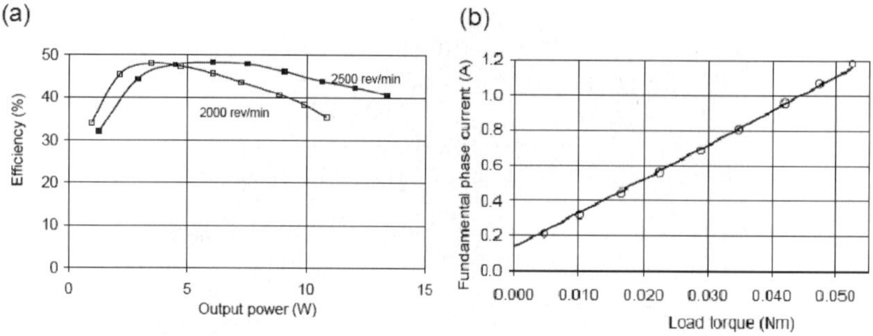

Fig. 9.39. Steady state performance characteristics of *VentrAssist*^{TM} BA2-4 pump driven by a six-step sensorless inverter: (a) efficiency; (b) fundamental phase current calculated on the basis of EMF (solid line) and obtained from laboratory tests (circles) [226].

A Japanese industry-academia group, called *frontier group*, which includes *Sumitomo Electric Industries, Ltd.*[2] (SEI), the manufacturer of HTS wires, has developed a 12.5 kW synchronous motor with HTS BiSrCaCuO (BSCCO) field excitation winding [249, 252]. This motor was integrated with a pod propulsor and completed in 2003.

The pod propulsor has outer diameter of 0.8 m and is 2 m long. The diameter of the propeller is 1 m. When a large conventional motor is used with a pod propulsor, the outside diameter of the propulsor becomes also large.

This HTS motor is an axial flux (disc type) three phase, eight-pole machine. The stationary field excitation system and rotating armature simplify the cooling system. Technical specifications of the HTS motor are given in Table 9.12. The expanded view of the motor is shown in Fig. 9.40. The main features include:

- machine is cooled by low-cost, easy-to-handle liquid nitrogen;
- uses the bismuth-based 1G BSCCO HTS wires that were fabricated in the lengths of several kilometers with the aid of SEI innovative manufacturing process;
- very low noise;
- very low leakage magnetic flux;
- the surface of the machine is kept at a room temperature and, therefore, can be installed in used in any place.

[2] *Sumitomo Electric Industries, Ltd.*, *Fuji Electric Systems Co., Ltd.*, *Hitachi, Ltd.*, *Ishikawajima-Harima Heavy Industries Co., Ltd.*, *Nakashima Propeller Co., Ltd.*, *Taiyo Nippon Sanso Corporation*, University of Fukui (H. Sugimoto)

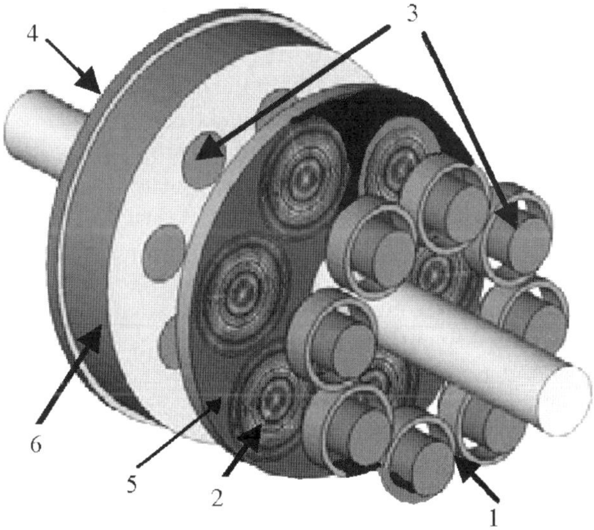

Fig. 9.40. Computer created 3D image of the axial flux 12.5 kW HTS motor for pod propulsor: 1 — HTS field excitation coils, 2 — armature excitation coil, 3 — cores, 4 — back yoke, 5 — fiberglass reinforced plate of cryostat, 6 — cryostat.

Table 9.12. Specifications of Japanese 12.5 kW HTS axial flux motor

Rated output power, kW	12.5
Rated torque, Nm	1194
Short time duty power, kW	62.5
Speed, rpm	100
Armature phase rms current, A	30
Armature current density, A/mm^2	5.0
Liquid nitrogen temperature, K	66
Diameter, m	0.65
Length, m	0.36
Number of turns of HTS field excitation coil	330
Dimensions of HTS BSC wire, mm	W – 4.3, T – 0.22
Diameters of HTS coil, mm	ID = 136, OD = 160
Filed excitation pole core diameter, mm	96
Number of turns of armature coil	850
Diameter of armature coil, mm	208
Diameter of armature copper wire, mm	2

The rotating armature is located in the center of the motor between twin stators. The coreless armature winding consists of 6 pancake coils wound with a copper wire. Coreless winding does not produce any cogging torque. The motor is furnished with slip rings and brushes to supply the heavy electric current to each of the armature phase windings, which is the main disadvantage.

The stationary field excitation system is placed in fiberglass reinforced plastic (FRP) cryostats. The FRP cryostat is lightweight, strong and does not conduct electric current (no eddy-current losses). The temperature of liquid nitrogen is 66 K in order to increase the field current. There are two field excitation systems (one on each side of the armature) with eight ferromagnetic poles and eight HTS coils each. The pole cores are concentric with respect to the HTS coils, which results in limited penetration of the magnetic flux into the coils. The pole cores protrude outside the cryostat where they are mechanically fixed to the disc-shaped ferromagnetic yokes, which create a back path for the magnetic flux. The cores are wound with ferromagnetic ribbon to reduce the eddy current losses. Since cores are outside the cryostat, the loss to cool cores is zero.

The maximum magnetic flux density in the core is only 1.8 T, so that the magnetic circuit is unsaturated. The maximum normal component of the magnetic flux density penetrating to HTS coils is 0.29T, i.e. below the value corresponding to critical current.

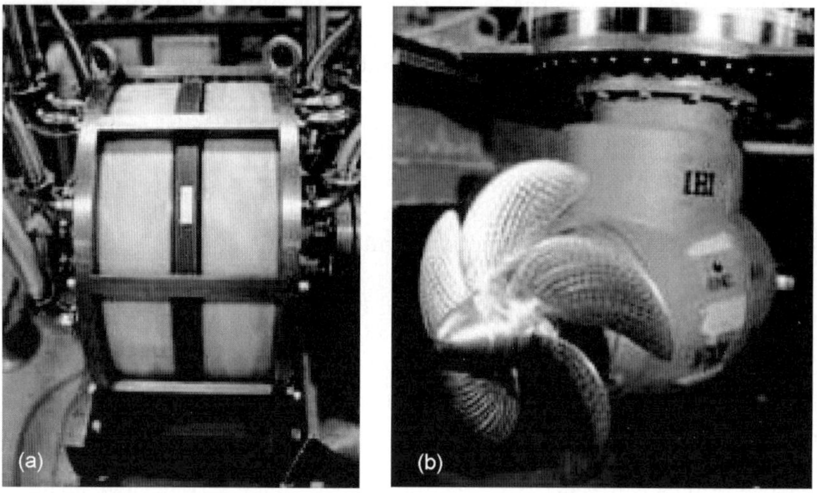

Fig. 9.41. Japanese 12.5 kW HTS motor for pod propulsor: (a) motor; (b) pod propulsor integrated with the motor.

The 12.5 kW HTS motor and pod propulsor are shown in Fig. 9.41. Performance characteristics obtained from laboratory tests are shown in Fig. 9.42. The maximum efficiency is 97.7 % and it remains practically constant for the load from 60 to 210 Nm. However, the cooling loss is not taken into account. The motor was loaded far below the rated torque, which is 1194 Nm (Table 9.12). The maximum shaft torque at laboratory tests was only 210 Nm, which corresponds to 2.2 kW shaft power at 100 rpm. The frontier group justifies it by the limit on the load equipment.

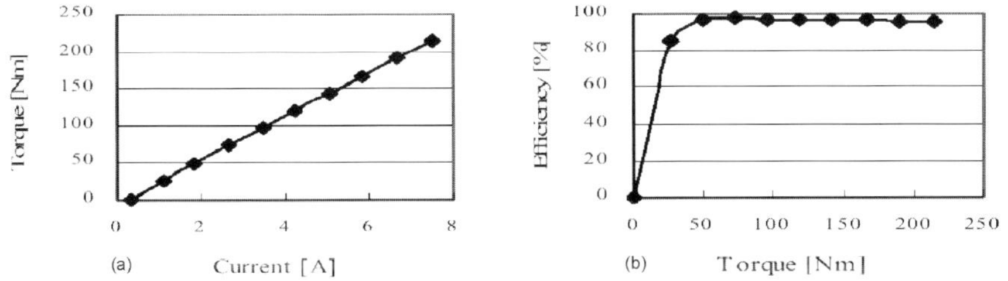

Fig. 9.42. Performance characteristics of the 12.5 kW HTS machine obtained from laboratory tests: (a) voltage versus speed (generating mode); (b) efficiency versus shaft torque (motoring mode).

The axial flux HTS motor has a small size and is characterized by low energy consumption. It is expected to find potential applications not only in pod propulsors, but also in railcar motors, wind generators and large industrial electromechanical drives, e.g. steel rolling mills.

The group has a plan to develop a commercial 400 kW, 250 rpm HTS motor. The development will include the test at a wharf the contra-rotating propeller pod propulsion system incorporating 2 units of a 400 kW full superconducting motor. The 400 kW motor is proposed for a coasting vessel. Further plans include development of larger motors 2.5 MW and 12.5 MW. Instead of BSCCO, coated YBCO wire will be the mainstream HTS wire. With YBCO wire the current density is considerably improved.

Another Japanese industry-academia group[3] has developed a 15 kW, 720 rpm, coreless axial flux HTS motor, similar to that shown in Figs 9.40 and 9.41. *Kitano Seiki* is marketing this motor [153].

The axial gap superconductive motor uses GdBaCuO bulk HTS for rotor excitation system and HTS tapes for the stator (armature) winding. The rotor system is in the central position of the machine and the twin stator (armature) is on both sides of the rotor (Fig. 9.43). The rotor has 8 poles 26-mm diameter each. The armature winding has 6 poles. The rotor assembly is 0.3 m long and is 0.5 m in diameter. The magnetic flux density in the air gap is 3 T at 77K under field cool magnetization. The bulk HTS poles are magnetized inside the machine using a pulsed-copper coils. The armature system is stationary so that there are no slip rings, only armature leads. The rotor field excitation system with bulk HTS does not require any leads. No electric power supply is needed to maintain the magnetic field. The machine can operate both as a motor and generator. The prototype has been designed for marine pod propulsors. Potential application to medical, biomedical, and environmental protection are under consideration.

[3] *Fuji Electric Systems*, *Fukui University*, *Hitachi*, *Ishikawa-hajima Heavy Industry* (IHI), *Kitano Seiki*, *Nakashima Propeller Co., Ltd.*, *Niigata Motors*, SEI, *Taiyou Nissan* and Tokyo University of Marine Science and Technology.

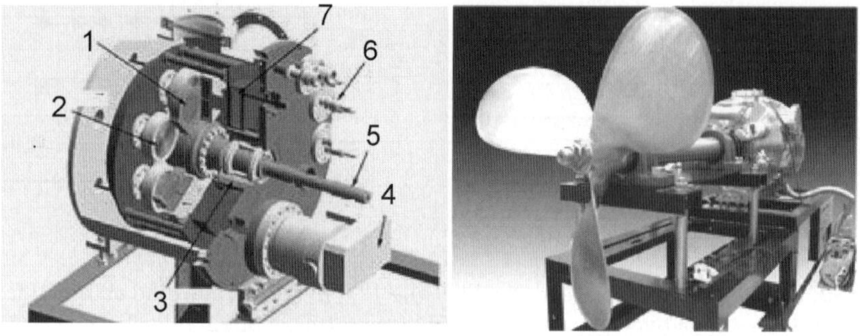

(a) (b)

Fig. 9.43. Axial flux brushless machine with bulk HTS rotor developed by *Kitano Seiki*, Japan: (a) computer 3D image; (b) motor and propeller on test bed. 1 — HTS bulk magnet, 2 — armature coil, 3 — magnetic seal unit, 4 — vacuum pump, 5 — liquid nitrogen inlet, 6 — armature winding terminal, 7 — coil cooling layer.

Numerical Example 9.1

Find the kinetic energy and tensile stress of the ring with dimensions $R_{out} = 0.142$ m, $R_{in} = 0.082$m, thickness $d = 0.022$m and density $\rho = 7800$ kg/m^3 rotating at the speed of 30 000 rpm.

Solution

The mass of the ring is

$$m = \pi\rho d(R_{out}^2 - R_{in}^2) = \pi 7800 \times 0.022(0.142^2 - 0.082^2) = 7.25 \text{ kg}$$

The moment of inertia according to eqn (9.2) is

$$J = \frac{1}{2} 7.25(0.142^2 + 0.082^2) = 0.097 \text{ kgm}^2$$

The rotational speed $n = 30\ 000/60 = 500$ rev/s and angular speed $\Omega = 2\pi 500 = 3141.6$ rad/s. Thus the kinetic energy according to eqn (9.4) is

$$E_k = \pi^3 7800 \times 0.022(0.142^4 - 0.082^4) \times 500^2 = 480.74 \text{ kJ}$$

The tensile stress according to eqn (9.2) is

$$\sigma = \frac{7800 \times 3141.6^2}{3}(0.082^2 + 0.082 \times 0.142 + 0.142^2) = 988.6 \text{ MPa}$$

The energy density according to eqn (9.6) is

$$e_k = \frac{480\ 740}{7.25} = 66.3 \text{ kJ/kg}$$

The shape factor of the ring [3] is

$$k_{sh} = e_k \frac{\rho}{\sigma} = 66\ 300 \frac{7800}{988.6 \times 10^6} = 0.523$$

Numerical Example 9.2

Find the main dimensions of a three-phase, 1200 W, 4-pole, 200 000 rpm AFPM synchronous generator for a mobile battery charger. The total nonmagnetic air gap including magnet retaining ring (Fig. 9.2) should be $g = 2$ mm, number of slots $s_1 = 12$, air gap magnetic flux density $B_g = 0.4$ T, line current density $A_m = 16\ 000$ A/m, PM inner–to–outer diameter ratio $k_d = 0.5$, PM temperature 350°C. Vacomax 225 SmCo PMs (*Vaccumschmelze*, Hanau, Germany) with $B_{r20} = 1.04$ T and $H_c = 760$ kA/m at 20°C are recommended. The temperature coefficient for B_r is $\alpha_B = -0.035$ %/°C, leakage flux coefficient $\sigma_{lM} = 1.3$, EMF–to–voltage ratio $\epsilon = 1.3$ ($E_f > V_1$), pole shoe width–to–pole pitch ratio $\alpha_i = 0.72$, efficiency $\eta = 0.8$ and power factor $\cos\phi = 0.8$.

Solution

The winding coefficients (2.8), (2.9) and (2.10) are $k_{d1} = 1$, $k_{p1} = 1$ and $k_{w1} = 1$ since $q_1 = 12/(4 \times 3) = 1$. The coefficient k_D according to eqn (2.27) is

$$k_D = \frac{1}{8}(1 + 0.5)(1 - 0.5^2) = 0.141$$

The outer diameter of PMs according to eqn (2.100) is

$$D_{out} = \sqrt[3]{\frac{1.3 \times 1200}{\pi^2 0.141 \times 1 \times (200\ 000/60) \times 0.4 \times 16\ 000 \times 0.8 \times 0.8}} = 0.044 \text{ m}$$

The remanent magnetic flux density at 350°C according to eqns (3.2) is

$$B_r = 1.04\left[1 + \frac{-0.035}{100}(350 - 20)\right] = 0.92 \text{ T}$$

Since demagnetization curves for different temperatures are parallel lines, the relative recoil magnetic permeability will be approximately the same as for 20°C (linear magnetization curve), i.e.

$$\mu_{rrec} = \frac{0.92}{0.4\pi \times 10^{-6} \times 760\ 000} = 0.963$$

The PM thickness with the leakage coefficient σ_{lm} included, can be calculated on the basis of eqns (3.13) and (3.9), i.e.

$$h_M \approx \mu_{rrec}\frac{\sigma_{lm}B_g}{B_r - \sigma_{lm}B_g}g = 1.089\frac{1.3 \times 0.4}{0.92 - 1.3 \times 0.4}0.002 = 0.0026 \ m$$

The inner diameter of PMs is $D_{in} = k_d D_{out} = 0.5 \times 44 = 22$ mm, radial length of PM $l_M = 0.5(44.0 - 22.0) = 11.0$ mm, average diameter of PMs $D = 0.5(44.0 + 22.0) = 33$ mm, average pole pitch $\tau = \pi 33/4 = 26$ and average PM circumferential width $b_p = \alpha_i\tau = 0.72 \times 26 = 18$ mm.

Numerical Example 9.3

A 10-kW, 2200-rpm electric motor operates at almost constant speed according to the torque profile given in Fig. 9.44. The overload capacity factor $k_{ocf} = T_{max}/T_{shr} = 1.8$. Find the thermal utilisation coefficient of the motor.

Solution

The required rated shaft torque is

$$T_{shr} = \frac{P_{out}}{2\pi n} = \frac{10,000}{2\pi \times (2200/60)} = 43.4 \text{ Nm}$$

The *rms* torque based on the given duty cycle is

$$T_{rms}^2(t_1 + t_2 + \ldots + t_n) = T_1^2 t_1 + T_2^2 t_2 + \ldots + T_n^2 t_n \quad \text{or} \quad T_{rms}^2 \sum t_i = \sum T_i^2 t_i$$

Thus

$$T_{rms} = \sqrt{\frac{\sum T_i^2 t_i}{\sum t_i}}$$
$$= \sqrt{\frac{25^2 \times 3 + 40^2 \times 8 + 65^2 \times 5 + 38^2 \times 30 + 15^2 \times 10}{3 + 8 + 5 + 30 + 10}}$$
$$= 38.1 \text{ Nm}$$

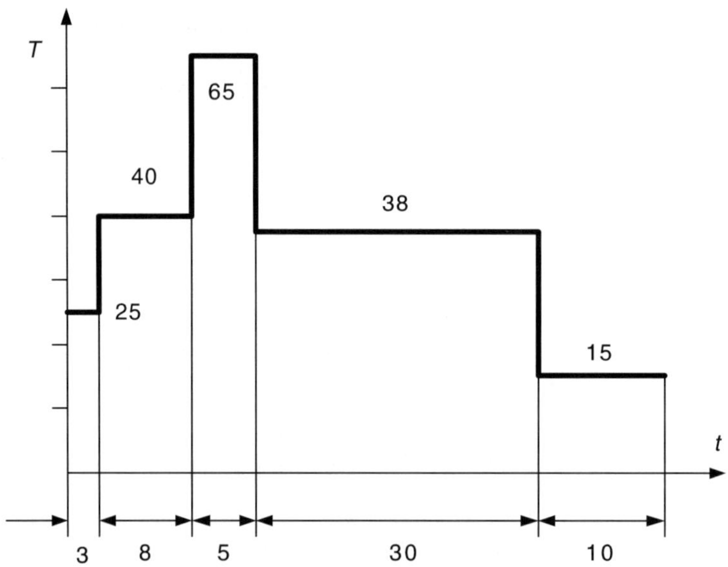

Fig. 9.44. Torque profile of the electric motor according to *Numerical example 9.3*.

The maximum torque in Fig. 9.44 cannot exceed the rated shaft torque times the overload capacity factor $k_{ocf}T_{shr} = 1.8 \times 43.4 = 78.12$ Nm. Also, the required T_{shr} should be greater than or equal to T_{rms}.

The coefficient of thermal utilisation of the motor is

$$\frac{T_{rms}}{T_{shr}} \times 100\% = \frac{38.1}{43.4} \times 100\% = 87.8\%$$

Symbols and Abbreviations

$\mathbf{A}$	magnetic vector potential
A	line current density; cross-section area
a	number of parallel current paths of the stator (armature) winding
$\mathbf{B}$	vector magnetic flux density
B	magnetic flux density; damping of the system
b	instantaneous value of the magnetic flux density; width of slot
b_p	pole shoe width
C_f	cost of frame
C_{ins}	cost of insulation
C_0	cost of all other components independent of the shape of the machine
C_{PM}	cost of PMs
C_{rc}	cost of the rotor core
C_{sh}	cost of shaft
C_w	cost of winding
c_{Cu}	cost of copper conductor per kg
c_E	armature constant (EMF constant)
c_{Fe}	cost of ferromagnetic core per kg
c_{ins}	cost of insulation per kg
c_p	specific heat at constant pressure
c_{PM}	cost of PMs per kg
c_{steel}	cost of steel per kg
c_v	heat capacity
D	diameter; duty cycle of power semiconductor switches
D_{in}	inner diameter of PMs equal to the inner diameter of stator bars
D_{out}	outer diameter of PMs equal to the outer diameter of stator bars
E	EMF (rms value); electric field intensity
E_f	EMF per phase induced by the rotor without armature reaction
E_i	internal EMF per phase
e	instantaneous EMF; eccentricity
F	force; number of winding sections
F_{12}	shape factor of two surfaces involved in radiation

$\mathcal{F}$	space and/or time distribution of the MMF
$\mathcal{F}_a$	armature reaction MMF
$\mathcal{F}_{exc}$	MMF of the rotor excitation system
f	frequency; friction factor
G	permeance; gap ratio g/R
g	air gap (mechanical clearance); gravitational acceleration
Gr	Grashof number
g'	equivalent air gap
$\mathbf{H}$	vector magnetic field intensity
H	magnetic field intensity
h	height; heat transfer coefficient
h_M	height of the PM
I	electric current
I_a	stator (armature) current
i	instantaneous value of current; enthalpy
$\mathbf{J}$	vector electric current density
J	moment of inertia
J_a	current density in the stator (armature) winding
K_c	current regulator gain
K_i	inverter gain
k	coefficient, general symbol; thermal conductivity
k_{1R}	skin effect coefficient for the stator conductor resistance
k_{ad}	reaction factor in d-axis
k_{aq}	reaction factor in q-axis
k_C	Carter's coefficient
k_d	inner-to-outer diameter ratio $k_d = D_{in}/D_{out}$
k_{d1}	distribution factor for fundamental
k_E	EMF constant $k_E = c_E\Phi_f$
k_f	form factor of the field excitation $k_f = B_{mg1}/B_{mg}$
k_i	stacking factor of laminations
k_{ocf}	overload capacity factor $k_{ocf} = T_{max}/T_{shr}$
k_{p1}	pitch factor for fundamental
k_{sat}	saturation factor of the magnetic circuit due to the main (linkage) magnetic flux
k_T	torque constant $k_T = c_T\Phi_f$
k_{w1}	winding factor $k_{w1} = k_{d1}k_{p1}$ for fundamental
L	inductance; length
l_{1e}	length of the one-sided end connection
L_i	armature stack effective length
l_M	axial length of PM
M	mutual inductance
M_o	momentum
m	number of phases; mass
$\dot{m}$	mass flow rate
m_a	amplitude modulation ratio

m_f	frequency modulation ratio
N	number of turns per phase; number of machines
Nu	Nusselt number
n	rotational speed in rpm; independent variables
n_0	no-load speed
P	active power
P_{elm}	electromagnetic power
P_{in}	input power
P_{out}	output power
Pr	Prandtl number
ΔP	active power losses
ΔP_{1Fe}	stator core losses
ΔP_{1w}	stator winding losses
ΔP_{2Fe}	rotor core losses
ΔP_e	eddy current losses in stator conductors
ΔP_{fr}	friction losses
ΔP_{PM}	losses in PMs
ΔP_{rot}	rotational (mechanical) losses
ΔP_{wind}	windage losses
Δp	specific core loss
p	number of pole pairs; pressure
p_r	radial force per unit area
$\wp$	wetted perimeter
Q	electric charge; reactive power; volumetric flow rate
Q_1	number of slots per pole
Q_c	number of stator coils
Q_{en}	enclosed electric charge
q_1	number of slots per pole per phase
R	radius; resistance
R_1	armature winding resistance of a.c. motors
R_{in}	inner radius of PMs equal to the inner radius of stator bars
R_{out}	outer radius of PMs equal to the outer radius of stator bars
Re	Reynolds number
$\Re_{\mu g}$	air gap reluctance
$\Re_{\mu la}$	external armature leakage reluctance
$\Re_{\mu M}$	permanent magnet reluctance
S	apparent power; surface
S_M	cross section area of PM; $S_M = w_M L_M$ or $S_M = b_p L_M$
s	cross section area of stator conductor
s_1	number of stator slots equal to the number of stator teeth
T	torque
T_d	electromagnetic torque developed by the machine
T_{drel}	reluctance torque
T_{dsyn}	synchronous or synchronizing torque
T_m	mechanical time constant

T_{sh}	shaft torque (output or load torque)
t	time; slot pitch
U	internal energy
u	tangential velocity
V	electric voltage; volume
v	instantaneous value of electric voltage; linear velocity
W	energy produced in outer space of PM; rate of change of the air gap energy
W_m	stored magnetic energy
w	energy per volume, $\mathrm{J/m^3}$; radial velocity
w_M	width of PM
X	reactance
X_1	stator winding leakage reactance
X_{ad}	d-axis armature reaction (mutual) reactance
X_{aq}	q-axis armature reaction (mutual) reactance
X_{sd}	d-axis synchronous reactance; $X_{sd} = X_1 + X_{ad}$
X_{sq}	q-axis synchronous reactance; $X_{sq} = X_1 + X_{aq}$
$\mathbf{Z}$	impedance $\mathbf{Z} = R + jX$; $\mid \mathbf{Z} \mid = Z = \sqrt{R^2 + X^2}$
z	number of coils in a phase group
α	complex attenuation constant of electromagnetic field
α_i	effective pole arc coefficient $\alpha_i = b_p/\tau$
γ	form factor of demagnetization curve of PM material
δ	power (load) angle
δ_i	inner torque angle
ϵ	eccentricity
ε	emissivity; surface spectral property
η	efficiency
γ	equivalent sand grain roughness
θ	rotor angular position for brushless motors
θ_m	slot pitch angle
θ_{re}	electrical angle corresponding to the coil layer width
ϑ	temperature; angle between $\mathbf{I}_a$ and $\mathbf{I}_{ad}$
λ	coefficient of leakage permeance (specific leakage permeance)
λ_T	turbulent parameter
μ	dynamic viscosity
μ_o	magnetic permeability of free space $\mu_o = 0.4\pi \times 10^{-6}$ H/m
μ_r	relative magnetic permeability
μ_{rec}	recoil magnetic permeability
μ_{rrec}	relative recoil permeability $\mu_{rrec} = \mu_{rec}/\mu_o$
ν	number of the stator νth harmonic; kinematic viscosity
ξ	reduced height of the stator conductor
ρ	specific mass density
σ	electric conductivity; Stefan-Boltzmann constant
σ_f	form factor to include the saturation effect
σ_p	output coefficient

σ_r	radiation factor
τ	average pole pitch; thermal time constant
Φ	magnetic flux
Φ_f	excitation magnetic flux
Φ_l	leakage flux
ϕ	power factor angle
Ψ	flux linkage $\Psi = N\Phi$; angle between $\mathbf{I}_a$ and $\mathbf{E}_f$
ψ	flux linkage
Ω	angular speed $\Omega = 2\pi n$
ω	angular frequency $\omega = 2\pi f$

Subscripts

a	armature (stator)
avg	average
c	conduction
cv	control volume
Cu	copper
d	direct axis; differential; developed
e	end connection; eddy-current
elm	electromagnetic
eq	equivalent
exc	excitation
ext	external
Fe	ferromagnetic
f	field; forced
fr	friction; free
g	air gap
h	hydraulic; hysteresis
in	inner
l	leakage
M	magnet
m	peak value (amplitude)
n, t	normal and tangential components
out	output, outer
q	quadrature axis
r	rated; remanent; radiation; rotor
r, θ, z	cylindrical coordinate system
rel	reluctance
rot	rotational
s	slot; synchronous; system; stator
sat	saturation
sh	shaft
st	starting

syn	synchronous or synchronizing
t	teeth; total
u	useful
v	convection
vent	ventilation
wind	windage
y	yoke
x, y, z	cartesian coordinate system
1	stator; fundamental harmonic; inlet
2	rotor; exit

Superscripts

inc	incremental
(sq)	square wave
(tr)	trapezoidal

Abbreviations

A/D	analog to digital
AFPM	axial flux permanent magnet
AIFI	American Iron and Steel Industry
a.c.	alternating current
AWG	American Wire Gauge
BPF	band pass filtering
BSCCO	bismuth-strontium-calcium-copper-oxygen
CAD	computer-aided design
CPU	central processor unit
DSP	digital signal processor
d.c.	direct current
EDM	electro-discharge machining
EMALS	electro-magnetic aircraft launch system
EMF	electromotive force
EMI	electromagnetic interference
EV	electric vehicle
FCEV	fuel cell electric vehicle
FDB	fluid dynamic bearing
FEM	finite element method
FPGA	field programmable gate array
FRP	fiberglass reinforced plastic
GCD	greatest common divisor
HDD	hard disk drive
HEV	hybrid electric vehicle
HTS	high temperature superconducting

IC	integrated circuit
IGBT	insulated-gate bipolar transistor
ISG	integrated starter-generator
LPF	low pass filter
LVAD	left ventricular assist device
MMF	magnetomotive force
MMT	moving magnet technologies
MOSFET	metal oxide semiconductor (MOS) field effect transistor
MTBF	mean time between failures
MVD	magnetic voltage drop
NdFeB	neodymium iron boron
PFM	pulse frequency modulation
PLD	programmable logic device
PM	permanent magnet
PWM	pulse width modulation
RFI	radio frequency interference
RFPM	radial flux permanent magnet
SEMA	segmented electro-magnetic array
SMC	soft magnetic composite
SmCo	samarium cobalt
SSC	solid state converter
YBCO	yttrium-barium-copper oxide ($YBa_2Cu_3O_{7-x}$)

References

1. Abdel-Razek A.A., Coulomb J.L., Feliachi M., and Sabonnadiere J.C. (1981). The calculation of electromagnetic torque in saturated electric machines within combined numerical and analytical solutions of the field equations, IEEE Trans. MAG-17(6):3250–3252.
2. Abdel-Razek A.A., Coulomb J.L., Feliachi M., and Sabonnadiere J.C. (1982). Conception of an air-gap element for the dynamic analysis of the electromagnetic field in electric machines, IEEE Trans. MAG-18(2):655–659.
3. Acarnley P.P., Mecrow B.C., Burdess J.S., Fawcett J.N., Kelly J.G., and Dickinson P.G. (1996). Design principles for a flywheel energy store for road vehicles, IEEE Trans. IA-32(6):1402–1408.
4. Accucore, TSC Ferrite International, Wadsworth, IL, U.S.A., (2000), www.tscinternational.com
5. Afonin A.A., and Cierznewski P. (1999). Electronically commutated disc-type permanent magnet motors (in Russian), Int. Conf. on Unconventional Electromechanical and Electr Systems UEES'99. St Petersburg, Russia, pp. 271–276.
6. Afonin A.A., Kramarz W., and Cierzniewski P. (2000) *Electromechanical Energy Converters with Electronic Commutation (in Polish).* Szczecin: Wyd Ucz PS.
7. Ahmed A.B., and de Cachan L.E. (1994). Comparison of two multidisc configurations of PM synchronous machines using an elementary approach, Int. Conf. on Electr. Machines ICEM'94, vol. 1, Paris, France, pp. 175–180.
8. Amaratunga G.A.J., Acarnley P.P., and McLaren P.G. (1985). Optimum magnetic circuit configurations for PM aerospace generators, IEEE Trans on AES, 21(2):230–255.
9. Anderson H.H. (1972). *Centrifugal pumps.* The Trade and Technical Press.
10. Angelo J.D., Chari M.V.K., and Campbell P. (1983). 3-D FE solution for a PM axial field machine, IEEE Trans. PAS-102(1):83–90.
11. Armensky E.V., and Falk G.B. (1978). *Fractional–Horsepower Electrical Machines.* Moscow: Mir Publishers.
12. Arnold D.P., Zana I., Herrault F., Galle P., Park J.W., Das, S., Lang, J. H., and Allen, M. G. (2005). Optimization of a microscale, axial-flux, permanent-magnet generator, Proc. 5th Int. Workshop Micro Nanotechnology for Power

Generation and Energy Conversion Apps. (PowerMEMS 2005), Tokyo, Japan, pp. 165–168.

13. Arnold D.P., Das S., Park, J.W., Zana, I., Lang, J.H., and Allen, M.G. (2006). Microfabricated high-speed axial-flux multiwatt permanent magnet generatorsPart II: Design, fabrication, and testing, Journal of Microelectromechanical Systems, 15(5):1351–1363.
14. Atallah K., Zhu Z.Q., Howe D., and Birch T.S. (1998). Armature Reaction Field and Winding Inductances of Slotless Permanent-Magnet Brushless Machines, IEEE Trans. MAG-34(8):3737–3744.
15. Balagurov V.A., Galtieev F.F., and Larionov A.N. (1964). *Permanent Magnet Electrical Machines (in Russian).* Moscow: Energia.
16. Baluja S. (1994). *Population-based incremental learning: A method for integrating genetic search based function optimization and competitive learning.* Tech. report No. CMU-CS-94-163, Carnegie Mellon University, Pittsburgh, U.S.A.
17. Barakat G., El-Meslouhi T., and Dakyo B. (2001). Analysis of the cogging torque behavior of a two-phase axial flux permanent magnet synchronous machine. IEEE Trans. MAG-37(4):2803–28005.
18. Baudot J.H. (1967). *Les Machines Éléctriques en Automatique Appliqueé (in French).* Paris: Dunod.
19. Becerra R.C., and Ehsani M. (1988). High-speed torque control of brushless PM motors, IEEE Trans. IE-35(3):402–405.
20. Berry C.H. (1954). *Flow and fan - principle of moving air through ducts.* The Industrial Press, New York.
21. Bertotti G., Boglietti A., Champi M., Chiarabaglio D., Fiorillo D., and Lazari M. (1991). An improved estimation of iron losses in rotating electrical machines, IEEE Trans. MAG-27(6):5007–5009.
22. Biwersi, S., Billet, L., Gandel, P, and Prudham, D. (2002). Low cost – high speed small size disk magnet synchronous motor, 8th Int. Conf. Actuator'2002, Bremen, Germany, pp. 196–200.
23. Bolognani S., Oboe R., and Zigliotto M. (1999). Sensorless full-digital PMSM drive with EKF estimation of speed and rotor position, IEEE Trans. IE-46(1): 184–191.
24. Bose B.K. (1988). A high-performance inverter-fed drive system of an interior PM synchronous machine, IEEE Trans. IA-24(6):987–998.
25. Braga G., Farini A., and Manigrasso R. (1991). Synchronous drive for motorized wheels without gearbox for light rail systems and electric cars, 3rd European Power Electronic Conf. EPE'91, vol. 4, Florence, Italy, pp. 78–81.
26. Brauer J.R., (ed.) (1988). *What Every Engineer Should Know about Finite Element Analysis.* New York: Marcel Dekker.
27. Bumby J.R., and Martin R. (2005). Axial-flux permanent-magnet air-cored generator for small-scale wind turbines IEE Proc.-Electr. Power Appl., 152(5):1065–1075
28. Campbell P. (1072). A new wheel motor for commuter cars, Electrical Review, (190), pp. 332–333.
29. Campbell P. (1974). Principle of a PM axial field DC machine, Proceedings of IEE, vol.121, no.1, pp. 1489–1494.
30. Campbell P. (1975). The magnetic circuit of an axial flux DC electrical machine, IEEE Trans. MAG-11(5):1541–1543.

31. Campbell P. (1979). Performance of a permanent magnet axial-field d.c. machine, IEE Proc Pt B 2(4):139–144.
32. Campbell P., Rosenberg D.J., and Stanton D.P. (1981). The computer design and optimization of axial field PM motor, IEEE Trans. PAS-100(4):1490–1497.
33. Campbell P. (1994). *Permanent magnet materials and their application.* Cambridge University Press, Cambridge, UK.
34. Capponi F., Terrigi R., Caricchi F., Del Ferraro L. (2007). Active output voltage regulation for an ironless axial-flux PM automotive alternator with electromechanical flux weakening, IEEE-IAS Annual Meeting, New Orleans, Paper no. 48P2.
35. Caricchi F., Crescimbini F., di Napoli A., Honorati O., Lipo T.A., Noia G., and Santini E. (1991). Development of a IGBT inverter driven axial flux PM synchronous motor drive, European Power Electronics Conf. EPE'91, Firenze, Italy, vol.3, pp. 482–487.
36. Caricchi F., Crescimbini F., Honorati O., and Santini E. (1992). Performance evaluation of an axial-flux PM generator, Int. Conf. on Electr. Machines ICEM'92, Manchester, U.K., vol. 2, pp. 761–765.
37. Caricchi F., Crescimbini F., di Napoli A., and Santini E. (1992). Optimum CAD-CAE design of axial flux permanent magnets motors, Int. Conf. on Electr. Machines ICEM'92, Manchester, U.K., vol. 2, pp. 637–641.
38. Caricchi F., Crescimbini F., Fedeli E., and Noia G. (1994). Design and construction of a wheel-directly-coupled axial-flux PM motor prototype for EVs, IEEE-IAS Annual Meeting, IAS-29, part 1, pp. 254–261.
39. Caricchi F., Crescembini F., and Santini E. (1995). Basic principle and design criteria of axial-flux PM machines having counterrotating rotors, IEEE Trans. IA-31(5):1062–1068.
40. Caricchi F., Crescimbini F., Mezzetti F., and Santini E. (1996). Multistage axial-flux PM machines for wheel-direct-drive, IEEE Trans. IA-32(4):882–888.
41. Caricchi F., Crescimbini F., di. Napoli A., and Marcheggiani M. (1996). Prototype of electric-vehicle-drive with twin water-cooled wheel direct-drive motors, IEEE Annual Power Electronics Specialists Conf. PESC'96, Part 2, pp. 1926–1932.
42. Caricchi F., Crescimbini F., Santini E., and Santucci C. (1997). Influence of the radial variation of the magnet pitches in slot-less PM axial flux motors, IEEE-IAS Annual Meeting, vol. 1, pp. 18–23.
43. Caricchi F., Crescimbini F., Santini E., and Santucci C. (1998). FEM evaluation of performance of axial flux slotted PM machines, IEEE IAS Annual Meeting, vol. 1, pp. 12–18.
44. Caricchi F., Crescimbini F., and Honorati O. (1998). Low-cost compact permanent magnet machine for adjustable-speed pump application, IEEE Trans. IA-34(1):109–116.
45. Caricchi F., Crescimbini F., Honorati O., Bianco G.L., and Santini E. (1998). Performance of core-less winding axial-flux PM generator with power output at 400Hz – 3000 rpm, IEEE Trans. IA-34(6):1263–1269.
46. Caricchi F., Santini E., Crescimbini F., and Solero L. (2000). High efficiency low volume starter/alternator for automotive applications, IEEE-IAS Annual Meeting, Rome, vol. 1, pp. 215–222.
47. Carter G.W. (1954). *Electromagnetic field in its engineering aspects.* Longmans.

48. Cascio A.M. (1997). Modeling, analysis and testing of orthotropic stator structures, Naval Symp. on Electr. Machines, Newport, RI, USA, pp. 91–99.
49. Chalmers B.J., Hamed S.A., and Baines G.D. (1985). Parameters and performance of a high-field permanent magnet synchronous motor for variable-frequency operation, Proc IEE Pt B 132(3):117–124.
50. Chalmers B.J., Spooner E., Honorati O., Crescimbini F., and Caricch F. (1997). Compact permanent magnet machines, Electr. Machines and Power Systems, 25(6): 635–648.
51. Chalmers B.J., Wu W., and Spooner E. (1999). An axial-flux permanent-magnet generator for a gearless wind energy system, IEEE Trans. EC-14(2): 251–257.
52. Chan C.C. (1982). *Axial-field electrical machines with yokeless armature core.* PhD Thesis, University of Hong Kong.
53. Chan C.C. (1987). Axial-field electrical machines: design and application, IEEE Trans. EC-2(2): 294–300.
54. Chandler P.L., and Patterson D.J. (1999). Counting the losses in very high efficiency machine design, World Renewable Energy Congress, Perth, Australia.
55. Chang L. (1994). Comparison of a.c. drives for electric vehicles — a report on experts' opinion survey, IEEE AES Systems Magazine 8: 7–11.
56. Chari M.V.K., and Silvester P.P. (ed.) (1980). *Finite element in electrical and magnetic field problems.* John Wiley & Sons, New York.
57. Chen J., and Chin K. (2003). Minimum copper loss flux-weakening control of surface mounted permanent magnet synchronous motors, IEEE Trans. PE-18(4): 929–936.
58. Chen S.X., Low T.S., Lin H., and Liu Z.J. (1996). Design trends of spindle motors for high performance hard disk drives. IEEE Trans. MAG-32(5): 3848–3850.
59. Chillet C., Brissonneau P., and Yonnet J.P. (1991). Development of a water cooled permanent magnet synchronous machine. Int. Conf. on Synchronous Machines SM100, vol 3, Zürich, Switzerland, pp. 1094–1097.
60. Chin Y.K., Nordlund E., and Staton D.A. (2003). Thermal analysis - lumped circuit model and finite element analysis, The 6th Int. Power Engineering Conf., Singapore, pp.1067–1072.
61. Chisholm D. (1983). *Two-phase flow in pipelines and heat exchangers.* George Godwin, New York.
62. Chung S.U., Hwang, G.Y., Hwang, S.M., Kang, B.S., and Kim H.G. (2002) Development of brushless and sensorless vibration motor used for mobile phones, IEEE Trans. MAG-18(5): 3000–3002.
63. Cistelecan M.V., Popescu M., and Popescu M. (2007). Study Of the number of slots / pole combinations for low speed permanent magnet synchronous generators, IEEE-IEMDC Conference, Turkey, pp. 1616–1620.
64. Coilgun research spawns mighty motors and more. Machine Design 9(Sept 24):24–25, (1993).
65. Coulomb J.L., and Meunier G. (1984). Finite element implementation of virtual work principle for magnetic or electric force and torque computation, IEEE Trans. MAG-20(5):1894–1896.
66. Crescimbini F., and Solero L. (2005). Advances in Propulsion Systems of Hybrid and Electric Vehicles, The 8th Brazilian Power Electronics Conference COBEP'05, Recife, Brazil.

67. Cros J., and Viarouge P. (2002). Synthesis of high performance PM motors with concentrated windings, IEEE Trans. EC-17(2):248–253.
68. Cvetkovski G., Petkovska L., Cundev M., and Gair S. (1998). Mathematical model of a PM axial field synchronous motor for a genetic algorithm optimisation, Int. Conf. on Electr. Machines ICEM'98, Istanbul, vol. 2, pp. 1172–1177.
69. Dąbrowski M., and Gieras J.F. (1977). *Induction motors with solid rotors (in Polish).* PWN, Warsaw-Poznan.
70. Dąbrowski M. (1977). *Construction of Electrical Machines (in Polish).* Warsaw: WNT.
71. Dąbrowski M. (1988). *Magnetic fields and circuits of electrical machines (in Polish).* Warsaw, WNT.
72. Dąbrowski M. (1980). Joint action of permanent magnets in an electrical machine (in Polish), Zeszyty Nauk. Polit. Pozn. Elektryka 21:7–17.
73. Day A.J., and Hirzel A. (2002). Redefining power generation, Gorham Conference, Cincinnati, OH, U.S.A. www.lightengineering.com
74. De Angelo C., Bossio G., Solsona J., García G., and Valla M. (2002). Sensorless speed control of permanent magnet motors with torque ripple minimization, 28th Annual Conf. of the IEEE Industrial Electronics Society (IECON'02), Sevilla, Spain, vol. 1, pp. 680–685.
75. De Angelo C., Bossio, G., Solsona J., García G., and Valla M. (2002). A rotor position and speed observer for permanent magnet motors with non-sinusoidal EMF waveform, 28th Annual Conf. of the IEEE Industrial Electronics Society (IECON'02), Sevilla, Spain, vol. 1, pp. 756–761.
76. De Bruijn F. (2005). The current status of fuel cell technology for mobile and stationary applications, Green Chemistry, 2005, 7, 132-150.
77. Dote Y., and Kinoshita S. (1990). *Brushless Servomotors. Fundamentals and Applications.* Oxford: Clarendon Press.
78. Douglas J.F., Gasiorek J.M., and Swaffield J.A. (1995). *Fluid mechanics.* 3rd ed., Longman Scientific & Technical.
79. Doyle M.R., Samuel D.J., Conway T., and Klimowski R.R. (2001). Electromagnetic aircraft launch system, Int. Electr. Machines and Drives Conf. (IEMDC'2001), Boston, MA, U.S.A.
80. Dunkerley S. (1893). On the whirling and vibration of shafts, Proc. of the Royal Society of London, vol. 54, pp. 365–370.
81. Eastham, J.F., Profumo, F., Tenconi, A., Hill-Cottingham R., Coles, P., and Gianolio, G. (2002). Novel axial flux machine for aircraft drive: design and modeling, IEEE Trans. MAG-38(5):3003–3005
82. El-Hasan T.S., Luk, P.C.K., Bhinder, F.S., and Ebaid M.S. (2000). Modular design of high-speed permanent-magnet axial-flux generators. IEEE Trans. MAG-36(5):3558–3561.
83. El-Hasan T., and Luk P.C.K. (2003). Magnet topology optimization to reduce harmonics in high speed axial flux generators, IEEE Trans. MAG-39(5):3340–3342.
84. Engelmann R.H., and Middendorf W.H. (ed.) (1995). *Handbook of electric motors.* Marcel Dekker, Inc., New York.
85. Engstrom J. (2000). Inductance of slotless machines, in *Proc. of the IEEE Nordic Workshop on Power and Industrial Electronics*, Aalborg, Denmark.
86. Evans P.D., and Easham J.F. (1980). Disc-geometry homopolar synchronous machine, Proc. of IEE, pt. B, 127(6): 299–307.

87. Evans P.D., and Easham J.F. (1983). Slot-less alternator with ac-side excitation, Proc. of IEE, 130(6): 399–406.
88. Evans P.D., and Easham J.F. (1983). Double-disc alternator with ac-side excitation, IEE Proc. Part B, EPA-130(2), pp. 95–102.
89. Ertugrul N., and Acarnley P.P. (1992). Analytical solution of the system equations of the axial field permanent magnet synchronous motor drive, Proceedings of ICEM'92, vol. 2, pp. 785–789.
90. Ertugrul N., and Acarnley P. (1994). A new algorithm for sensorless operation of permanent magnet motors, IEEE Trans. IA-30(1):126–133.
91. Favre E., Cardoletti L., and Jufer M. (1993). Permanent magnet synchronous motors: a comprehensive approach to cogging torque suppression, IEEE Trans. IA-29(6):1141–1149.
92. Ficheux R., Caricchi F., Honorati O., and Crescimbini F. (2000). Axial flux permanent magnet motor for direct drive elevator systems without machine room, IEEE-IAS Annual Meeting, Rome, vol. 1, pp. 190–197.
93. *Film coil motor.* EmBest, Seoul, Korea, (2001), www.embest.com
94. Fitzgerald A.E., and Kingsley C. (1961). *Electric Machinery.* 2nd ed., New York: McGraw-Hill.
95. Flack T.J., and Volschenk A.F. (1994). Computational aspects of time-stepping finite element analysis using an air-gap element, Int. Conf. on Electr. Machines ICEM'94, Paris, France, vol. 3, pp. 158–163.
96. Fracchia M., and Sciutto G. (1994). Cycloconverter Drives for Ship Propulsion, Symp. on Power Electronics, Electr. Drives, Advanced Electr. Motors SPEEDAM'94, Taormina, Italy, pp. 255–260.
97. Fratta A., Villata F., and Vagati A. (1991). Extending the voltage saturated performance of a DC brushless drive, European Power Electronic Conf. EPE'91, vol. 4, Florence, Italy, pp. 134–138.
98. World Fuel Cell Council, www.fuelcellworld.org.
99. Furlani E.P. (1992). A method for predicting the field in axial field motors, IEEE Trans. MAG-28(5):2061–2066.
100. Furlani E.P. (1994). Computing the field in permanent magnet axial-field motors, IEEE Trans. MAG-30(5):3660–3663.
101. Gair S., Eastham J.F., and Profumo F. (1995). Permanent magnet brushless d.c. drives for electric vehicles, Int. Aeagean Conf. on Electr. Machines and Power Electronics ACEMP'95, Kuşadasi, Turkey, pp. 638–643.
102. Gieras I.A., and Gieras J.F. (2006). Recent advancements in permanent magnet motors technology for medical applications, (invited paper), 25th Int. Symp. on Micromachines and Servo Drives MIS'06, Bialowieza, Poland, pp.7–13.
103. Gieras J.F. (1981). Electrodynamic levitation forces – theory and small-scale test results, Acta Technica CSAV, No. 4, pp. 389–414.
104. Gieras J.F. (1983). Simplified theory of double-sided linear induction motor with squirrel-cage elastic secondary, IEE Proc. Part B 130(6):424–430.
105. Gieras J.F., Santini E., and Wing M. (1998). Calculation of synchronous reactances of small permanent magnet alternating-current motors: comparison of analytical approach and finite element method with measurements, IEEE Trans. MAG-34(5):3712–3720.
106. Gieras J.F., and Wing M. (2002). *Permanent magnet motor technology: design and applications.* 2nd ed., Marcel Dekker, New York.

107. Gieras J.F., and Gieras I.A. (2002). Performance analysis of a coreless permanent magnet brushless motor, IEEE 37th IAS Meeting, Pittsburgh, PA, U.S.A.
108. Glinka T. (1995). *Electrical Micromachines with Permanent Magnet Excitation (in Polish).* Gliwice (Poland): Silesian Techn University, Gliwice.
109. Goetz J., and Takahashi, T. (2003). A design platform optimized for inner loop control, presented at 24th Int. Exhibition and Conf. on Power Electronics, Intelligent Motion and Power Quality (PCIM 2003), Nuremburg, Germany.
110. Gottvald A., Preis K., Magele C., Biro O., and Savini A. (1992). Global optimization methods for computational electromagnetics, IEEE Trans. MAG-28(2):1537–1540.
111. Grovor L. (1962). *Inductance calculations - working formulas and tables.* Dover, New York.
112. Gu C., Wu W., and Shao K. (1994). Magnetic field analysis and optimal design of d.c. permanent magnet core-less disk machines, IEEE Trans. MAG-30(5) Pt 2, pp. 3668–3671.
113. Hakala H. (2000). Integration of motor and hoisting machine changes the elevator business, Int. Conf. on Electr. Machines ICEM'2000, vol 3, Espoo, Finland, pp. 1242–1245.
114. Halbach K. (1980). Design of permanent multipole magnets with oriented rare earth cobalt material, Nuclear Instruments and Methods, vol. 169, pp. 1–10.
115. Halbach K. (1981). Physical and optical properties of rare earth cobalt magnets, Nuclear Instruments and Methods, vol. 187, pp. 109–117.
116. Halbach K. (1985). Application of permanent magnets in accelerators and electron storage rings, J. Appl. Physics, vol. 57, pp. 3605–3608.
117. Hamarat S., Leblebicioglu K., and Ertan H.B. (1998). Comparison of deterministic and non-deterministic optimization algorithms for design optimization of electrical machines, Int. Conf. on Electr. Machines ICEM'98, Istanbul, Turkey, vol.3, pp. 1477–1482.
118. Hanitsch R., Belmans R., and Stephan R. (1994). Small axial flux motor with permanent magnet excitation and etched air gap winding, IEEE Trans. MAG-30(2):592–594.
119. Hanselman D.C. (2003). *Brushless permanent-magnet motor design.* Cranston, RI: The Writers' Collective.
120. *Hardware interfacing to the TMS320C25.* Texas Instruments.
121. Heller B., and Hamata V. (1977). *Harmonic Field Effect in Induction Machines.* Prague: Academia (Czechoslovak Academy of Sciences).
122. Hendershot J.H., and Miller T.J.E. (1994). *Design of Brushless Permanent Magnet Motors.* Oxford: Clarendon Press.
123. Holland J.H. (1994). *Adaption in nature and artificial systems.* Bradford Books, U.S.A.
124. Holman J.P. (1992). *Heat transfer.* 7th ed., McGraw-Hill (UK), London.
125. Honorati O., Solero L., Caricchi F., and Crescimbini F. (1998). Comparison of motor drive arrangements for single-phase PM motors, Int. Conf. on Electr. Machines, ICEM'98, vol. 2, Istanbul, Turkey, pp. 1261–1266.
126. Honsinger V.B. (1980). Performance of polyphase permanent magnet machines, IEEE Trans. PAS-99(4):1510–1516.

127. Hoole S.R. (1989). *Computer-aided analysis and design of electromagnetic devices.* Elsevier Science Publishing, New York.
128. Hrabovcová V., and Bršlica V. (1990). Disk synchronous machines with permanent magnets – electric and thermal equivalent circuits, Electr. Drives Symp., Capri, Italy, pp. 163–169.
129. Hredzak B., and Gair S. (1996). Elimination of torque pulsation in a direct drive EV wheel motor, IEEE Trans. MAG-32(5) Pt 2, pp. 5010–5012.
130. Huang M., Ye Y., Fan C., Clancy T. (2007). Application of disc permanent magnet linear synchronous motor in novel oil-pumping machines, 6th Int. Symp. on Linear Drives for Industrial Applications, LDIA'07, Lille France, paper number: 122.
131. Huang S., Luo J., Leonardi F., and Lipo T.A. (1996). A general approach to sizing and power density equations for comparison of electrical machines, IEEE-IAS Annual Meeting, San Diego, CA, U.S.A., pp. 836–842.
132. Huang S., Luo J., Leonardi F., and Lipo T.A. (1999). A comparison of power density for axial flux machines based on general purpose sizing equations, IEEE Trans. EC-14(2):185–192.
133. Hughes A., and Miller T.J. (1977). Analysis of fields and inductances in air-cored and iron-cored synchronous machines, Proceedings of IEE, 124(2): 121–126.
134. Incropera F.P., and DeWitt D.P. (2001). *Fundamentals of heat and mass transfer.* 5th ed., John Wiley & Sons, New York.
135. Ivanuskin V.A., Sarapulov F.N., and Szymczak P. (2000). *Structural Simulation of Electromechanical Systems and Their Elements (in Russian).* Szczecin: Wyd Ucz PS.
136. Jahns T.M. (1984). Torque production in permanent magnet synchronous motor drives with rectangular current excitation, IEEE Trans. IA-20(4):803–813.
137. Jahns T.M. (1987). Flux weakening regime operation of an interior PM synchronous motor drive, IEEE Trans. IA-23(4):681–689.
138. Jang J., Sul S.K., Ha J., Ide K., and Sawamura M. (2003). Sensorless drive of surface-mounted permanent-magnet motor by high-frequency signal injection based on magnetic saliency, IEEE Trans. IA-39(4): 1031–1039.
139. Jang G.H., and Chang J.H. (1999). Development of dual air gap printed coil BLDC motor, IEEE Trans. MAG-35(3): 1789–1792.
140. Jang G.H., and Chang J.H. (2002). Development of an axial-gap spindle motor for computer hard disk drives using PCB winding and dual air gaps, IEEE Trans. MAG-38(5): 3297–3299.
141. Jensen C.C., Profumo F., and Lipo T.A. (1992). A low loss permanent magnet brushless d.c. motor utilizing tape wound amorphous iron, IEEE Trans. IA-28(3):646–651.
142. Jones B.L., and Brown J.E. (1987). Electrical variable-speed drives, IEE Proc. Pt A 131(7):516–558.
143. Kamper M.J., and Mackay A.T. (1995). Optimum control of the reluctance synchronous machine with a cageless flux barrier rotor, Trans. of SAIEE, 86(2): 49–56.
144. Kamper M.J., Van der Merwe F.S., and Williamson S. (1996). Direct finite element design optimisation of cageless reluctance synchronous machine, IEEE Trans. EC-11(3):547–555.

145. Kamper M.J., and Jackson S. (1998). Performance of small and medium power flux barrier rotor reluctance synchronous machine drives, Proc. of ICEM'98, Istanbul, Turkey, vol. 1, pp. 95-99.
146. Kamper M.J., Wang R-J, and Rossouw F.G. (2007). Analysis and performance evaluation of axial flux air-cored stator permanent magnet machine with concentrated coils, IEEE-IEMDC Conference, Turkey, pp. 13-20.
147. Kenjo T., and Nagamori S. (1985). *Permanent magnet and brushless d.c. motors.* Oxford: Clarendon Press.
148. Kenjo T. (1990). *Power electronics for the microprocessor era.* Oxford: OUP.
149. Kenjo T. (1991). *Electric motors and their control.* Oxford: OUP.
150. Kessinger R.L., and Robinson S. (1997). SEMA-based permanent magnet motors for high-torque, high-performance, Naval Symp. on Electr. Machines, Newport, RI, U.S.A., pp. 151-155
151. Kessinger R.L. (2002). *Introduction to SEMA motor technology.* Kinetic Art and Technology, Greenville, IN, U.S.A.
152. King R.D., Haefner K.B., Salasoo L., and Koegl R.A. (1995). Hybrid electric transit bus pollutes less, conserves fuel, IEEE Spectrum 32(7): 26–31.
153. Kitano Seiki Co., Ltd., www.kitano-seiki.co.jp
154. Kleen S., Ehrfeld W., Michel F., Nienhaus M., and Stölting H.D. (2000). Penny-motor: A family of novel ultraflat electromagnetic micromotors, Int. Conf. Actuator'2000, Bremen, Germany, pp. 193–196.
155. Klug L. (1990). Axial field a.c. servomotor, Electr. Drives and Power Electronics Symp. EDPE'90, Košice, Slovakia, pp. 154–159.
156. Klug L. (1991). Synchronous servo motor with a disc rotor (in Czech), Elektrotechnický Obzor 80(1-2):13–17.
157. Klug L., and Guba R. (1992). Disc rotor a.c. servo motor drive, Electr. Drives and Power Electronics Symp. EDPE'92, Košice, Slovakia, pp. 341–344.
158. Komęza K., Pelikant A., Tegopoulos J. and Wiak S. (1994). Comparative computation of forces and torques of electromagnetic devices by means of different formulae, IEEE Trans. MAG-30(5):3475–3478.
159. Kostenko M., and Piotrovsky L. (1974). *Electrical Machines.* vol.1: Direct Current Machines and Transformers. Moscow: Mir Publishers.
160. Kubzdela S., and Węgliński B. (1988). Magnetodielectrics in induction motors with disk rotors, IEEE Trans. MAG-24(1):635–638.
161. Lammeraner J., and Štafl M. (1964). *Eddy Currents.* London: Iliffe Books.
162. Lange A., Canders W.R., Laube F., and Mosebach H. (2000). Comparison of different drive systems for a 75 kW electrical vehicles drive, Int. Conf. on Electr. Machines ICEM'2000, vol. 3, Espoo, Finland, pp. 1308–1312.
163. Lange A., Canders W.R., and Mosebach H. (2000). Investigation of iron losses of soft magnetic powder components for electrical machines, Int. Conf. on Electr. Machines ICEM'2000, vol. 3, Espoo, Finland, pp. 1521–1525.
164. Leihold R., Bossio G., Garcia G., and Valla M. (2001). A new strategy to extend the speed range of a permanent magnet a.c. motor, The 6th Brazilian Power Electronics Conference (COBEP'2001), Florianópolis, SC, Brazil.
165. Leung W.S., and Chan C.C. (1980). A new design approach for axial-field electrical machine, IEEE Trans. PAS-99(4):1679–1685.
166. Libert F., and Soulard J. (2004). Investigation on pole-slot combinations for permanent magnet machines with concentrated windings, Proceedings of ICEM, Poland, pp. 530–535.

167. Linke M., Kennel R., and Holtz J. (2002). Sensorless position control of permanent magnet synchronous machine without limitation at zero speed, IEEE-IECON Annual Conf., Spain, pp. 674–679.
168. Linke M., Kennel R., and Holtz J. (2003). Sensorless speed and position control of synchronous machines using alternating carrier injection, IEEE Electr. Mach. and Drives Conf. (IEMDC'03) (Madison/Wisconsin, USA), pp. 1211–1217.
169. Liu C.T., Chiang T.S., Zamora J.F., and Lin S.C. (2003). Field-oriented control evaluations of a single-sided permanent magnet axial-flux motor for an electric vehicle, IEEE Trans. MAG-39(5):3280–3282.
170. Lombard N.F., and Kamper M.J. (1999). Analysis and performance of an ironless stator axial flux PM machine, IEEE Trans. EC-14(4):1051–1056.
171. Lovatt H.C., Ramdenand V.S., and Mecrow B.C. (1998). Design of an in-wheel motor for a solar-powered electric vehicle, Proc. of IEE: Electric Power Applications, vol. 145, No. 5, pp. 402–408.
172. Lowther D.A., and Silvester P.P. (1986). *Computer-aided design in magnetics.* Berlin: Springer Verlag.
173. Lukaniszyn M., Wróbel R., Mendrela A., and Drzewoski R. (2000). Towards optimisation of the disc-type brushless d.c. motor by changing the stator core structure, Int. Conf. on Electr. Machines ICEM'2000, Vol. 3, Espoo, Finland, pp. 1357–1360.
174. Lukaniszyn M., Mendrela E., Jagiello M., and Wróbel R. (2002). Integral parameters of a disc-type motor with axial stator flux, Zesz. Nauk.Polit. Slaskiej, vol. 200, Elecktryka no. 177, pp. 255–262.
175. Magnetfabrik Schramberg GmbH & Co, Schramberg–Sulgen, (1989).
176. Magnussen F., and Sadarangani C. (2003). Winding factors and joule losses of permanent magnet machines with concentrated windings, Proc. IEEE International Electrical machines and drives conference (IEMDC), Madison (USA), vol. 1, pp. 333–339.
177. Magnussen F., Thelin P., and Sadarangani C. (2004). Performance evaluation of permanent magnet synchronous machines with concentrated and distributed windings including the effect of field-weakening, Proc. IEE International Conference on Power Electronics, Machines and Drives (PEMD), Edinburgh (UK), vol. 2, pp. 679-685.
178. Mangan J., and Warner A. (1998). *Magnet wire bonding.* Joyal Product Inc., Linden, NJ, U.S.A., www.joyalusa.com
179. Maric D.S., Hiti S., Stancu C.C., Nagashima J.M., and Rutledge D.B. (1999). Two flux weakening schemes for surface-mounted permanent-magnet synchronous drives – design and transient response consideration. IEEE-IAS 34th Annual Meeting, Phoenix, AZ, pp. 673–678.
180. Marignetti F., and Scarano M. (2000). Mathematical modeling of an axial-flux PM motor wheel, Int. Conf. on Electr. Machines ICEM'2000, vol. 3, Espoo, Finland, pp. 1275–1279.
181. *Maxon Motor.* Sachseln, Switzerland: Interelectric AG, (1991/92).
182. Mbidi D.N., Van der Westhuizen K., Wang R-J, Kamper M.J., and Blom J. (2000). Mechanical design considerations of double stage axial-flux permanent magnet machine, IEEE-IAS 35th Annual Meeting, Rome, vol. 1, pp. 198–201.
183. McFee S., and Lowther D.A. (1987). Towards accurate and consistent force calculation in finite element based computational magnetostatics, IEEE Trans. MAG-23(5):3771–3773.

184. Mellara B., and Santini E. (1994). FEM computation and optimization of L_d and L_q in disc PM machines," 2nd Int. Workshop on Electr. and Mag. Fields, Leuven, Belgium, paper No. 89.
185. Mendrela E., Lukaniszyn M., and Macek-Kaminska K. (2002). *Electronically commutated d.c. brushless disc motors (in Polish).* Warsaw: Gnome.
186. Metzger D.E., and Afgan N.H. (1984). *Heat and mass transfer in rotating machinery.* Hemisphere Publishing Corporation, Washington DC, U.S.A.
187. Miller T.J.E. (1989). *Brushless Permanent–Magnet and Reluctance Motor Drives.* Oxford: Clarendon Press.
188. Miller D.S. (1990). *Internal flow systems.* 2nd ed., BHRA Information Services, The Fluid Eng. Centre, Cranfield, Bedford, U.K.
189. Millner A.R. (1994). Multi-hundred horsepower permanent magnet brushless disk motors, IEEE Appl. Power Electronics Conf. and Exp. APEC'94, pp. 351–355.
190. Mills A.F. (1995). *Basic heat and mass transfer.* Richard D. Irwin, U.S.A.
191. Mishler W.R. (1981). Test results on a low amorphous iron induction motor, IEEE Trans. PAS-100(6):860–866.
192. Miti G.K., and Renfrew A.C. (1998). Field weakening performance of the TORUS motor with rectangular current excitation, Int. Conf. on Electr. Machines ICEM'98, Istanbul, Vol. 1, pp. 630–633.
193. Mizia J., Adamiak K., Eastham A.R., and Dawson G.E. (1988). Finite element force calculation: comparison of methods for electric machines, IEEE Trans. MAG-24(1):447–450.
194. Mohan N., Undeland T.M., Robbins W.P. (1989). *Power Electronics Converters, Applications, and Design.* New York: J Wiley & Sons.
195. Mongeau P. (1997). High torque/high power density permanent magnet motors, Naval Symp. on Electr. Machines, Newport, RI, USA, pp. 9–16.
196. Morimoto S., Sanada, M., Takeda, Y. (1994). Wide-speed operation of interior permanent magnet synchronous motors with high performance current regulator, IEEE Trans. IA-30(4):920–926.
197. Muljadi E., Butterfield C.P., and Wan Y.H. (1999). Axial flux, modular, permanent magnet generator with a toroidal winding for wind turbine applications, IEEE Trans. IA-35(4):831–836.
198. Munson B.R., Young D.F., and Okiishi T.H. (1994). *Fundamentals of fluid mechanics.* 2nd ed., New York, John Wiley & Sons.
199. Nagashima J. (2005). Wheel Hub Motors for Automotive Applications, Proceedings of the 21st Electric Vehicles Symposium, Monaco, April 2-5.
200. Nasar S.A., Boldea I., and Unnewehr L.E. (1993). *Permanent magnet, reluctance, and self-synchronous motors.* Boca Raton: CRC Press.
201. Neyman L.R. (1949). *Skin effect in ferromagnetic bodies (in Russian).* GEI, Leningrad–Moskva.
202. Ofori-Tenkorang J., and Lang J.H. (1995). A comparative analysis of torque production in Halbach and conventional surface-mounted permanent magnet synchronous motors, IEEE-IAS Annual Meeting, Orlando, CA, U.S.A., pp. 657–663.
203. O'Neil S.J. (1997). Advances in motor technology for the medical industry, Medical Device and Diagnostic Journal Ind. Magazine, No. 5.
204. Osin I.L., Kolesnikov V.P., and Yuferov F.M. (1976). *Permanent Magnet Synchronous Micromotors (in Russian).* Moscow: Energia.

205. Ovrebo S. (2002). Comparison of excitation signals for low and zero speed estimation of the rotor position in a axial flux PMSM, Nordic Workshop on Power and Industrial Electronics NORPIE 2002, Stockholm, Sweden.
206. Owen J.M. (1988). Air cooled gas turbine discs: a review of recent research, Int. Journal of Heat and Fluid Flow, vol. 9, No. 4, pp. 354–365.
207. Owen J.M. (1989). An approximate solution for the flow between a rotating and a stationary disk, ASME Journal of Turbomachinery, Vol. 111, No. 4, pp. 323–332.
208. Owen J.M. (1971). The Reynolds analogy applied to flow between a rotating and a stationary disc, Int. Journal of Heat and Mass Transfer, Vol.14, pp.451–460.
209. Owen J.M. (1971). The effect of forced flow on heat transfer from a disc rotating near a stator, Int. Journal of Heat and Mass Transfer, vol. 14, pp. 1135–1147.
210. Owen J.M., and Rogers R.H. (1989). *Flow and heat transfer in rotating-disc system, Vol. 1: Rotor-stator systems.* Research Studies Press, Taunton, UK.
211. Owen J.M, and Rogers R.H. (1995). *Flow and heat transfer in rotating-disc system, Vol. 2: Rotating cavities.* Research Studies Press, Taunton, UK.
212. Parker R.J. (1990). *Advances in Permanent Magnetism.* New York: J Wiley & Sons.
213. Parviainen A., Pyrhonen J., and Kontkanen P. (2005). Axial flux permanent magnet generator with concentrated winding for small wind power applications, IEEE Int'l Conf. on Electric Machines and Drives, pp. 1187–1191.
214. Parviainen A., Niemela M., and Pyrhonen J. (2004). Modeling of Axial Flux Permanent-Magnet Machines, IEEE Trans. IA-40(5):1333–1340.
215. Patterson D., and Spée R. (1995). The design and development of an axial flux permanent magnet brushless d.c. motor for wheel drive in a solar powered vehicle, IEEE Trans. IA-31(5):1054–1061.
216. Perho J., and Ritchie E. (1996). A motor design using reluctance network, 2nd Int. Conf. on Unconventional Electr. and Electromechanical Systems, UEES'96, Alushta, Ukraine, pp. 309–314.
217. Plat D. (1989). Permanent magnet synchronous motor with axial flux geometry, IEEE Trans. MAG-25(4):3076–3079.
218. Powell M.J.D. (1964). An efficient method for finding the minimum of a function of several variables without calculating derivatives, Computer Journal, Vol. 7, pp. 155–162.
219. Profumo F., Zhang Z., and Tenconi A. (1997). Axial flux machines drives: a new viable solution for electric cars, IEEE Trans. IE-44(1):39–45.
220. Profumo F., Tenconi A., Zhang Z., and Cavagnino A. (1998). Design and realization of a novel axial flux interior PM synchronous motor for wheel-motors applications, Int. Conf. on Electr. Machines ICEM'98, Istanbul, Vol. 3, pp. 1791–1796.
221. Profumo F., Tenconi A., Zhang Z., and Cavagnino A. (1998). Novel axial flux interior PM synchronous motor realized with powdered soft magnetic materials, IEEE-IAS Annual Meeting, vol. 1, pp. 152–159.
222. Pullen K., Etemad M.R., and Fenocchi A. (1996). The high speed axial flux disc generator unlocking the potential of the automotive gas turbine, in Proc. IEE Colloq. Machines and Drives for Electric and Hybrid Vehicles, IEE Colloq. Dig., London, U.K., pp. 8/1-4.

223. Rahman K., Ward T., Patel N., Nagashima J., Caricchi F., and Crescimbini F. (2006). Application of Direct Drive Wheel Motor for Fuel Cell Electric and Hybrid Electric Vehicle Propulsion Systems, IEEE Trans. IA-42(5):1185–1192.
224. Rajashekara K., Kawamura A., and Matsuse K., (ed.) (1996). *Sensorless Control of AC Motor Drives.* New York: IEEE Press.
225. Ramsden V.S., Mecrow B.C., and Lovatt H.C. (1997). Design of an in wheel motor for a solar-powered electric vehicle, Proc. of EMD'97, pp. 192–197.
226. Ramsden V.S., Watterson P.A., Holliday W.M., Tansley G.D., Reizes J.A., and Woodard J.C. (2000). A rotary blood pump, Journal IEEE Australia, vol. 20, pp.17–22
227. Rao J.S. (1983). Rotor dynamics, 3rd edition, New Age Int. Publishers, New Delhi.
228. Ratnajeevan S., Hoole H., and Haldar M.K. (1995). Optimisation of electromagnetic devices: circuit models, neural networks and gradient methods in concept, IEEE Trans. MAG-31(3):2016–2019.
229. Reichert, K. (2004). Motors with concentrated, non overlapping windings, some characteristics, Int. Conf. on Electr. Machines ICEM'2004, Cracow, Poland, Paper No. 541.
230. Salon S.J. (1995). *Finite element analysis of electrical machines.* Kluwer Academic Publishers, Norwell, MA, U.S.A.
231. Say M.G. (1992). *Alternating Current Machines.* Singapore: ELBS with Longman.
232. Sayers A.T. (1990). *Hydraulic and compressible flow turbomachines.* McGraw-Hill (U.K.), London.
233. Schaefer G. (1991). Field weakening of brushless PM servomotors with rectangular current, European Power Electronic Conf. EPE'91, vol. 3, Florence, Italy, pp. 429–434.
234. Schroedl M. (1996). Sensorless control of AC machines at low speed and standstill based on the INFORM method, IEEE-IAS 31st Annual Meeting, San Diego, CA, pp. 270–277.
235. Scowby S.T., Dobson R.T., and Kamper M.J. (2004). Thermal modelling of an axial flux permanent magnet machine, Applied Thermal Engineering, vol. 24, No. 2-3, pp. 193–207.
236. SGS–Thomson Motion Control Applications Manual (1991).
237. Sidelnikov B., and Szymczak P. (2000). Areas of application and appraisal of control methods for converter-fed disc motors (in Polish), Prace Nauk. IMNiPE, Technical University of Wroclaw, 48: Studies and Research 20:182–191.
238. Silvester P.P., and Ferrari R.L. (1990). *Finite Elements for Electrical Engineers.* 2nd ed. Cambridge: Cambridge University Press.
239. Sitapati K., and Krishnan R. (2001). Performance comparisons of radial and axial field permanent magnet brushless machines, IEEE Trans. IA-37(5):1219–1226.
240. Skaar S.E., Krovel O., and Nilssen R. (2006). Distribution, coil span and winding factors for PM machines with concentrated windings, Int. Conf. on Electr. Machines ICEM-2006, Chania, Greece, paper 346.
241. Soderlund L., Koski A., Vihriala H., Eriksson J.T., and Perala R. (1997). Design of an axial flux PM wind power generator, Int. Conf. on Electr. Machine and Drives ICEMD'97, Milwaukee, WI, U.S.A., pp. 224–228.

242. Somaloy^{TM}500, *Höganäs*, Höganäs, Sweden, (1997), www.hoganas.com
243. Spooner E., and Chalmers B.J. (1988). Toroidally-wound, slotless, axial-flux PM brushless d.c. motor, Int. Conf. on Electr. Machines ICEM'88, Pisa, Italy, Vol. 3, pp. 81–86.
244. Spooner E., Chalmers B., and El-Missiry M.M. (1990). "A compact brushless d.c. machine," Electr. Drives Symp. EDS'90, Capri, Italy, pp. 239–243.
245. Spooner E., and Chalmers B.J. (1992). TORUS: a slotless, toroidal-stator, permanent-magnet generator, Proc. of IEE, Pt. B EPA vol. 139, pp. 497–506.
246. Stec T.F. (1994). Amorphous magnetic materials Metgkass 2605S-2 and 2605TCA in application to rotating electrical machines, NATO ASI Modern Electrical Drives, Antalya, Turkey.
247. Stefani P., and Zandla G. (1992). Cruise liners diesel electric propulsion. Cyclo- or synchroconverter? The shipyard opinion, Int. Symp. on Ship and Shipping Research vol. 2, Genoa, Italy, pp. 6.5.1–6.5.32.
248. Strachan P.J., Reynaud F.P., and von Backström T.W. (1992), The hydrodynamic modeling of torque converters, R&D Journal, South African Inst. of Mech. Eng., vol. 8, No. 1, pp. 21–29.
249. Sugimoto H., Nishikawa T., Tsuda T., Hondou Y., Akita Y., Takeda T., Okazaki T., Ohashi S., and Yoshida Y. (2006). Trial manufacture of liquid nitrogen cooling high temperature superconductivity rotor, Journal of Physics: Conference Series, (43):780-783
250. Sullivan C.R. (2001). Computationally efficient winding loss calculation with multiple windings, arbitrary waveforms, and two-dimensional or three-dimensional field geometry, IEEE Trans. PE-16(1):142–150.
251. Takano H., Itoh T., Mori K., Sakuta A., and Hirasa T. (1992). Optimum values for magnet and armature winding thickness for axial-field PM brushless DC motor, IEEE Trans. IA-28(2):350–357.
252. Takeda T., Togawa H., and Oota T. (2006). Development of liquid nitrogen-cooled full superconducting motor, IHI Engineering Review, 39(2):89–94
253. Timar P.L., Fazekas A., Kiss J., Miklos A., and Yang S.J. (1989). *Noise and Vibration of Electrical Machines.* Amsterdam: Elsevier.
254. Toliyat H.A., and Campbell S. (2003). *DSP-based electromechanical motion control.* Boca Raton, FL, CRC Press.
255. Tomassi G., Topor M., Marignetti F., and Boldea I. (2006). Characterization of an axial-flux machine with non-overlapping windings as a generator, Electromotion, vol. 13, pp. 73–79.
256. Upadhyay P.R., and Rajagopal K.R. (2005). Comparison of performance of the axial-field and radial-field permanent magnet brushless direct current motors using computer aided design and finite element methods, Journal of Applied Physics 97, 10Q506, pp. 1–3.
257. Varga J.S. (1992). A breakthrough in axial induction and synchronous machines, Int. Conf. on Electr. Machines ICEM'1992, vol. 3, Manchester, UK, pp. 1107–1111.
258. Voldek A.I. (1974). *Electrical Machines (in Russian).* St Petersburg: Energia.
259. Wallace R., Lipo T.A., Moran L.A., and Tapia J.A. (1997). Design and construction of a PM axial flux synchronous generator, Int. Conf. on Electr. Machines and Drives ICEMD'97, Milwaukee, WI, U.S.A., pp. MA1 4.1–4.3.
260. Wang R-J, Kamper M.J., and Dobson R.T. (2005). Development of a thermofluid model for axial field permanent magnet machines, IEEE Trans EC-20(1):80–87.

261. Wang R-J, Mohellebi H., Flack T., Kamper M., Buys J., and Feliachi M. (2002). Two-dimensional Cartesian air-gap element (CAGE) for dynamic finite element modeling of electrical machines with a flat air gap, IEEE Trans. MAG-38(2):1357–1360.
262. Wang R-J, and Kamper M.J. (2002). Evaluation of eddy current losses in axial flux permanent magnet (AFPM) machine with an ironless stator, 37th IEEE-IAS Meeting, Pittsburgh, USA, 2:1289–1294.
263. Wang R-J, and Kamper M.J. (2004). Calculation of eddy current loss in axial field permanent magnet machine with coreless stator, IEEE Trans EC-19(3):532–538.
264. Wang R-J, Kamper M.J., Van der Westhuizen K., and Gieras J.F. (2005). Optimal design of a coreless stator axial field permanent magnet generator, IEEE Trans MAG-41(1):55–64.
265. Węgliński, B. (1990). Soft magnetic powder composites — dielectromagnetics and magnetodielectrics, Reviews on Powder Metallurgy and Physical Ceramics, Vol. 4, No. 2, Freund Publ. House Ltd., London, UK.
266. White F.M. (1994). *Fluid mechanics.* McGraw-Hill Book Company, New York.
267. Wiak S., and Welfle H. (2001). *Disc type motors for light electric vehicles (in Polish).* Lodz: Technical University of Lodz.
268. Wijenayake A.H., Bailey J.M., and McCleer P.J. (1995). Design optimization of an axial gap PM brushless d.c. motor for electric vehicle applications, IEEE-IAS Annual Meeting, pp. 685–691.
269. Williamson S., and Smith J.R. (1980). The application of minimisation algorithms in electrical engineering, Proc. of IEE, vol.127, Pt. A, No. 8, pp. 528–530.
270. Wong W.Y. (1977). *Heat transfer for engineers.* Longmans.
271. Wu R., and Slemon G.R. (1991). A permanent magnet motor drive without a shaft sensor, IEEE Trans. IA-27(5): 1005–1011.
272. Wurster R. (1999). PEM fuel cells in stationary and mobile applications: pathways to commercialization, 6th International Technical Congress (BIEL'99), 13-19 September.
273. Wurster R. (2005). Status quo fuel cell development in europe, Fuel Cell Congress (F-Cell 2005), Workshop: 'Germany meets Canada', Stuttgart, 28 September.
274. Xu L., Xu X., Lipo T.A., and Novotny D.W. (1991). Vector control of a synchronous reluctance motor including saturation and iron loss, IEEE Trans. IA-27(5):977–987.
275. Zangwill W.I. (1967). Nonlinear programming via penalty functions, Management Science, Vol. 13, pp. 344–358.
276. Zhang Z., Profumo F., and Tenconi A. (1994). Axial flux interior PM synchronous motors for electric vehicle drives, Symp. on Power Electronics, Electr. Drives, Advanced Electr. Motors SPEEDAM'94, Taormina, Italy, pp. 323–328.
277. Zhang Z., Profumo F., and Tonconi A. (1996). Axial flux wheel machines for electric vehicles, Electr. Machines and Power Systems, vol.24, no.8, pp. 883–896.
278. Zhang Z., Profumo F., and Tonconi A. (1996). Axial flux versus radial flux permanent magnet motors, Electromotion, vol. 3, pp. 134–140.

279. Zhang Z., Profumo F., and Tenconi A. (1997). Analysis and experimental validation of performance for an axial flux PM brushless d.c. motor with powder iron metallurgy cores, IEEE Trans. MAG-33(5):4194–4196.
280. Zhu Z.Q, Chen Y.S., and Howe D. (2000). Online optimal flux-weakening control of permanent-magnet brushless AC drives, IEEE Trans. IA-36(6):1661–1668.
281. Zhilichev Y.N. (1996). Calculation of 3D magnetic field of disk-type micromotors by integral transformation method,, IEEE Trans. MAG-32(1):248–253.
282. Zhilichev Y.N. (1998). Three-dimensional analytic model of permanent magnet axial flux machine, IEEE Trans. MAG-34(6):3897–3901.

Patents

[P1] US132. Davenport T. (1837). Improvement in propelling machinery by magnetism and electro-magnetism.

[P2] US405858. Tesla N. (1889). Electro-Magnetic Motor.

[P3] US3069577. Lee R. (1962). Disc rotor for induction motor.

[P4] US3230406. Henry-Baudot J. (1966). High frequency electromechanical generator.

[P5] US3231774. Henry-Baudot J. (1966). A.C. rotating electric machines with printed circuit armatures.

[P6] US3239702. Van de Graaff R.J. (1966). Multi-disk electromagnetic power machinery.

[P7] US3280353. Haydon A.W., et al (1966). Electric rotating machine.

[P8] US3293466. Henry-Baudot J. (1966). Axial airgap electric rotary machines.

[P9] US3315106. Reynst M.F. (1967). Disk shaped electric motor.

[P10] US3375386. Hayner P.F., et al (1968). Printed circuit motor.

[P11] US3383535. Lohr T.E. (1968). Electric motor.

[P12] US3407320. McLean W.B. (1968). Wafer type submersible motor for underwater devide.

[P13] US3440464. Tolmie R.J. (1969). Electromagnetic apparatus.

[P14] US3443133. Henry-Baudot J. (1969). Stator for axial airgap electric rotary machines.

[P15] US3487246. Long B.E. (1969). Electric machine.

[P16] US3524250. Burr R.P. (1970). Method of manufacturing electrical wire wound machines.

[P17] US3524251. Burr R.P. (1970). Method of manufacturing disc-type wire wound electrical machines.

[P18] US3525007. Henry-Baudot J. (1970). Axial airgap machines having means to reduce eddy current losses in the armature windings.

[P19] US3529191. Henry-Baudot J. (1970). Homopolar machine having disklike rotor.

[P20] US3534469. Keogh R.J. (1970). Method of manufacturing windings for disc-type dc machine armatures.

[P21] US3538704. Kawanaka H., et al (1970). Balance wheel motor in a timepiece.

[P22] US3550645. Keogh R.J. (1970). Wire wound armature, method and apparatus for making same.

[P23] US3555321. Gruener R., et al (1971). Eddy current brake for use in business machines or the like.
[P24] US3558947. Burr R.P. (1971). Discoidal wire wound armatures.
[P25] US3575624. Keogh R.J. (1971). Wire wound disc armature.
[P26] US3678314. Carter A.H. (1972). Discoidal electric motor.
[P27] US3745388. Frederick D.M. (1973). Axial air gap motor.
[P28] US3922574. Whiteley E. (1975). Permanent magnet hermetic synchronous motor.
[P29] US3979619. Whiteley E. (1976). Permanent magnet field structure for dynamoelectric machines.
[P30] US3999092. Whiteley E. (1976). Permanent magnet synchronous dynamoelectric machine.
[P31] US4007387. Rustecki R.Z. (1977). Electrical synchronous machines.
[P32] US4020372. Whiteley E. (1977). Cooling of discoidal dynamoelectric machines.
[P33] US4059777. Whiteley E. (1977). Cooling of discoidal dynamoelectric machines.
[P34] US4068143. Whiteley E. (1978). Discoidal winding for dynamoelectric machines.
[P35] US4076340. Meinke P., et al (1978). Drive and bearing support for a disc-shaped rotor.
[P36] US4100443. Kuwako T. (1978). Electrical rotary machine.
[P37] US4187441. Oney W.R. (1980). High power density brushless dc motor.
[P38] US4188556. Hahn J.H. (1980). Electro-mechnical machine.
[P39] US4237396. Blenkinsop P.T., et al (1980). Electromagnetic machines with permanent magnet excitation.
[P40] US4253031. Frister M. (1981). Directly driven dynamo electric machine-gas turbine generator structure.
[P41] US4297604. Tawse I.S. (1981). Axial air gap alternators/generators of modular construction.
[P42] US4319152. van Gils A.W. (1982). Laminated winding for electric machines.
[P43] US4363988. Kliman G.B. (1982). Induction disk motor with metal tape components.
[P44] US4371801. Richter E. (1983). Method and apparatus for output regulation of multiple disk permanent magnet machines.
[P45] US4390805. Hahn J.H. (1983). Electromechnical machine.
[P46] US4398112. van Gils A.W. (1983). Aminated winding for electric machines.
[P47] US4435662. Tawse I.S. (1984). Axial air gap alternators/generators of modular construction.
[P48] US4508998. Hahn J.H. (1985). Brushless disc-type dc motor or generator.
[P49] US4536672. Kanayama K., et al (1985). Flat type rotary electric machine.
[P50] US4567391. Tucker H.F., et al (1986). Permanent magnet disc rotor machine.
[P51] US4578610. Kliman G.B., et al (1986). Synchronous disk motor with amorphous metal stator and permanent magnet rotor and flywheel.
[P52] US4605873. Hahn J.H. (1986). Electromechanical machine.
[P53] US4644207. Catterfeld F.C., et al (1987). Integrated dual pump system.
[P54] US4701655. Schmider F. (1987). D.C. machine, with mechanical and electrical connections among collector segments.
[P55] US4710667. Whiteley E. (1987). Brushless d.c. dynamoelectric machine with decreased magnitude of pulsations of air gap flux.

[P56] US4720640. Anderson B.M., et al (1988). Fluid powered electrical generator.
[P57] US4823039. Lynch C. (1989). Electrical machines.
[P58] US4841393. MacLeod D.J., et al (1989). Spindle motor for a disc drive.
[P59] US4864175. Rossi L. (1989). Rotor for an electric motor.
[P60] US4866321. Blanchard H.J., et al (1989). Brushless electrical machine for use as motor or generator.
[P61] US4868443. Rossi L. (1989). Tachogenerator for electric machines.
[P62] US4959578. Varga J.S. (1990). Dual rotor axial air gap induction motor.
[P63] US4978878. Dijken R.H. (1990). Electric multipolar machine.
[P64] US4996457. Hawsey R.A., et al (1991). Ultra-high speed permanent magnet axial gap alternator with multiple stators.
[P65] US5001412. Carter I.W., et al (1991). Alternator starter.
[P66] US5021698. Pullen K.R., et al (1991). Axial field electrical generator.
[P67] US5097140. Crall F.W. (1992). Alternator starter.
[P68] US5130595. Arora R.S. (1992). Multiple magnetic paths machine.
[P69] US5194773. Clarke P.W. (1993). Adjustable speed split stator rotary machine.
[P70] US5200659. Clarke P.W. (1993). Axial and radial field electric rotating machines having relatively rotatable first and second stators.
[P71] US5334898. Skybyk D. (1994). Polyphase brushless dc and ac synchronous machines.
[P72] US5334899. Skybyk D. (1994). Polyphase brushless dc and ac synchronous machines.
[P73] US5394321. McCleer P.J., et al (1995). Quasi square-wave back-emf permanent magnet ac machines wih five or more phases.
[P74] US5396140. Goldie J.H., et al (1995). Parallel air gap serial flux a.c. electrical machine.
[P75] US5535582. Paweletz A. (1996). Drive for a shaftless spinning rotor of an open end spinning machine.
[P76] US5548950. Paweletz A. (1996). Motor for a shaftless spinning rotor for an open-end spinning machine.
[P77] US5615618. Berdut E. (1997). Orbital and modular motors using permanent magnets and interleaved iron or steel magnetically permeable members.
[P78] US5619087. Sakai K. (1997). Axial-gap rotary-electric machine.
[P79] US5637941. Paweletz A. (1997). Shaftless spinning rotor for an open-end spinning machine.
[P80] US5642009. McCleer P.J., et al (1997). Quasi square-wave back-emf permanent magnet ac machines with five or more phases.
[P81] US5710476. Ampela M.J. (1998). Armature design for an axial-gap rotary electric machine.
[P82] US5731645. Clifton D.B., et al (1998). Integrated motor/generator/flywheel utilizing a solid steel rotor.
[P83] US5744896. Kessinger R.L., Jr., et al (1998). Interlocking segmented coil array.
[P84] US5777421. Woodward R.C., Jr. (1998). Disc-type electrical machine.
[P85] US5801473. Helwig A. (1998). Open stator axial flux electric motor.
[P86] US5877578. Mitcham A.J., et al (1999). Rotor disc construction for use in an electrical machine.
[P87] US5905321. Clifton D.B., et al (1999). Energy storage flywheel apparatus and methods.

[P88] US5920138. Clifton D.B., et al (1999). Motor/generator and axial magnetic bearing utilizing common magnetic circuit.
[P89] US5925965. Li Y., et al (1999). Axial flux reluctance machine with two stators driving a rotor.
[P90] US5932935. Clifton D.B., et al (1999). Energy storage flywheel emergency power source and methods.
[P91] US5955808. Hill W. (1999). Multi-phase electric machine with offset multi-polar electric pole units.
[P92] US5955809. Shah M.J. (1999). Permanent magnet generator with auxiliary winding.
[P93] US5955816. Clifton D.B., et al (1999). Energy storage flywheel apparatus and methods.
[P94] US5982069. Rao D.K. (1999). Axial gap machine phase coil having tapered conductors with increasing width in radial direction.
[P95] US5982074. Smith S.H., et al (1999). Axial field motor/generator.
[P96] US6002193. Canini J-M, et al (1999). Basic module for a discoidal electric machine, and corresponding electric machine.
[P97] US6037696. Sromin A., et al (2000). Permanent magnet axial air gap electric machine.
[P98] US6040650. Rao D.K. (2000). Stator with coplanar tapered conductors.
[P99] US6046518. Williams M.R. (2000). Axial gap electrical machine.
[P100] US6064135. Hahn J.H. (2000). Electromechanical machine and armature structure therefore.
[P101] US6155969. Schima H., et al (2000). Centrifugal pump for pumping blood and other shear-sensitive liquids.
[P102] US6163097. Smith S.H., et al (2000). Motor generator including interconnected stators and stator laminations.
[P103] US6166472. Pinkerton J.F., et al (2000). Airgap armature coils and electric machines using same.
[P104] US6194802B1. Rao D.K. (2001). Axial gap motor with radially movable magnets to increase speed capablity.
[P105] US6199666B1. Aulanko E., et al (2001). Elevator drive machine.
[P106] US6404097B1. Pullen K.R. (2002). Rotary electrical machines.
[P107] US6445105B1. Kliman G.B., et al (2002). Axial flux machine and method of fabrication.
[P108] US6476534B1. Vanderbeck W.E., et al (2002). Permanent magnet phase-control motor.
[P109] US6515390B1. Lopatinsky E.L., et al (2003). Electric drive apparatus with a rotor having two magnetizied disks.
[P110] US6633106B1. Swett D.W. (2003). Axial gap motor-generator for high speed operation.
[P111] US6674214B1. Knorzer K-H, et al (2004). Electric axial flow machine.
[P112] US6700271B2. Detela A. (2004). Hybrid synchronous motor equipped with toroidal winding.
[P113] US6700288B2. Smith J.S. (2004). High speed rotor.
[P114] US6707221B2. Carl R.J. (2004). Axial flux machine, stator and fabrication method.
[P115] US6750584B2. Smith J.S. (2004). High speed rotor.
[P116] US6750588B1. Gabrys C.W. (2004). High performance axial gap alternator motor.

[P117] US6762523B1. Lisowski L. (2004). Continuously variable electromagnetic transmission.
[P118] US6768239B1. Kelecy P., et al (2004). Electromotive devices using notched ribbon windings.
[P119] US6794791B2. Ben Ahmed A.H., et al (2004). Motor/generator with energized reluctance and coil in the air gap.
[P120] US6803694B2. Decristofaro N.J., et al (2004). Unitary amorphous metal component for an axial flux electric machine.
[P121] US6828710B1. Gabrys C.W. (2004). Airgap armature.
[P122] US6833647B2. Saint-Michel J., et al (2004). Discoid machine.
[P123] US6844656B1. Larsen K.D., et al (2005). Electric multipole motor/generator with axial magnetic flux.
[P124] US6891301B1. Hsu J.S. (2005). Simplified hybrid-secondary uncluttered machine and method.
[P125] US6940200B2. Lopatinsky E., et al (2005). Electric drive.
[P126] US6967554B2. Eydelie A., et al (2005). Coil for a rotary electric machine.
[P127] US6977454B2. Hsu J.S. (2005). Hybrid-secondary uncluttered permanent magnet machine and method.
[P128] US6995494B2. Haugan O., et al (2006). Axial gap brushless dc motor.
[P129] US7034422B2. Ramu K. (2006). Radial-axial electromagnetic flux electric motor, coaxial electromagnetic flux electric motor, and rotor for same.
[P130] US7034425B2. Detela A. (2006). Hybrid synchronous electric machine.
[P131] US7034427B2. Hirzel A.D. (2006). Selective alignment of stators in axial airgap electric devices comprising low-loss materials.
[P132] US7042109B2. Gabrys C.W. (2006). Wind turbine.
[P133] US7064469B2. Jack A., et al (2006). Core back of an electrical machine and method for making the same.
[P134] US7067950B2. Hirzel A.D., et al (2006). Efficient high-speed electric devices using low-loss materials.
[P135] US7084548B1. Gabrys C.W. (2006). Low cost high speed electrical machine.
[P136] US7098566B2. Rajasingham A.I. (2006). Axial gap electrical machine.
[P137] US7105979B1. Gabrys C.W. (2006). Compact heteropolar hybrid alternator-motor.
[P138] US7109625B1. Jore L.M., et al (2006). Conductor optimized axial field rotary energy device.
[P139] US7144468B2. Decristofaro N.J., et al (2006). Method of constructing a unitary amorphous metal component for an electric machine.
[P140] US7148594B2. Rajasingham A.I (2006). Axial gap electrical machine.
[P141] US7157826B2. Rajasingham A.I (2006). Axial gap electrical machine.
[P142] US7157829B2. Van Tichelen P., et al (2007). Axial flux permanent magnet generator/motor.
[P143] US7170212B2. Balson J.C., et al (2007). Synchronous axial field electrical machine.
[P144] US7187098B2. Hasebe M., et al (2007). Axial gap rotating electrical machine.
[P145] US7190101B2. Hirzel A.D. (2007). Stator coil arrangement for an axial airgap electric device including low-loss materials.
[P146] US7226018B2. Sullivan S. (2007). Landing gear method and apparatus for braking and maneuvering.
[P147] US7230361B2. Hirzel A.D. (2007). Efficient high-speed electric devices using low-loss materials.

[P148] US7237748B2. Sullivan S. (2007). Landing gear method and apparatus for braking and maneuvering.
[P149] US7262536B2. Rahman K.M., et al (2007). Gearless wheel motor drive system.

Index